普通高等教育“十三五”规划教材

应用化学专业英语

李 杰 王 俊 主编

中国石化出版社

内 容 提 要

《应用化学专业英语》是根据大学英语教学大纲专业阅读部分的要求，结合石油石化行业需要编写的，旨在为应用化学及相关专业提供一本比较系统、特色鲜明的专业英语教学用书。

本书共分为8个单元，内容涵盖无机化学、有机化学、物理化学、精细化学品化学、高分子化学、表面活性剂化学、油田化学和石油化学等专业知识。介绍了从事化学工作的基础知识，同时还简单介绍了一般科技论文的写法，并附有实验室常用仪器名称和化学常见缩略语。

本书可供应用化学本科生的专业英语教学，也可作为相关专业的教师、研究生以及从事石油石化及相关专业的科技人员的参考用书。

图书在版编目（CIP）数据

应用化学专业英语 / 李杰，王俊主编．—北京：中国石化出版社，2017.9（2019.7 重印）
普通高等教育“十三五”规划教材
ISBN 978-7-5114-4631-2

Ⅰ.①应… Ⅱ.①李… ②王… Ⅲ.①应用化学-英语-高等学校-教材 Ⅳ.①069

中国版本图书馆 CIP 数据核字（2017）第 218339 号

中国石化出版社出版发行
地址：北京市朝阳区吉市口路 9 号
邮编：100020　电话：(010)59964500
发行部电话：(010)59964526
http://www.sinopec-press.com
E-mail:press@sinopec.com
北京科信印刷有限公司印刷
全国各地新华书店经销
*
787×1092 毫米 16 开本 16 印张 420 千字
2019 年 7 月第 1 版　第 2 次印刷
定价：52.00 元

前　言

高等学校理工科本科《大学英语教学大纲》规定，专业英语是大学英语教学中一个不可缺少的组成部分，是继大学英语之后为提高大学生英文文献的读写能力而开设的一门专业必修课，通过本课程的学习，可以增加学生对专业英语词汇、语法和结构的了解与掌握。

随着我国石油化学工业的发展，同国外的科技交流和合作日趋增多，各类应用化学信息的交流日益广泛，尤其需要提高大学生通过英文文献熟练获取和共享信息资源的能力。目前，不同高校的应用化学专业方向不尽相同，石油石化行业急需特色鲜明、实用性较强的应用化学专业英语教材。

本书信息涵盖面大，既包括化学基本知识、基本理论，又重点编排了精细化工、油田化学和石油化学等专业方向的英文科技书籍文献，能够反映石油、石化行业发展趋势，并初步涉及了英文写作方法。内容选自近年来国外原版的教学用书及学术论文，有较强的系统性和完整性，内容编排由浅入深，符合教学规律，能够体现现代科技英语的篇章结构特点和词汇特点。本书第1、第4单元由李杰编写，第7、第8单元由王俊编写，第3、第5单元由李翠勤编写，第2、第6单元由牛瑞霞编写，全书由李杰统稿。本书可供化学和应用化学本科生的专业英语教学，也可作为相关专业的教师、研究生以及从事石油石化及相关专业科技人员的参考用书。

本书编写过程中参阅了大量国内外书籍和科技文献，在此谨向这些书籍和科技文献的作者致以诚挚的谢意！另外，由于编者时间和水平有限，书中难免存在疏漏和不足之处，敬请读者批评指正。

目　　录

Unit 1 Foundations of Chemistry

1.1 The History of Inorganic Chemistry

Even before alchemy became a subject of study, many chemical reactions were used and the products applied to daily life. For example, the first metals used were probably gold and copper, which can be found in the metallic state. Copper can also be readily formed by the reduction of malachite-basic copper carbonate, $Cu_2(CO_3)(OH)_2$—in charcoal fires. Silver, tin, antimony, and lead were also known as early as 3000 BC. Iron appeared in classical Greece and in other areas around the Mediterranean Sea by 1500 BC. At about the same time, colored glasses and ceramic glazes, largely composed of silicon dioxide (SiO_2, the major component of sand) and other metallic oxides, which had been melted and allowed to cool to amorphous solids, were introduced.

Alchemists were active in China, Egypt, and other centers of civilization early in the first centuries AD. Although much effort went into attempts to "transmute" base metals into gold, the treatises of these alchemists also described many other chemical reactions and operations. Distillation, sublimation, crystallization, and other techniques were developed and used in their studies. Because of the political and social changes of the time, alchemy shifted into the Arab world and later (about 1000 to 1500 AD) reappeared in Europe. Gunpowder was used in Chinese fireworks as early as 1150, and alchemy was also widespread in China and India at that time. Alchemists appeared in art, literature, and science until at least 1600, by which time chemistry was beginning to take shape as a science. Roger Bacon (1214-1294), recognized as one of the first great experimental scientists, also wrote extensively about alchemy.

By the 17th century, the common strong acids (nitric, sulfuric, and hydrochloric) were known, and more systematic descriptions of common salts and their reactions were being accumulated. The combination of acids and bases to form salts was appreciated by some chemists. As experimental techniques improved, the quantitative study of chemical reactions and the properties of gases became more common, atomic and molecular weights were determined more accurately, and the groundwork was laid for what later became the periodic table. By 1869, the concepts of atoms and molecules were well established, and it was possible for Mendeleev and Meyer to describe different forms of the periodic table.

The chemical industry, which had been in existence since very early times in the form of factories for the purification of salts and the smelting and refining of metals, expanded as methods for the preparation of relatively pure materials became more common. In 1896, Becquerel discovered radioactivity, and another area of study was opened. Studies of subatomic particles, spectra, and electricity finally led to the atomic theory of Bohr in 1913, which was soon modified by the quantum mechanics of

Schrödinger and Heisenberg in 1926 and 1927.

Inorganic chemistry as a field of study was extremely important during the early years of the exploration and development of mineral resources. Qualitative analysis methods were developed to help identify minerals and, combined with quantitative methods, to assess their purity and value. As the industrial revolution progressed, so did the chemical industry. By the early 20th century, plants for the production of ammonia, nitric acid, sulfuric acid, sodium hydroxide, and many other inorganic chemicals produced on a large scale were common.

In spite of the work of Werner and Jørgensen on coordination chemistry near the beginning of the 20th century and the discovery of a number of organometallic compounds, the popularity of inorganic chemistry as a field of study gradually declined during most of the first half of the century. The need for inorganic chemists to work on military projects during World War Ⅱ rejuvenated interest in the field. As work was done on many projects (not least of which was the Manhattan Project, in which scientists developed the fission bomb that later led to the development of the fusion bomb), new areas of research appeared, old areas were found to have missing information, and new theories were proposed that prompted further experimental work. A great expansion of inorganic chemistry started in the 1940s, sparked by the enthusiasm and ideas generated during World War Ⅱ.

In the 1950s, an earlier method used to describe the spectra of metal ions surrounded by negatively charged ions in crystals (crystal field theory) was extended by the use of molecular orbital theory to develop ligand field theory for use in coordination compounds, in which metal ions are surrounded by ions or molecules that donate electron pairs. This theory gave a more complete picture of the bonding in these compounds. The field developed rapidly as a result of this theoretical framework, the new instruments developed about this same time, and the generally reawakened interest in inorganic chemistry.

In 1955, Ziegler and associates and Natta discovered organometallic compounds that could catalyze the polymerization of ethylene at lower temperatures and pressures than the common industrial method used up to that time. In addition, the polyethylene formed was more likely to be made up of linear rather than branched molecules and, as a consequence, was stronger and more durable. Other catalysts were soon developed, and their study contributed to the rapid expansion of organometallic chemistry, still one of the fastest growing areas of chemistry today.

The study of biological materials containing metal atoms has also progressed rapidly. Again, the development of new experimental methods allowed more thorough study of these compounds, and the related theoretical work provided connections to other areas of study. Attempts to make model compounds that have chemical and biological activity similar to the natural compounds have also led to many new synthetic techniques.

One current problem that bridges organometallic chemistry and bioinorganic chemistry is the conversion of nitrogen to ammonia:

$$N_2+3H_2 \longrightarrow 2NH_3$$

This reaction is one of the most important industrial processes, with over 120 million tons of ammonia produced in 1990 worldwide. However, in spite of metal oxide catalysts introduced in the Haber-Bosch process in 1913 and improved since then, it is also a reaction that requires temperatures near 400℃ and 200 atm pressure and that still result in a yield of only 15% ammonia. Bacteria, however, manage to fix nitrogen (convert it to ammonia and then to nitrite and nitrate) at 0.8 atm at room temperature in nodules on the roots of legumes. The nitrogenase enzyme that catalyzes this reaction is a complex iron-molybdenum-sulfur protein. The structures of the active sites have been determined by X-ray crystallography.

From *Inorganic chemistry* by *Gary L. Miessler*

New Words and Expressions

alchemy[ˈælkəmɪ] 炼金术;炼丹术
malachite [ˈmæləkaɪt] [矿] 孔雀石
antimony [ˈæntɪmənɪ] [化学] 锑(符号 Sb)
BC abbr.公元前(Before Christ)
Mediterranean Sea 地中海
ceramic glazes 陶瓷釉
amorphous solid [材料学] 非晶形固体
exploration and development 勘探与开发
qualitative analysis [化]定性分析
rejuvenated [rɪˈdʒuːvəneɪtɪd] 更生的,恢复活力的
quantum mechanic 量子力学
ligand field theory 配位场理论
nitrogenase[ˈnaɪtrədʒəneɪs] 固氮酶;定氮酶
crystallography[krɪstəˈlɒgrəfɪ] 结晶学;晶体学

Notes

(1) **Roger Bacon** 是英国早期的一位哲学家和科学家,方济各会修士,提倡经验主义,主张通过实验获得知识。

(2) **Becquerel**(1852~1908)法国物理学家,放射性的发现者,标志着原子核物理学的开始,获得 1903 年的诺贝尔物理学奖。

(3) **Bohr**(1885~1962) 丹麦物理学家,在卢瑟福模型的基础上,提出了电子在核外的量子化轨道,描绘出了完整而令人信服的原子结构学说,成功地说明了原子的稳定性和氢原子光谱线规律。

(4) **Coordination chemistry** 配位化学,是研究配位化合物的结构、成键、反应、分类和制备的学科。配位化学的理论包括配位化合物的价键理论、晶体场理论及配位场理论。1893 年瑞士化学家 A.韦尔纳首先提出这类化合物的正确化学式及配位理论,于 1913 年获得诺贝尔化学奖。

(5) **Haber-Bosch process** 哈伯-博施法，它是在20世纪初发展出来，由大气中氮制氨的化学方法。是化学方法方面最重要的发明之一，因为它使大气中氮的固定成为可能，从而还能由将转化为硝酸来生产肥料(和炸药)所需的硝酸盐。哈伯(F.Haber)在理论的实验上证明，如何维持来自空气的氮和来自水中的氢在适当的温度和压力，并在有催化剂的情况下反应。博施(C.Bosch)还证明如何在工业规模上实现这种方法。

1.2 The Structure of an Atom and Distribution of Electrons

1.2.1 The Structure of an atom

An atom consists of a tiny dense nucleus surrounded by electrons that are spread throughout a relatively large volume of space around the nucleus. The nucleus contains positively charged protons and neutral neutrons, so it is positively charged. The electrons are negatively charged. Because the amount of positive charge on a proton equals the amount of negative charge on an electron, a neutral atom has an equal number of protons and electrons. Atoms can gain electrons and thereby become negatively charged, or they can lose electrons and become positively charged. However, the number of protons in an atom does not change.

Protons and neutrons have approximately the same mass and are about 1800 times more massive than an electron. This means that most of the mass of an atom is in its nucleus. However, most of the volume of an atom is occupied by its electrons, and that is where our focus will be because it is the electrons that form chemical bonds.

The atomic number of an atom equals the number of protons in its nucleus. The atomic number is also the number of electrons that surround the nucleus of a neutral atom. For example, the atomic number of carbon is 6, which means that a neutral carbon atom has six protons and six electrons. Because the number of protons in an atom does not change, the atomic number of a particular element is always the same—all carbon atoms have an atomic number of 6.

The mass number of an atom is the sum of its protons and neutrons. Not all carbon atoms have the same mass number, even though they all have the same number of protons, they do not all have the same number of neutrons. For example, 98.89% of naturally occurring carbon atoms have six neutrons—giving them a mass number of 12—and 1.11% have seven neutrons—giving them a mass number of 13. These two different kinds of carbon atoms (^{12}C and ^{13}C) are called isotopes. Isotopes have the same atomic number (i.e., the same number of protons), but different mass numbers because they have different numbers of neutrons. The chemical properties of isotopes of a given element are nearly identical.

Naturally occurring carbon also contains a trace amount of ^{14}C, which has six protons and eight neutrons. This isotope of carbon is radioactive, decaying with a half-life of 5730 years. (The half-life is the time it takes for one-half of the nuclei to decay.) As long as a plant or animal is alive, it takes in as much as ^{14}C it excretes or exhales. When it dies, it no longer takes in ^{14}C, so the ^{14}C in the organism slowly decreases. Therefore, the age of an organic substance can be determined by its ^{14}C content.

The atomic weight of a naturally occurring element is the average weighted mass of its atoms. Because an atomic mass unit (amu) is defined as exactly 1/12 of the mass of ^{12}C, the atomic mass of ^{12}C is 12.0000 amu; the atomic mass of ^{13}C is 13.0034 amu. Therefore, the atomic weight of carbon is 12.011 amu ($0.9889 \times 12 + 0.01111 \times 13.0034 = 12.011$). The molecular weight is the sum of the atomic weights of all the atoms in the molecule.

1.2.2 The Distribution of electrons in an atom

Electrons are moving continuously. Like anything that moves, electrons have kinetic energy, and this energy is what counters the attractive force of the positively charged protons that would otherwise pull the negatively charged electrons into the nucleus. For a long time, electrons were perceived to be particles—infinitesimal "planets" orbiting the nucleus of an atom. In 1924, however, a French physicist named Louis de Broglie showed that electrons also have wavelike properties. He did this by combining a formula developed by Einstein that relates mass and energy with a formula developed by Planck relating frequency and energy. The realization that electrons have wavelike properties spurred physicists to propose a mathematical concept known as quantum mechanics.

Quantum mechanics uses the same mathematical equations that describe the wave motion of a guitar string to characterize the motion of an electron around a nucleus. The version of quantum mechanics most useful to chemists was proposed by Erwin Schrödinger in 1926. According to Schrödinger, the behavior of each electron in an atom or a molecule can be described by a wave equation. The solutions to the Schrödinger equation are called wave functions or orbitals. They tell us the energy of the electron and the volume of space around the nucleus where an electron is most likely to be found.

According to quantum mechanics, the electrons in an atom can be thought of as occupying a set of concentric shells that surround the nucleus. The first shell is the one closest to the nucleus. The second shell lies farther from the nucleus, and even farther out lie the third and higher numbered shells. Each shell contains subshells known as atomic orbitals. Each atomic orbital has a characteristic shape and energy and occupies a characteristic volume of space, which is predicted by the Schrödinger equation. An important point to remember is that the closer the atomic orbital is to the nucleus, the lower is its energy.

The first shell consists of only an s atomic orbital; the second shell consists of s and p atomic orbitals; the third shell consists of s, p, and d atomic orbitals; and the fourth and higher shells consist of *s*, *p*, *d*, and *f* atomic orbitals (Table 1.1).

Table 1.1 Distribution of electrons in the first four shells that surround the nucleus

	First shell	Second shell	Third shell	Fourth shell
Atomic orbitals	*s*	*s*, *p*	*s*, *p*, *d*	*s*, *p*, *d*, *f*
Number of atomic orbitals	1	1, 3	1, 3, 5	1, 3, 5, 7
Maximum number of electrons	2	8	18	32

Each shell contains one s atomic orbital. The second and higher shells—in addition to their *s* orbital—each contain three degenerate *p* atomic orbitals. Degenerate orbitals are orbitals that have the same energy. The third and higher shells—in addition to their s and *p* orbitals—also contain five degenerate *d* atomic orbitals, and the fourth and higher shells also contain seven degenerate *f* atomic orbitals. Because a maximum of two electrons can coexist in an atomic orbital (see the Pauli exclusion principle, below), the first shell, with only one atomic orbital, can contain no more than two electrons. The second shell, with four atomic orbitals—one s and three p—can have a total of eight electrons. Eighteen electrons can occupy the nine atomic orbitals—one *s*, three *p*, and five *d* —of the third shell, and 32 electrons can occupy the 16 atomic orbitals of the fourth shell. In studying organic chemistry, we will be concerned primarily with atoms that have electrons only in the first and second shells.

The ground-state electronic configuration of an atom describes the orbitals occupied by the atom's electrons when they are all in the available orbitals with the lowest energy. If energy is applied to an atom in the ground state, one or more electrons can jump into a higher energy orbital. The atom then would be in an excited-state electronic configuration. The following principles are used to determine which orbitals electrons occupy:

1. The aufbau principle (aufbau is German for "building up") tells us the first thing we need to know to be able to assign electrons to the various atomic orbitals. According to this principle, an electron always goes into the available orbital with the lowest energy. The relative energies of the atomic orbitals are as follows:

$$1s<2s<2p<3s<3p<4s<3d<4p<5s<4d<5p<6s<4f<5d<6p<7s<5f$$

Because a 1*s* atomic orbital is closer to the nucleus, it is lower in energy than a 2*s* atomic orbital, which is lower in energy—and is closer to the nucleus—than a 3*s* atomic orbital. Comparing atomic orbitals in the same shell, we see that an s atomic orbital is lower in energy than a p atomic orbital, and a *p* atomic orbital is lower in energy than a *d* atomic orbital.

2. The Pauli exclusion principle states that (a) no more than two electrons can occupy each atomic orbital, and (b) the two electrons must be of opposite spin. It is called an exclusion principle because it states that only so many electrons can occupy any particular shell.

From these first two rules, we can assign electrons to atomic orbitals for atoms that contain one, two, three, four, or five electrons. The single electron of a hydrogen atom occupies a 1*s* atomic orbital, the second electron of a helium atom fills the 1*s* atomic orbital, the third electron of a lithium atom occupies a 2*s* atomic orbital, the fourth electron of a beryllium atom fills the 2*s* atomic orbital, and the fifth electron of a boron atom occupies one of the 2*p* atomic orbitals. (The subscripts *x*, *y*, and *z* distinguish the three 2*p* atomic orbitals.) Because the three p orbitals are degenerate, the electron can be put into any one of them. Before we can continue to larger atoms—those containing six or more electrons—we need Hund's rule:

3. Hund's rule states that when there are degenerate orbitals—two or more orbitals with the same energy—an electron will occupy an empty orbital before it will pair up with another electron. In this way, electron repulsion is minimized. The sixth electron of a carbon atom, therefore, goes into an empty 2*p* atomic orbital, rather than pairing up with the electron already occupying a 2*p* atomic orbital. The seventh electron of a nitrogen atom goes into an empty 2*p* atomic orbital, and the eighth elec-

tron of an oxygen atom pairs up with an electron occupying a 2*p* atomic orbital rather than going into a higher energy 3*s* orbital.

From *organic chemistry* by *Paula Yurkanis Bruice*

New Words and Expressions

proton [ˈprəʊˌtɔn] [物] 质子
nucleus [ˈnjuːklɪəs] 核,核心;原子
neutron [ˈnjuːtrɔn] [物] 中子
infinitesimal [ɪnfɪnɪˈtesɪml] 无穷小的;无限小的;极小的
quantum mechanics 量子力学
isotope [ˈaɪsətəʊp] [化] 同位素
wavefunction 波(动)函数
ground state 基态
excited state 受激状态;激发态
lithium [ˈlɪθiəm] (化)锂
electronic configuration [物化] 电子构型;电子排布
beryllium [bəˈrɪliəm] 铍(元素符号 Be)
degenerate orbital 简并轨道
subscript [ˈsʌbskrɪpt] 下标,脚注

Notes

(1) **The aufbau principle** 构造原理(aufbau 是德文"构造"),是指随核电荷数递增,大多数元素的电中性基态原子的电子按顺序填入核外电子运动轨道。

(2) **Pauli exclusion principle** 泡利不相容原理,是微观粒子运动的基本规律之一。它指出:在费米子组成的系统中,不能有两个或两个以上的粒子处于完全相同的状态。在原子中完全确定一个电子的状态需要四个量子数,所以泡利不相容原理在原子中就表现为:不能有两个或两个以上的电子具有完全相同的四个量子数,这成为电子在核外排布形成周期性从而解释元素周期表的准则之一。

(3) **Hund's rule** 洪特规则,是指在能量相等的轨道上,自旋平行的电子数目最多时,原子的能量最低。所以在能量相等的轨道上,电子尽可能自旋平行地多占不同的轨道。例如碳原子核外有 6 个电子,按能量最低原理和泡利不相容原理,首先有 2 个电子排布到第一层的 1s 轨道中,另外 2 个电子填入第二层的 2s 轨道中,剩余 2 个电子排布在 2 个 p 轨道上,具有相同的自旋方向,而不是两个电子集中在一个 p 轨道,自旋方向相反。作为洪特规则的补充,能量相等的轨道全充满、半满或全空的状态比较稳定。

1.3 Ionic, Covalent, and Polar Bonds

In trying to explain why atoms form bonds, G.N.Lewis proposed that an atom is most stable if its outer shell is either filled or contains eight electrons and it has no electrons of higher energy. According to Lewis's theory, an atom will give up, accept, or share electrons in order to achieve a filled outer shell or an outer shell that contains eight electrons. This theory has come to be called the octet rule.

Lithium (Li) has a single electron in its 2*s* atomic orbital. If it loses this electron, the lithium atom ends up with a filled outer shell—a stable configuration. Removing an electron from an atom takes energy—called the ionization energy. Lithium has relatively low ionization energy—the drive to achieve a filled outer shell with no electrons of higher energy causes it to lose an electron relatively easily. Sodium (Na) has a single electron in its 3*s* atomic orbital. Consequently, sodium also has relatively low ionization energy, when it loses an electron, it is left with an outer shell of eight electrons. Elements (such as lithium and sodium) that have low ionization energies are said to be electropositive—they readily lose an electron and thereby become positively charged. The elements in the first column of the periodic table are all electropositive—each readily loses an electron because each has a single electron in its outermost shell.

Electrons in inner shells (those below the outermost shell) are called core electrons. Core electrons do not participate in chemical bonding. Electrons in the outermost shell are called valence electrons, and the outermost shell is called the valence shell. Carbon, for example, has two core electrons and four valence electrons.

Lithium and sodium each have one valence electron. Elements in the same column of the periodic table have the same number of valence electrons, and because the number of valence electrons is the major factor determining an element's chemical properties, elements in the same column of the periodic table have similar chemical properties. Thus, the chemical behavior of an element depends on its electronic configuration.

When we draw the electrons around an atom, as in the following equations, core electrons are not shown; only valence electrons are shown. Each valence electron is shown as a dot. Notice that when the single valence electron of lithium or sodium is removed, the resulting atom—now called an ion—carries a positive charge.

Fluorine has seven valence electrons. Consequently, it readily acquires an electron in order to have an outer shell of eight electrons. When an atom acquires an electron, energy is released. Elements in the same column as fluorine (e.g., chlorine, bromine, and iodine) also need only one electron to have an outer shell of eight, so they, too, readily acquire an electron. Elements that readily acquire an electron are said to be electronegative—they acquire an electron easily and thereby become negatively charged.

1.3.1 Ionic Bonds

Because sodium gives up an electron easily and chlorine acquires an electron readily, when sodium metal and chlorine gas are mixed, each sodium atom transfers an electron to a chlorine atom, and crystalline sodium chloride (table salt) is formed as a result. The positively charged sodium ions and negatively charged chloride ions are independent species held together by the attraction of opposite charges. A bond is an attractive force between two atoms. Attractive forces between opposite charges are called electrostatic attractions. A bond that is the result of only electrostatic attractions is called an ionic bond. Thus, an ionic bond is formed when there is a transfer of electrons, causing one atom to become a positively charged ion and the other to become a negatively charged ion.

Sodium chloride is an example of an ionic compound. Ionic compounds are formed when an element on the left side of the periodic table (an electropositive element) transfers one or more electrons to an element on the right side of the periodic table (an electronegative element).

1.3.2 Covalent Bonds

Instead of giving up or acquiring electrons, an atom can achieve a filled outer shell by sharing electrons. For example, two fluorine atoms can each attain a filled shell of eight electrons by sharing their unpaired valence electrons. A bond formed as a result of sharing electrons is called a covalent bond.

Two hydrogen atoms can form a covalent bond by sharing electrons. As a result of covalent bonding, each hydrogen acquires a stable, filled outer shell (with two electrons).

Similarly, hydrogen and chlorine can form a covalent bond by sharing electrons. In doing so, hydrogen fills its only shell and chlorine achieves an outer shell of eight electrons.

A hydrogen atom can achieve a completely empty shell by losing an electron. Loss of its sole electron results in a positively charged hydrogen ion. A positively charged hydrogen ion is called a proton because when a hydrogen atom loses its valence electron, only the hydrogen nucleus—which consists of a single proton—remains. A hydrogen atom can achieve a filled outer shell by gaining an electron, thereby forming a negatively charged hydrogen ion, called a hydride ion.

$$H\cdot \longrightarrow H^{+} + e^{-}$$

a hydrogen atom   a proton

$$H\cdot + e^{-} \longrightarrow H:^{-}$$

a hydrogen atom   a hydride ion

Because oxygen has six valence electrons, it needs to form two covalent bonds to achieve an outer shell of eight electrons. Nitrogen, with five valence electrons, must form three covalent bonds, and carbon, with four valence electrons, must form four covalent bonds to achieve a filled outer shell. Notice that all the atoms in water, ammonia, and methane have filled outer shells.

1.3.3 Polar Covalent Bonds

In the F—F and H—H covalent bonds shown previously, the atoms that share the bonding electrons are identical. Therefore, they share the electrons equally; that is, each electron spends as much time in the vicinity of one atom as in the other. An even (nonpolar) distribution of charge results. Such a bond is called a nonpolar covalent bond.

In contrast, the bonding electrons in hydrogen chloride, water, and ammonia are more attracted to one atom than another because the atoms that share the electrons in these molecules are different and have different electronegativities. Electronegativity is the tendency of an atom to pull bonding electrons toward itself. The bonding electrons in hydrogen chloride, water, and ammonia molecules are more attracted to the atom with the greater electronegativity. This results in a polar distribution of charge. A polar covalent bond is a covalent bond between atoms of different electronegativities. Notice that electronegativity increases as you go from left to right across a row of the periodic table or up any of the columns.

A polar covalent bond has a slight positive charge on one end and a slight negative charge on the other. Polarity in a covalent bond is indicated by the symbols δ^+ and δ^-, which denote partial positive and partial negative charges, respectively. The negative end of the bond is the end that has the more electronegative atom. The greater the difference in electronegativity between the bonded atoms, the more polar the bond will be.

The direction of bond polarity can be indicated with an arrow. By convention, the arrow points in the direction in which the electrons are pulled, so the head of the arrow is at the negative end of the bond; a short perpendicular line near the tail of the arrow marks the positive end of the bond.

You can think of ionic bonds and nonpolar covalent bonds as being at the opposite ends of a continuum of bond types. An ionic bond involves no sharing of electrons. A nonpolar covalent bond involves equal sharing. Polar covalent bonds fall somewhere in between, and the greater the difference in electronegativity between the atoms forming the bond, the closer the bond is to the ionic end of the continuum. C-H Bonds are relatively nonpolar, because carbon and hydrogen have similar electronegativities (electronegativity difference = 0.4). N—H Bonds are relatively polar (electronegativity difference = 0.9), but not as polar as O—H bonds (electronegativity difference = 1.4). The bond between sodium and chloride ions is closer to the ionic end of the continuum (electronegativity difference = 2.1), but sodium chloride is not as ionic as potassium fluoride (electronegativity difference = 3.2).

Understanding bond polarity is critical to understanding how organic reactions occur, because a central rule that governs the reactivity of organic compounds is that electron-rich atoms or molecules are attracted to electron-deficient atoms or molecules. Electrostatic potential maps (often simply called potential maps) are models that show how charge is distributed in the molecule under the map. Therefore, these maps show the kind of electrostatic attraction an atom or molecule has for another atom or molecule, so you can use them to predict chemical reactions.

A polar bond has a dipole—it has a negative end and a positive end. The size of the dipole is indicated by the dipole moment, which is given the Greek letter μ. The dipole moment of a bond is equal to the magnitude of the charge (e) on the atom (either the partial positive charge or the partial negative charge, because they have the same magnitude) times the distance between the two charges (d).

A dipole moment is reported in a unit called a debye (D) (pronounced de-bye). Because the charge on an electron is 4.80×10^{-10} electrostatic units (esu) and the distance between charges in a polar bond is on the order of 10^{-8} cm, the product of charge and distance is on the order of 10^{-18} esu cm. A dipole moment of 1.5×10^{-18} esu cm can be more simply stated as 1.5 D.

In a molecule with only one covalent bond, the dipole moment of the molecule is identical to the dipole moment of the bond. For example, the dipole moment of hydrogen chloride (HCl) is 1.1 D because the dipole moment of the single bond H—Cl is 1.1 D. The dipole moment of a molecule with more than one covalent bond depends on the dipole moments of all the bonds in the molecule and the geometry of the molecule.

From *organic chemistry* by *Paula Yurkanis Bruice*

New Words and Expressions

lithium [ˈlɪθiəm] 锂(符号 Li)
core electron 核心电子
electrostatic attraction 静电引力
fluorine [ˈfluəriːn] [化] 氟
iodine [ˈaɪədiːn] [化]碘
outer shell 外层,外电子层
hydride ion 氢阴离子
electronegativity [elektrəʊnɪɡəˈtɪvɪtɪ] 阴电性,负电性,电负性
dipole [ˈdaipəul] 偶极子
by convention 根据惯例,按约定,依照通例
perpendicular line 垂线;[数] 垂直线;正交线
continuum [kənˈtɪnjʊəm] [数] 连续统;[经] 连续统一体
dipole moment 偶极矩
valence electron [物化] 价电子

Notes

(1) **G.N.Lewis** 路易斯(1875~1946)美国化学家,主要成就有三方面:创立电子对理论;提出了新的酸、碱理论和开辟了热力学非理想体系的研究。

(2) **Octet rule** 八隅体规则(或称八电子规则)是化学中一个简单的规则,即原子间的组合趋向令各电子的价层都拥有八个电子,与惰性气体拥有相同的电子排列。主族元素,如碳、氮、氧、卤素族、钠、镁都依从这个规则。简单而言,当组成离子或分子的组成原子的最外电子层有八个电子,它们便会趋向稳定,而若不满 8 个时,原子间会互相共享或交换电子达到平衡稳定。例如 Cl 与 Na 形成 NaCl 的结构。

(3) **Debye** 德拜(1884-1966),荷兰物理学家与物理化学家,由于在 X 射线衍射和分子偶极矩理论方面的杰出贡献,德拜获得了 1936 年诺贝尔化学奖。

1.4 Hybridization Theory

It is sometimes convenient to label the atomic orbitals that combine to form molecular orbitals as hybrid orbitals, or hybrids. In this method, the orbitals of the central atom are combined into equivalent hybrids. These hybrid orbitals are then used to form bonds with other atoms whose orbitals overlap properly. This approach is not essential in describing bonding, but was developed as part of the valence bond approach to bonding to describe equivalent bonds in a molecule.

1.4.1 SP^3 Hybridizaton

Definition

In sp^3 hybridization, the 2s orbital is mixed with all three of the 2p orbitals to give a set of four sp^3 hybrid orbitals. (The number of hybrid orbitals must equal the number of original atomic orbitals used for mixing.) The hybrid orbitals will each have the same energy but will be different in energy from the original atomic orbitals. That energy difference will reflect the mixing of the respective atomic orbitals. The energy of each hybrid orbital is greater than the original s orbital but less than the original p orbitals (Fig.1.1).

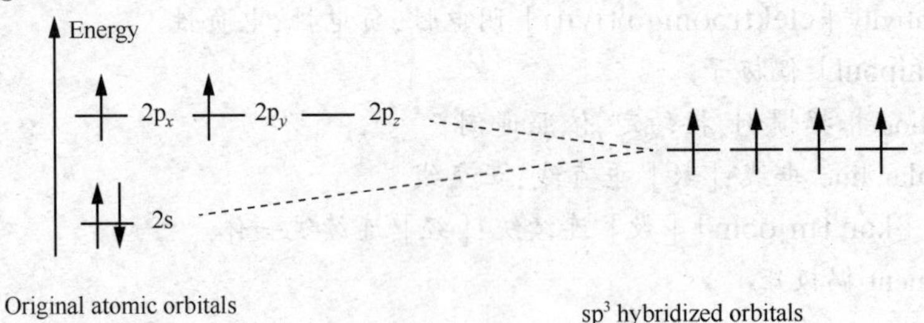

Fig.1.1 sp^3 Hybridization

Electronic Configuration

The valence electrons for carbon can nowbe fitted into the sp^3 hybridized orbitals. There were a total of four electrons in the original 2s and 2p orbitals. The s orbital was filled and two of the p orbitals were half filled. After hybridization, there is a total of four hybridized sp^3 orbitals all of equal energy. By Hund's rule, they are all half filled with electrons which means that there are four unpaired electrons. Four bonds are now possible.

Geometry

Each of the sp^3 hybridized orbitals has the same shape—a rather deformed looking dumbbell (Fig.1.2). This deformed dumbbell looks more like a p orbital than an s orbital since more p orbitals were involved in the mixing process.

Each sp^3 orbital will occupy a space as far apart from each other as possible by pointing to the corners of a tetrahedron (Fig.1.3). Here, only the major lobe of each hybridized orbital has been shown and the angle between each of these lobes is 109.5°. This is what is meant

Minor lobe Major lobe

Fig.1.2 sp^3 Hybridized orbital

by the expression tetrahedral carbon. The three-dimensional shape of the tetrahedral carbon can be represented by drawing a normal line for bonds in the plane of the page. Bonds going behind the page are represented by a hatched wedge, and bonds coming out the page are represented by a solid wedge.

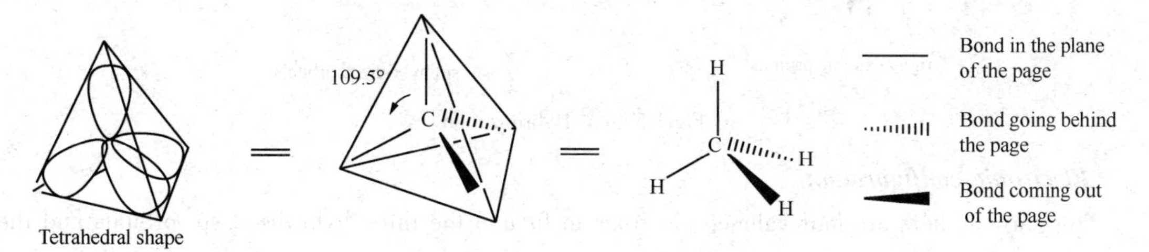

Fig.1.3 Tetrahedral shape of an sp^3 hybridized carbon

Sigma Bonds

A half-filled sp^3 hybridized orbital from one carbon atom can be used to form a bond with a half-filled sp^3 hybridized orbital from another carbon atom. In Fig.1.4a, the major lobes of the two sp^3 orbitals overlap directly leading to a strong σ bond. It is the ability of hybridized orbitals to form strong σ bonds that explains why hybridization takes place in the first place. The deformed dumbbell shapes allow a much better orbital overlap than would be obtained from a pure s orbital or a pure p orbital. A σ bond between an sp^3 hybridized carbon atom and a hydrogen atom involves the carbon atom using one of its half-filled sp^3 orbitals and the hydrogen atom using its half-filled 1s orbital (Fig.1.4b).

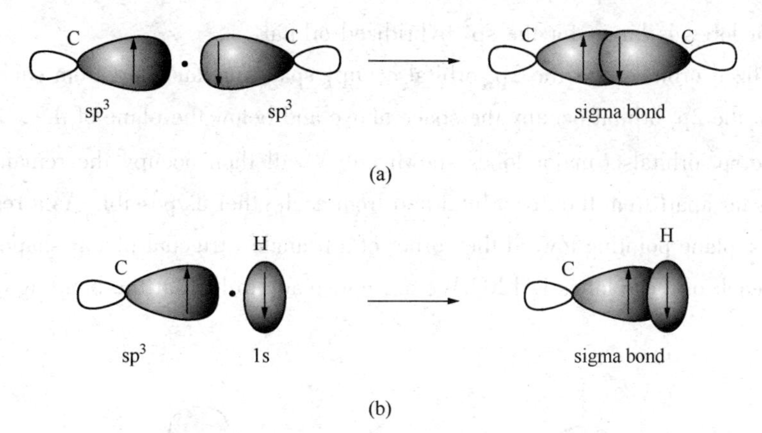

Fig.1.4 (a) σ Bond between two sp^3 hybridized carbons; (b) σ Bond between an sp^3 hybridized carbon and hydrogen

1.4.2 SP² Hybridization

Definition

In sp^2 hybridization, the s orbital is mixed with two of the 2p orbitals (e.g. $2p_x$ and $2p_z$) to give three sp^2 hybridized orbitals of equal energy. The remaining $2p_y$ orbital is unaffected. The energy of each hybridized orbital is greater than the original s orbital but less than the original p orbitals. The remaining 2p orbital (in this case the $2p_y$ orbital) remains at its original energy level (Fig.1.5).

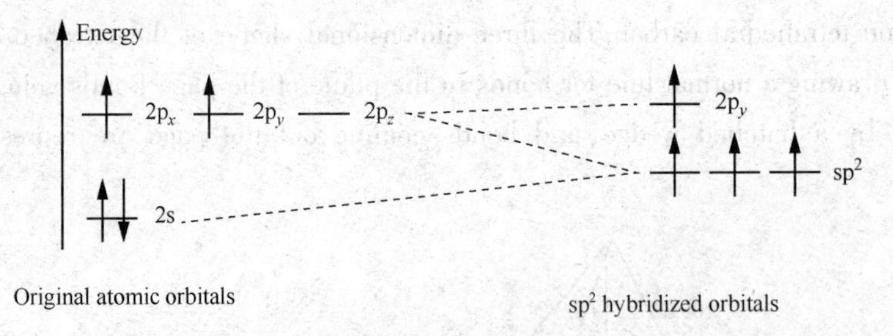

Fig.1.5 sp^2 Hybridization

Electronic configuration

For carbon, there are four valence electrons to fit into the three hybridized sp^2 orbitals and the remaining 2p orbital. The first three electrons are fitted into each of the hybridized orbitals according to Hund's rule such that they are all half-filled. This leaves one electron still to place. There is a choice between pairing it up in a half-filled sp^2 orbital or placing it into the vacant 2p$_y$ orbital. The usual principle is to fill up orbitals of equal energy before moving to an orbital of higher energy. However, if the energy difference between orbitals is small (as here) it is easier for the electron to fit into the higher energy 2p$_y$ orbital resulting in three half-filled sp^2 orbitals and one half-filled p orbital. Four bonds are possible.

Geometry

The 2p$_y$ orbital has the usual dumbbell shape. Each of the sp^2 hybridized orbitals has a deformed dumbbell shape similar to a sp^3 hybridized orbital. However, the difference between the sizes of the major and minor lobes is larger for the sp^2 hybridized orbital.

The hybridized orbitals and the 2p$_y$ orbital occupy spaces as far apart from each other as possible. The lobes of the 2p$_y$ orbital occupy the space above and below the plane of the x and z axes (Fig. 1.6a). The three sp^2 orbitals (major lobes shown only) will then occupy the remaining space such that they are as far apart from the 2p$_y$ orbital and from each other as possible. As a result, they are all placed in the x-z plane pointing toward the corner of a triangle (trigonal planar shape; Fig. 1.6b). The angle between each of these lobes is 120°. We are now ready to look at the bonding of an sp^2 hybridized carbon.

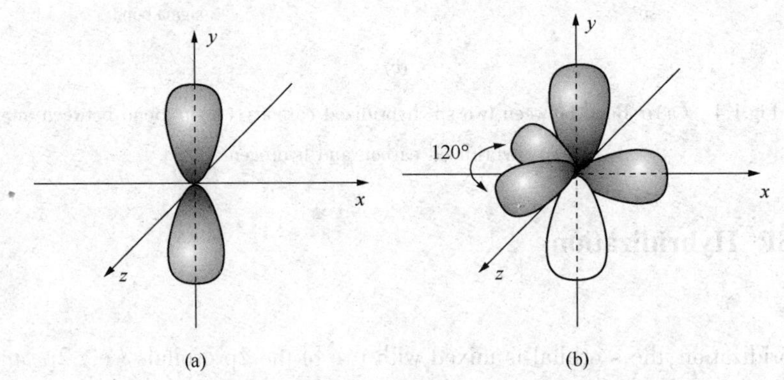

Fig.1.6 (a) Geometry of the 2p$_y$ orbital; (b) Geometry of the 2p$_y$ orbital and the sp^2 hybridized orbitals

1.4.3 sp Hybridization

Definition

In sp hybridization, the 2s orbital is mixed with one of the 2p orbitals (e.g.$2p_x$) to give two sp hybrid orbitals of equal energy. This leaves two 2p orbitals unaffected ($2p_y$ and $2p_z$) with slightly higher energy than the hybridized orbitals (Fig.1.7).

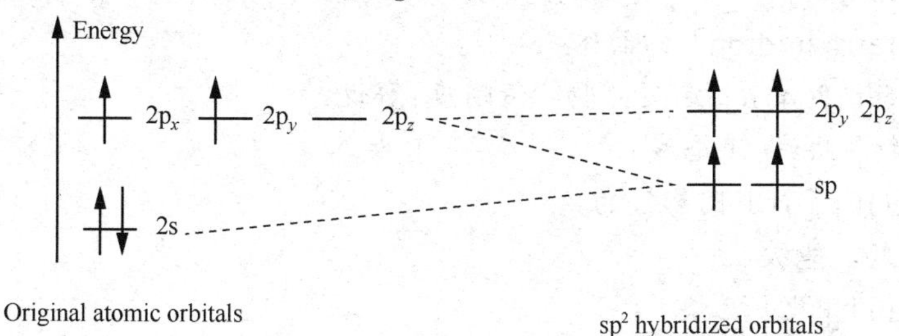

Fig.1.7 sp Hybridization

Electronic Configuration

For carbon, the first two electrons fit into each *sp* orbital according to Hund's rule such that each orbital has a single unpaired electron. This leaves two electrons which can be paired up in the half-filled sp orbitals or placed in the vacant $2p_y$ and $2p_z$ orbitals. The energy difference between the orbitals is small and so it is easier for the electrons to fit into the higher energy orbitals than to pair up. This leads to two half-filled sp orbitals and two half-filled 2p orbitals, and so four bonds are possible.

Geometry

The 2p orbitals are dumbbell in shape while the sp hybridized orbitals are deformed dumbbells with one lobe much larger than the other. The $2p_y$ and $2p_z$ orbitals are at right angles to each other (Fig.1.8a). The sp hybridized orbitals occupy the space left over and are in the *x* axis pointing in opposite directions (only the major lobe of the sp orbitals are shown in black; Fig.1.8b).

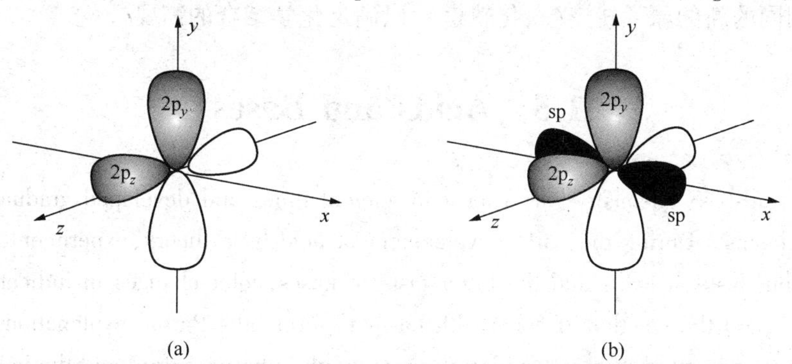

Fig.1.8 (a) $2p_y$ and $2p_z$ orbitals of an sp hybridized carbon;
(b) $2p_y$, $2p_z$ and sp hybridized orbitals of an sp hybridized carbon

A molecule using the two sp orbitals for bonding will be linear in shape. There are two common functional groups where such bonding takes place-alkynes and nitriles.

From *organic chemistry* by *G.Patrick*

New Words and Expressions

hybrid orbital 杂化轨道

dumbbell [ˈdʌmbel] 哑铃;<美俚>蠢人;笨蛋;傻瓜

tetrahedron [tetrəˈhiːdrən] 四面体

definition[defɪˈnɪʃ(ə)n] 定义;[物] 清晰度;解说

normal line[数] 法线;正垂线

hatched [ˈhætʃt] [计] 阴影线的

energy level[物] 能级

ethyne [ˈiːθaɪn] [化] 乙炔

nitrile [ˈnaɪtrɪl] [有化] 腈,腈类

trigonal [ˈtrig(ə)n(ə)l] 三角形的;三角的

Notes

Hybridization theory 杂化轨道理论。价键理论简明地阐明了共价键的形成过程和本质,成功解释了共价键的方向性和饱和性,但在解释一些分子的空间结构方面却遇到了困难。1931 年鲍林提出了杂化轨道理论,获 1954 年诺贝尔化学奖,丰富和发展了现代价键理论。杂化轨道理论是从电子具有波动性、波可以叠加的观点出发,认为一个原子和其他原子形成分子时,中心原子所用的原子轨道(波函数)不是原来纯粹的 s 轨道或 p 轨道,而是若干不同类型、能量相近的原子轨道经叠加混杂、重新分配轨道的能量和调整空间伸展方向,组成了同等数目的能量完全相同的新的原子轨道(杂化轨道),以满足化学结合的需要。

1.5 Acids and Bases

Practical acid-base chemistry was known in ancient times and developed gradually during the time of the alchemists. During the early development of acid-base theory, experimental observations included the sour taste of acids and the bitter taste of bases, color changes in indicators caused by acids and bases, and the reaction of acids with bases to form salts. Partial explanations included the idea that all acids contained oxygen (oxides of nitrogen, phosphorus, sulfur, and the halogens all form acids in water), but by the early 19th century, many acids that do not contain oxygen were known. By 1838, Liebig defined acids as "compounds containing hydrogen, in which the hydrogen can be replaced by a metal," a definition that still works well in many instances.

Although many other acid-base definitions have been proposed and have been useful in particular types of reactions, only a few have been widely adopted for general use. Among these are the ones attributed to Arrhenius (based on hydrogen and hydroxide ion formation), Bronsted-Lowry (hydrogen ion donors and acceptors), and Lewis (electron pair donors and acceptors). Others have received less attention or are useful only in a narrow range of situations.

1.5.1 Arrhenius Theory

An early attempt to provide a framework to observations on the chemistry of substances that react in water to produce acids or bases was provided by S.A. Arrhenius. At that time, the approach was limited to aqueous solutions, and the definitions of an acid and a base were given in these terms. Of course we now know that acid-base behavior is not limited to these cases, but it applies much more broadly. If we consider the reaction between gaseous HCl and water,

$$HCl(g) + H_2O(l) \longrightarrow H_3O^+(aq) + Cl^-(aq)$$

We see that the solution contains H_3O^+, the hydronium ion or, as it is perhaps more generally known, the oxonium ion. In aqueous solution, HNO_3 also ionizes, as illustrated in the reaction

$$HNO_3(aq) + H_2O(l) \longrightarrow H_3O^+(aq) + NO_3^-(aq)$$

In studying the properties of solutions of substances such as HCl and HNO_3, Arrhenius was led to the idea that the acidic properties of the compounds were due to the presence of an ion that we now write as H_3O^+ in the solutions. He therefore proposed that an acid is a substance whose water solution contains H_3O^+. The properties of aqueous solutions of acids are the properties of the H_3O^+ ion, a solvated proton (hydrogen ion) that is known as the hydronium ion in much of the older chemical literature but also referred to as the oxonium ion.

1.5.2 Brønsted-Lowry Theory

J.N. Brønsted and T.M. Lowry independently arrived at definitions of an acid and a base that do not involve water. They recognized that the essential characteristic of an acid-base reaction was the transfer of a hydrogen ion (proton) from one species (the acid) to another (the base). According to these definitions, an acid is a proton donor and a base is a proton acceptor. The proton must be donated to some other species so there is no acid without a base. According to Arrhenius, HCl is an acid because its water solution contains H_3O^+, which indicates that an acid can exist independently with no base being present. When examined according to the Brønsted-Lowry theory,

$$HCl(g) + H_2O(l) \longrightarrow H_3O^+(aq) + Cl^-(aq)$$

the above reaction is an acid-base reaction not because the solution contains H_3O^+ but rather because a proton is transferred from HCl (the acid) to H_2O (the base).

Once a substance has functioned as a proton donor, it has the potential to accept a proton (react as a base) from another proton donor. For example, acetate ions are produced by the reaction

$$HC_2H_3O_2(aq) + NH_3 \longrightarrow NH_4^+(aq) + C_2H_3O_2^-(aq)$$

The acetate ion can now function as a proton acceptor from a suitable acid. For example,

$$HNO_3(aq) + C_2H_3O_2^-(aq) \longrightarrow HC_2H_3O_2(aq) + NO_3^-(aq)$$

In this reaction, the acetate ion is functioning as a base. On the other hand, Cl^- has very little tendency to function as a base because it comes from HCl, which is a very strong proton donor. According to the Brønsted-Lowry theory, the species remaining after a proton is donated is called the conjugate base of that proton donor.

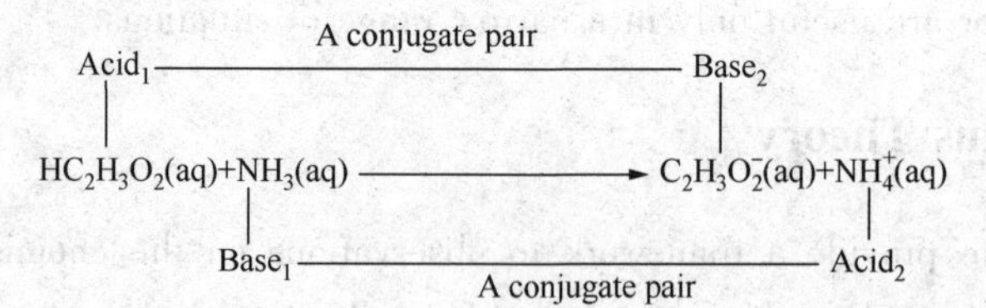

In this reaction, acetic acid donates a proton to produce its conjugate, the acetate ion, which is able to function as a proton acceptor. Ammonia accepts a proton to produce its conjugate, the ammonium ion, which can function as a proton donor. Two species that differ by the transfer of a proton are known as a conjugate pair. The conjugate acid of H_2O is H_3O^+, and the conjugate base of H_2O is OH^-.

Characteristics of the reactions described so far lead to several conclusions regarding acids and bases according to the Brønsted-Lowry theory.

1. There is no acid without a base. The proton must be donated to something else.

2. The stronger an acid is, the weaker its conjugate will be as a base. The stronger a base is, the weaker its conjugate will be as an acid.

3. A stronger acid reacts to displace a weaker acid. A stronger base reacts to displace a weaker base.

4. The strongest acid that can exist in water is H_3O^+. If a stronger acid is placed in water, it will donate protons to water molecules to produce H_3O^+.

5. The strongest base that can exist in water is OH^-. If a stronger base is placed in water, it will accept protons from water to produce OH^-.

1.5.3 Lewis Theory

Up to this point, we have dealt with the subject of acid-base chemistry in terms of proton transfer. If we seek to learn what it is that makes NH_3 a base that can accept a proton, we find that it is because there is an unshared pair of electrons on the nitrogen atom where the proton can attach. Conversely, it is the fact that the hydrogen ion seeks a center of negative charge that makes it leave an acid such as HCl and attach to the ammonia molecule. In other words, it is the presence of an unshared pair of electrons on the base that results in proton transfer. Sometimes known as the electronic theory of acids and bases, this shows that the essential characteristics of acids and bases do not always depend on the transfer of a proton. This approach to acid-base chemistry was first developed by G.N.Lewis in the 1920s.

When the reaction

$$HCl(g)+NH_3(g)\longrightarrow NH_4Cl(s)$$

is considered from the standpoint of electrons, we find that the proton from HCl is attracted to a center of negative charge on the base. The unshared pair of electrons on the nitrogen atom in ammonia is just that type of center. When the proton attaches to the NH_3 molecule, both of the electrons used in the bond come from the nitrogen atom. Thus, the bond is a coordinate covalent bond (or simply a coordinate bond) that is formed as the result of the reaction between an acid and a base.

Lewis acid-base chemistry provides one of the most useful tools ever devised for systematizing an enormous number of chemical reactions. Because the behavior of a substance as an acid or a base has nothing to do with proton transfer, many other types of reactions can be considered as acid-base reactions. For example,

$$BCl_3(g) + :NH_3(g) \longrightarrow H_3N:BCl_3(s)$$

involves the BCl_3 molecule attaching to the unshared pair of electrons on the NH_3 molecule to form a coordinate bond. Therefore, this reaction is an acid-base reaction according to the Lewis theory. The product of such a reaction results from the addition of two complete molecules, and as a result, it is often referred to as an acid-base adduct or complex. The BCl_3 is an electron pair acceptor, a Lewis acid. The NH_3 is an electron pair donor, a Lewis base. According to these definitions, it is possible to predict what types of species will behave as Lewis acids and bases.

The following types of species are Lewis acids:

(1) Molecules that have fewer than eight electrons on the central atom (e.g., BCl_3, $AlCl_3$)

(2) Ions that have a positive charge (e.g., H^+, Fe^{3+}, Cr^{3+}).

(3) Molecules in which the central atom can add additional pairs of electrons even though it already has an octet or more of electrons (e.g., $SbCl_3$, PCl_5, SF_4).

Lewis bases include the following types of species:

(1) Anions that have an unshared pair of electrons (e.g., OH^-, H^-, F^-, PO_4^{3-}).

(2) Neutral molecules that have unshared pairs of electrons (e.g., NH_3, H_2O, R_3N, ROH, PH_3).

We can now write equations for many reactions that involve a Lewis acid reacting with a Lewis base. The following are a few examples:

$$SbF_5 + F^- \longrightarrow SbF_6^-$$

$$AlCl_3 + R_3N \longrightarrow R_3N:AlCl_3$$

$$H^+ + PH_3 \longrightarrow PH_4^+$$

$$Cr^{3+} + 6\ NH_3 \longrightarrow Cr(NH_3)_6^{3+}$$

Acid-base behavior according to the Lewis theory has many of the same aspects as does acid-base theory according to the Brønsted-Lowry theory.

(1) There is no acid without a base. An electron pair must be donated to one species (the acid) by another (the base).

(2) An acid (or base) reacts to displace a weaker acid (or base) from a compound.

(3) The interaction of a Lewis acid with a Lewis base is a type of neutralization reaction because the acidic and basic characters of the reactants are removed.

From in *organic chemistry* by *James E. House*

New Words and Expressions

alchemist [ˈælkimist] 炼金术士

halogen [ˈhælədʒən] 卤素,卤化物

terminology[ˌtɜːmɪˈnɒlədʒɪ] 专门名词;术语,术语学;用辞

hydroxide ion 氢氧离子,羟离子

hydronium ion 水合氢离子

oxonium ion[无化]水合氢离子

solvated proton 溶剂化质子

proton donor 质子给(予)体

proton acceptor 质子接受体

acetate ion 醋酸根离子

conjugate[ˈkɒndʒəgeɪt] 结合的;成对的;配合的;共轭的

coordinate covalent bond 配位共价键

adduct[ˈædʌkt] [化]加合物

Notes

(1) **Arrhenius Theory** 阿伦尼乌斯酸碱理论,酸、碱是一种电解质,它们在水溶液中会离解,能离解出氢离子的物质是酸;能离解出氢氧根离子的物质是碱。第一次从定量的角度来描写酸碱的性质和它们在化学反应中的行为。该理论只适用于水溶液中的情况,而不能解释水溶液中不含氢氧根的物质显碱性及非水溶液中不含氢离子和氢氧根离子的物质也会表现出酸性或碱性的现象(如乙醇钠在乙醇溶液中显强碱性)。

(2) **Brønsted-Lowry Theory** 布仑斯特-劳里质子理论, 丹麦化学家布朗斯特和英国化学家劳里于 1923 年分别提出酸碱质子理论,也称为布朗斯特-劳里酸碱理论。该理论认为,凡是能给出质子(H^+)的物质都是酸,凡能接受质子的物质都是碱,而既能给出质子,也能接受质子的物质称为两性物质。酸和碱不是孤立的,它们通过质子互相联系,用通式可以表示为:酸 → 碱+质子,这样的一对酸碱称为共轭酸碱对,其中的酸和碱分别称为相应物质的共轭酸及共轭碱。与阿伦尼乌斯酸碱理论不同的是,布朗斯特酸碱不仅限于电中性的分子,也包括带电的阴阳离子。而该理论之下的酸碱反应则是两对共轭酸碱对之间传递质子的反应,不一定生成盐和水。

(3) **Lewis Theory** 路易斯酸碱理论,由吉尔伯特·牛顿·路易斯在 1923 年提出,结合了布朗斯特-劳里和酸碱溶剂理论的特点,在水溶液和非水溶剂中都有很广的应用。该理论着重探讨电子的给予与获得,路易斯酸被定义为电子接受体,而路易斯碱则是电子给予体。

1.6 Laws of Thermodynamics

Thermodynamics involves work and heat. It began in the 19th century with the efforts of engineers to increase the efficiency of steam engines, but it has become the general theory of the macroscopic behavior of matter at equilibrium. It is based on empirical laws, as is classical mechanics. Although classical mechanics has been superseded by relativistic mechanics and quantum mechanics, thermodynamics is an unchallenged theory. No exceptions have been found to the laws of thermodynamics.

1.6.1 The first law of thermodynamics

Mechanical Work

The quantitative measurement of work was introduced by Carnot, who defined an amount of work done on an object as the height it is lifted times its weight. This definition was extended by Coriolis, who provided the presently used definition of work: The amount of work done on an object equals the force exerted on it times the distance it is moved in the direction of the force. If a force Fz is exerted on an object in the z direction, the work done on the object in an infinitesimal displacement dz in the z direction is

$$dw = Fzdz \text{ (definition of work)}$$

where dw is the quantity of work. The SI unit of force is the newton), and the SI unit of work is the joule.

Heat

Joseph Black was the first to distinguish between the quantity of heat and the "intensity" of heat (temperature) and to recognize latent heat absorbed or given off in phase transitions. However, Black believed in the caloric theory of heat, which incorrectly asserted that heat was an "imponderable" fluid called "caloric." This incorrect theory was not fully discredited until several decades after Black's death.

A small amount of heat added to a system is proportional to its change in temperature of the system if there is no phase change or chemical reaction:

$$dq = CdT$$

where dq is an infinitesimal amount of heat transferred to the object and dT is a resulting infinitesimal change in temperature. The proportionality constant C is called the heat capacity of the system.

Equation does not indicate that C is a derivative of q with respect to T. We will see that dq is an inexact differential so that the heat capacity C depends on the way in which the temperature of the system is changed. If the temperature is changed at constant pressure, the heat capacity is denoted by C_P and is called the heat capacity at constant pressure. If the temperature is changed at constant volume, the heat capacity is denoted by C_V and is called the heat capacity at constant volume. These two heat capacities are not generally equal to each other.

If an object is heated from temperature T_1 to temperature T_2 without any chemical reaction or phase change occurring, the quantity of heat transferred to the object is given by

$$q = \int_c dq = \int_{T_1}^{T_2} CdT$$

If the heat capacity is independent of temperature

$$q = C\int_{T_1}^{T_2} dT = C(T_2 - T_1) = C\Delta T \quad \text{(if } C \text{ is independent of } T\text{)}$$

Apositive value of q indicates heat transferred to the system and a negative value indicates heat transferred from the system to its surroundings.

The calorie was the first metric unit of heat and was originally defined as the amount of heat required to raise the temperature of 1 gram of liquid water by 1℃ at 15℃. The specific heat capacity of liquid water equals 1.00 cal · K^{-1} · g^{-1} at 15℃. The heat capacity of liquid water is nearly temperature-independent and we can use this value for any temperature with good accuracy. The calorie is now defined to equal 4.184 J (exactly) so that heat capacities can be expressed in joules as well as in calories.

Internal Energy: The First Law of Thermodynamics

Although Lavoisier discredited the phlogiston theory of combustion, which held that combustion was the loss of an "imponderable" fluid called phlogiston, he was one of the principal promoters of the equally incorrect caloric theory of heat espoused by Black, which asserted that heat was an imponderable fluid called "caloric." The first experimental studies that discredited the caloric theory were done by Count Rumford. Rumford was at one time in charge of manufacturing cannons for the Elector of Bavaria, the ruler who made him a count. Rumford noticed that when a cannon was bored with a dull boring tool, more heat was produced than when a sharp tool was used. He carried out a systematic set of experiments and was able to show by using a very dull tool that there was no apparent limit to the amount of heat that could be generated by friction. He immersed the cannon in water to transfer the heat from the cannon and even generated enough heat to have melted the cannon. Rumford's results showed that "caloric" was not simply being extracted from the cannon, because if caloric existed only a definite amount could be stored in a cannon without melting it. Work must have been converted to heat.

Rumford calculated an approximate value for the "mechanical equivalent of heat," or the amount of heat to which a joule of work could be converted. Better values were obtained by Mayer in 1842 and Joule in 1847. Joule carried out experiments in which changes of temperature were produced either by doing work on a system or by heating it. A falling mass turned a stirring paddle in a sample of water, doing work on the liquid. The rise in temperature of the water was measured and the amount of work done by the falling mass was compared with the amount of heat required to produce the same change in temperature. Joule found that the ratio of the work to the amount of heat was always the same, approximately 4.18 J of work for 1.00 cal of heat. The calorie is now defined to be exactly 4.184 J.

Based on the experiments of Rumford, Mayer, Helmholtz, Joule, and many others since the time of Joule, we now state the first law of thermodynamics as it applies to a system whose kinetic and potential energy do not change: For a closed system and for any process that begins and ends with equilibrium states

$$\Delta U = q + w$$

where q is the amount of heat transferred to the system and w is the work done on the system and where ΔU is the change in the value of U, the internal energy, which is a state function:

$$\Delta U = U(final) - U(initial)$$

In spite of the work of Rumford, Mayer, and Joule the credit for announcing the first law of thermodynamics went to Helmholtz.

We accept the first law of thermodynamics as an experimentally established law and accept the internal energy as a state function. This law is a version of the law of conservation of energy, which is a general law of physics to which there are no known exceptions. Apparent violations of energy conservation led particle physicists to search for previously unknown particles that could be transferring energy to or from a system, leading to the discovery of the neutrino. Occasionally an unknown inventor in search of gullible investors announces a machine that will produce more energy than it takes in, violating the first law of thermodynamics. Such nonexistent machines are known as perpetual motion machines of the first kind. It is the total energy of a system that is governed by conservation of energy. The first law of thermodynamics as stated in above Equation applies to a closed system whose center of mass is not accelerated and whose gravitational potential energy does not change. If work is done to change the kinetic or potential energy of the system as a whole this amount of work must be subtracted from the total work done to obtain the amount of work that changes the internal energy.

1.6.2 The second law of thermodynamics

The statement of the first law of thermodynamics defines the internal energy and asserts as a generalization of experiment fact that it is a state function. The second law of thermodynamics establishes the entropy as a state function, but in a less direct way.

There are two important physical statements of the second law of thermodynamics. The Kelvin statement involves cyclic processes, which are processes in which the final state of the system is the same as its initial state: It is impossible for a system to undergo a cyclic process whose sole effects are the flow of an amount of heat from the surroundings to the system and the performance of an equal amount of work on the surroundings. In other words, it is impossible for a system to undergo a cyclic process that turns heat completely into work done on the surroundings.

The Clausius statement is: It is impossible for a process to occur that has the sole effect of removing a quantity of heat from an object at a lower temperature and transferring this quantity of heat to an object at a higher temperature. In other words, heat cannot flow spontaneously from a cooler to a hotter object if nothing else happens. The Clausius statement of the second law is closely related to ordinary experience. The Kelvin statement is less closely related, and it is remarkable that the statements are equivalent to each other and to the mathematical statement of the second law, which establishes that the entropy is a state function.

The Clausius statement applies only to cyclicprocesses. Heat can be completely turned into work done on the surroundings without violating the second law if the system undergoes a process that is not cyclic. For example, in an isothermal reversible expansion of an ideal gas $\Delta U = 0$, so that

$$W_{surr} = -w = \Delta U - q = q \quad \text{(ideal gas, isothermal volume change)}$$

Heat transferred to the system has been completely turned into work done on the surroundings. However, the process is not cyclic and there is no violation of the second law of thermodynamics.

1.6.3 The third law of thermodynamics

The third law of thermodynamics was first stated by Nernst: For certain isothermal chemical reactions between solids, the entropy changes approach zero as the thermo dynamic temperature approaches zero. Nernst based this statement on his analysis of experimental data obtained by T. W. Richards, who studied the entropy changes of chemical reactions between solids as the temperature was made lower and lower. The statement of Nernst was sometimes called Nernst's heat theorem, although it is a statement of experimental fact and not a mathematical theorem.

In 1911 Planck proposed extending Nernst's statement to assert that the entropies of individual substances actually approach zero as the temperature approaches zero. However, there is no experimental justification for this assertion. In 1923 Lewis proposed the following statement of the third law: "If the entropy of each element in some crystalline state be taken as zero at the absolute zero of temperature, every substance has a finite positive entropy—but at the absolute zero of temperature the entropy may become zero, and does so become in the case of perfect crystalline substances." We base our entropy calculations on this statement.

The restriction to perfect crystals was made necessary by the discoveries of Simon and Giauque, who found that substances such as CO and NO fail to obey the third law in their ordinary solid forms. These substances easily form metastable crystals with some molecules in positions that are the reverse of the equilibrium positions, and ordinary crystals are in metastable states such that their entropies do not approach zero at 0 K. We will return to this topic later in this section.

From *Physical Chemistry* by *Robert G.Mortimer*

New Words and Expressions

thermodynamics [ˌθɜːməʊdaɪˈnæmɪks] 热力学

macroscopic behavior 宏观行为

empirical law 实证定律;[统计] 经验法则;经验定律

relativistic mechanics 相对论力学

quantum mechanics 量子力学

infinitesimal displacement [数] 无穷小位移;微小位移

latent heat 潜热

cyclic process 循环过程

phlogiston theory 燃素学说

entropy [ˈentrəpɪ] [热] 熵(热力学函数)

caloric theory 热质说

heat capacity 热容量;热容

neutrino [njuːˈtriːnəʊ] 中微子

isothermal [ˈaɪsəʊˈθɜːməl] 等温的,等温线的

metastable [metəˈsteɪbəl] 亚稳的,相对稳定的

Notes

（1）**Joseph Black** 约瑟夫·布莱克（1728~1799），英国化学家，物理学家。在化学上的主要贡献是首先用天平来研究化学变化，创造了定量化学分析方法。在物理学上，布莱克澄清了热量和温度这两个不同的概念。他由此提出了比热容的理论，创立了测定热量的方法量热术。他还发现并解释了物态相变中的"潜热"，并定量地确定了冰熔解潜热和估计了水汽化潜热的大小。

（2）**Lavoisier** 拉瓦锡（1743~1794）法国著名化学家，近代化学的奠基人之一，"燃烧的氧学说"的提出者。与他人合作制定出化学物种命名原则，创立了化学物种分类新体系。拉瓦锡根据化学实验的经验，用清晰的语言阐明了质量守恒定律和它在化学中的运用。这些工作，特别是他所提出的新观念、新理论、新思想，为近代化学的发展奠定了重要的基础，因而后人称拉瓦锡为近代化学之父。

1.7 Coordination Chemistry

1.7.1 Coordination Compounds

Coordination compounds contain coordinate covalent bonds formed by the reactions of metal ions with groups of anions or polar molecules. The metal ion in these kinds of reactions acts as a Lewis acid, accepting electrons, whereas the anions or polar molecules act as Lewis bases, donating pairs of electrons to form bonds to the metal ion. Thus, a coordinate covalent bond is a covalent bond in which one of the atoms donates both of the electrons that constitute the bond. Often a coordination compound consists of a complex ion and one or more counter ions. In writing formulas for such coordination compounds, we use square brackets to separate the complex ion from the counter ion.

Some coordination compounds, such as $Fe(CO)_5$, do not contain complex ions. Most but not all of the metals in coordination compounds are transition metals. Our understanding of the nature of coordination compounds stems from the classic work of Alfred Werner, who prepared and characterized many coordination compounds. In 1893, at the age of 26, Werner proposed what is now commonly referred to as Werner's coordination theory.

Nineteenth-century chemists were puzzled by a certain class of reactions that seemed to violate valence theory. For example, the valences of the elements in cobalt(Ⅲ) chloride and those in ammonia seem to be completely satisfied, and yet these two substances react to form a stable compound having the formula $CoCl_3 \cdot 6NH_3$. To explain this behavior, Werner postulated that most elements exhibit two types of valence: primary valence and secondary valence. In modern terminology, primary valence corresponds to the oxidation number and secondary valence to the coordination number of the element. In $CoCl_3 \cdot 6NH_3$, according to Werner, cobalt has a primary valence of 3 and a secondary valence of 6.

Today we use the formula $[Co(NH_3)_6]Cl_3$ to indicate that the ammonia molecules and the cobalt atom form a complex ion; the chloride ions are not part of the complex but are counter ions, held to the complex ion by Coulombic attraction.

1. Properties of Transition Metals

Transition metals are those that either have incompletely filled d subshells or form ions with incompletely filled d subshells. Incompletely filled d subshells give rise to several notable properties, including distinctive colors, the formation of paramagnetic compounds, catalytic activity, and the tendency to form complex ions. The most common transition metals are scandium through copper, which occupy the fourth row of the periodic table. Most of the transition metals exhibit a close-packed structure in which each atom has a coordination number of 12. Furthermore, these elements have relatively small atomic radii. The combined effect of closest packing and small atomic size results in strong metallic bonds. Therefore, transition metals have higher densities, higher melting points and boiling points, and higher heats of fusion and vaporization than the main group and Group 2B metals.

Transition metals exhibit variable oxidation states in their compounds. Note that all these metals can exhibit the oxidation state +3 and nearly all can exhibit the oxidation state +2. Of these two, the +2 oxidation state is somewhat more common for the heavier elements. The highest oxidation state for a transition metal is +7, exhibited by manganese ($4s^2 3d^5$). Transition metals exhibit their highest oxidation states in compounds that contain highly electronegative elements such as oxygen and fluorine—for example, V_2O_5, CrO_3, and Mn_2O_7.

2. Ligands

The molecules or ions that surround the metal in a complex ion are called ligands. The formation of covalent bonds between ligands and a metal can be thought of as a Lewis acid-base reaction. (Recall that a Lewis base is a species that donates a pair of electrons.) To be a ligand, a molecule or ion must have at least one unshared pair of valence electrons, as these examples illustrate.

Therefore, ligands play the role of Lewis bases. The transition metal, on the other hand, acts as a Lewis acid, accepting (and sharing) pairs of electrons from the Lewis bases. The atom in a ligand that is bound directly to the metal atom is known as the donor atom. For example, nitrogen is the donor atom in the $[Cu(NH_3)_4]^{2+}$ complex ion.

The coordination number in a coordination compound refers to the number of donor atoms surrounding the central metal atom in a complex ion. The coordination number of Cu^{2+} in $[Cu(NH_3)_4]^{2+}$ is 4. The most common coordination numbers are 4 and 6, although coordination numbers of 2 and 5 are also known.

Depending on the number of donor atoms a ligand possesses, it is classified as monodentate (1 donor atom), bidentate (2 donor atoms), or polydentate (> 2 donor atoms). Bidentate and polydentate ligands are also called chelating agents because of their ability to hold the metal atom like a claw (from the Greek chele, meaning "claw"). One example is EDTA, a polydentate ligand used to treat metal poisoning. Six donor atoms enable EDTA to form a very stable complex ion with lead. This stable complex enables the body to remove lead from the blood.

The oxidation state of a transition metal in a complex ion is determined using the known charges of the ligands and the known overall charge of the complex ion. In the complex ion $[PtCl_6]^{2-}$, for example, each chloride ion ligand has an oxidation number of−1.For the overall charge of the ion to be 2,the Pt must have an oxidation number of+4.

3.Nomenclature of Coordination Compounds

Now that we have discussed the various types of ligands and the oxidation numbers of metals, our next step is to learn how to name coordination compounds.The rules for naming ionic coordination compounds are as follows:

(1) The cation is named before the anion, as in other ionic compounds. The rule holds regardless of whether the complex ion bears a net positive or a net negative charge.In the compounds $K_2[Fe(CN)_6]$ and $[Co(NH_3)_4]Cl$,for example,we name the K^+ and $[Co(NH_3)_4]^+$ cations first, respectively.

(2) Within a complex ion,the ligands are named first,in alphabetical order,and the metal ion is named last.

(3) The names of anionic ligands end with the letter *o*, whereas neutral ligands are usually called by the names of the molecules.The exceptions are H_2O (aqua), CO (carbonyl), and NH_3 (ammine).

(4) When two or more of the same ligand are present,use Greek prefixes *di*,*tri*,*tetra*,*penta*, and *hexa*,to specify their number.Thus,the ligands in the cation $[Co(NH_3)_4Cl_2]^+$ are "tetraammin-edichloro." (Note that prefixes are not used for the purpose of alphabetizing the ligands.)

(5)The oxidation number of the metal is indicated in Roman numerals immediately following the name of the metal.For example,the Roman numeral Ⅲ is used to indicate the +3 oxidation state of chromium in $[Cr(NH_3)_4Cl_2]^+$,which is called tetraamminedichlorochromium(Ⅲ) ion.

(6) If the complex is an anion,its name ends in *-ate*. In $K_4[Fe(CN)_6]$,for example,the anion $[Fe(CN)_6]^{4-}$ is called hexacyanoferrate (Ⅱ) ion.Note that the Roman numeral indicating the oxidation state of the metal follows the suffix *-ate*.

1.7.2 Structure of Coordination Compounds

The geometry of a coordination compound often plays a significant role in determining its properties.In studying the geometry of coordination compounds,we sometimes find that there is more than one way to arrange the ligands around the central atom.Such compounds in which ligands are arranged differently,known as stereoisomers,have distinctly different physical and chemical properties. Coordination compounds may exhibit two types of stereoisomerism:*geometric and optical.*

Geometrical isomers are stereoisomers that cannot be interconverted without breaking chemical bonds.Geometric isomers come in pairs.We use the terms cis and trans to distinguish one geometric isomer of a compound from the other.Cis means that two particular atoms (or groups of atoms) are adjacent to each other,and trans means that the atoms (or groups of atoms) are on opposite sides in the structural formula.The cis and trans isomers of coordination compounds generally have quite different colors,melting points,dipole moments,and chemical reactivities.Note that although the types of bonds are the same in both isomers (two Pt N and two Pt Cl bonds),the spatial arrangements are different.Another example is the tetraamminedichlorocobalt(Ⅲ) ion.

Optical isomers are nonsuperimposable mirror images. (Superimposable means that if one structure is laid over the other, the positions of all the atoms will match.) Like geometric isomers, optical isomers come in pairs. However, the optical isomers of a compound have identical physical and chemical properties, such as melting point, boiling point, dipole moment, and chemical reactivity toward molecules that are not themselves optical isomers. Optical isomers differ from each other, though, in their interactions with plane-polarized light, as we will see.

The structural relationship between two optical isomers is analogous to the relationship between your left and right hands. If you place your left hand in front of a mirror, the image you see will look like your right hand. Your left hand and right hand are mirror images of each other. They are nonsuperimposable, however, because when you place your left hand over your right hand (with both palms facing down), they do not match. This is why a right-handed glove will not fit comfortably on your left hand. Careful examination reveals that the trans isomer and its mirror image are superimposable, but the cis isomer and its mirror image are not. Thus, the cis isomer and its mirror image are optical isomers.

Optical isomers are described as chiral (from the Greek word for "hand") because, like your left and right hands, chiral molecules are nonsuperimposable. Isomers that are superimposable with their mirror images are said to be achiral. Chiral molecules play a vital role in enzyme reactions in biological systems. Many drug molecules are chiral, although only one of a pair of chiral isomers is biologically effective.

Chiral molecules are said to be optically active becauseof their ability to rotate the plane of polarization of polarized light as it passes through them. Unlike ordinary light, which vibrates in all directions, plane-polarized light vibrates only in a single plane. We use a polarimeter to measure the rotation of polarized light by optical isomers. A beam of unpolarized light first passes through a Polaroid sheet, called the polarizer, and then through a sample tube containing a solution of an optically active, chiral compound. As the polarized light passes through the sample tube, its plane of polarization is rotated either to the right (clockwise) or to the left (counterclockwise). This rotation can be measured directly by turning the analyzer in the appropriate direction until minimal light transmission is achieved. If the plane of polarization is rotated to the right, the isomer is said to be dextrorotatory and the isomer is labeled *d*; if the rotation is to the left, the isomer is levorotatory and the isomer is labeled *l*. The *d* and *l* isomers of a chiral substance, called enantiomers, always rotate the plane of polarization by the same amount, but in opposite directions. Thus, in an equimolar mixture of two enantiomers, called a racemic mixture, the net rotation is zero.

From *Chemistry* by *Julia Burdge*

New Words and Expressions

Coulombic attraction 库仑引力

paramagnetic [pærəmæg'netɪk] 顺磁性的；常磁性的

coordination number [物] 配位数

chelating agent [化学] 螯合剂;络合剂
alphabetical order 字母顺序
Stereoisomer [物化] 立体异构体
Geometricalisomer 几何异构体
Optical isomer [物化] 光学异构体
chiral 手性的;对掌性的
achiral [ə'kaɪərəl] 非手性的
polarimeter 旋光计;偏光计
polaroid sheet 偏振片
dextrorotatory 右旋的;右旋性的
levorotatory 左旋的
racemic mixture [化学] 外消旋混合物
polydentate 多配位基的

Notes

(1) **Alfred Werner**,阿尔弗雷德·韦尔纳(1866~1919),法国化学家,1893年在《无机化学领域中的新见解》一书中提出络合物的配位理论,提出了配位数这个重要概念。韦尔纳的理论可以说是现代无机化学发展的基础,韦尔纳因创立配位化学而获得1913年诺贝尔化学奖。著有《立体化学教程》。。

(2) **Transition metal**,是指元素周期表中d区的一系列金属元素,这一区域包括3到12一共十个族的元素,但不包括f区的内过渡元素。过渡金属由于具有未充满的价层d轨道,基於十八電子規則,性质与其他元素有明显差别。由于空的d轨道的存在,过渡金属很容易形成配合物。金属元素采用杂化轨道接受电子以达到16或18电子的稳定状态。当配合物需要价层d轨道参与杂化时,d轨道上的电子就会发生重排,有些元素重排后可以使电子完全成对,这类物质称为反磁性物质。相反,当价层d轨道不需要重排,或重排后还有单电子时,生成的配合物就是顺磁性的。反磁性的物质没有颜色,而顺磁性的物质有颜色,其颜色因物质而异,甚至两种异构体的颜色都是不同的。一些金属离子的颜色也是有单电子的缘故。

1.8 Nomenclature of Organic Chemistry

1.8.1 Alkanes nomenclature

1. Simple alkanes

The names of the simplest straight chain alkanes are shown in Fig.1.9.

2. Branched alkanes

Branched alkanes are alkanes with alkyl substituents branching off from the main chain. They are named by the following procedure:

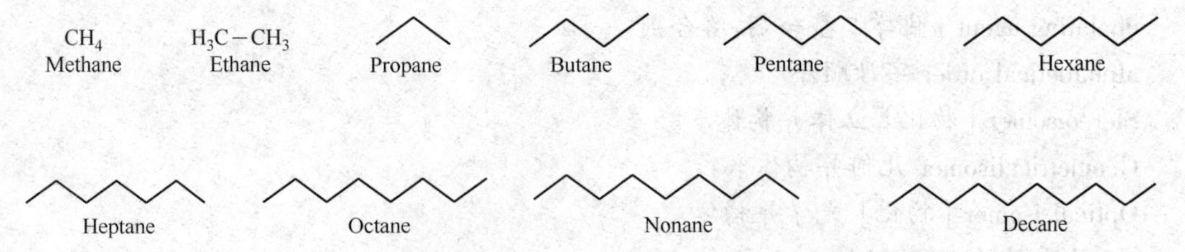

Fig.1.9 Nomenclature of simple alkanes

① identify the longest chain of carbon atoms. In the example shown (Fig.1.10a), the longest chain consists of five carbon atoms and a pentane chain;

② number the longest chain of carbons, starting from the end nearest the branch point (Fig.1.10b);

③ identify the carbon with the branching group (number 2 in Fig.1.10b);

④ identify and name the branching group. (In this example it is CH_3. Branching groups (or substituents) are referred to as alkyl groups (C_nH_{2n+1}) rather than alkanes (C_nH_{2n+2}). Therefore, CH_3 is called methyl and not methane.);

⑤ name the structure by first identifying the substituent and its position in the chain, then naming the longest chain. The structure in Fig. 1. 10 is called 2-methylpentane. Notice that the substituent and the main chain are one complete word, that is, 2-methylpentane rather than 2-methyl pentane.

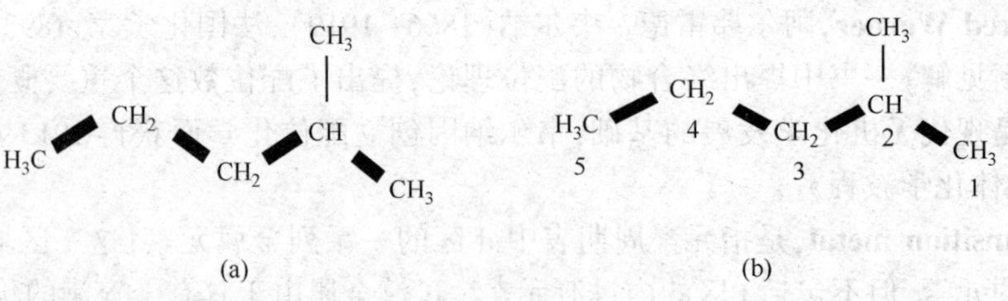

Fig.1.10 (a) identify the longest chain; (b) number the longest chain

3. Multi-branched alkanes

If there is more than one alkyl substituentpresent in the structure then the substituents are named in alphabetical order, numbering again from the end of the chain nearest the substituents. The structure in Fig. 1. 11 is 4-ethyl-3-methyloctane and not 3-methyl-4-ethyloctane. If a structure has identical substituents, then the prefixes di-, tri-, tetra-, *et cetera* are used to represent the number of substituents. For example, the structure in Fig.1.12 is called 2,2-dimethylpentane and not 2-methyl-2-methylpentane.

Fig.1.11 4-Ethyl-3-methyloctane

The prefixes di-, tri-, tetra-, etc. are used for identical substituents, but the order in which they are written is still dependent on the alphabetical order of the substituents themselves (i.e. ignore the

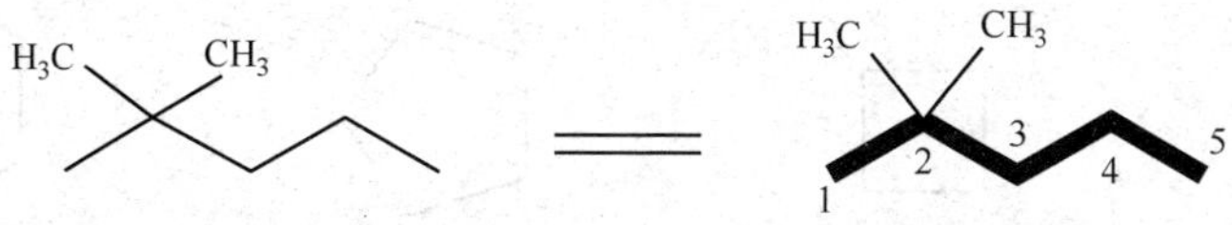

Fig.1.12　2,2-Dimethylpentane

di-, tri-, tetra-, et cetera). For example, the structure in Fig. 1. 13 is called 5-ethyl-2, 2-dimethyldecane and not 2,2-dimethyl-5-ethyldecane.

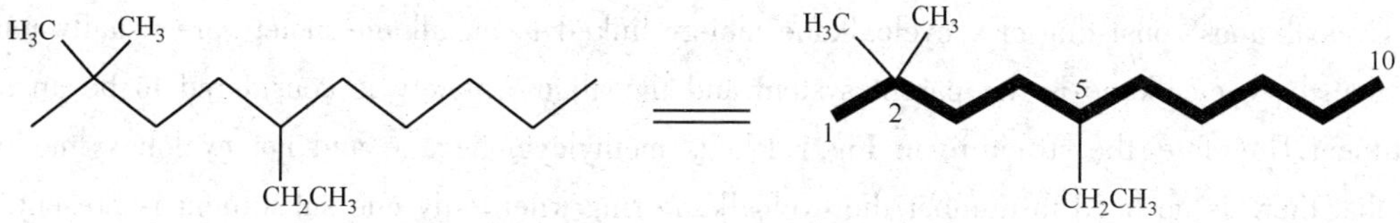

Fig.1.13　5-Ethyl-2,2-dimethyldecane

Identical substituents can be in different positions on the chain, but the same rules apply. For example, the structure in Fig.1.14 is called 5-ethyl-2,2,6-trimethyldecane.

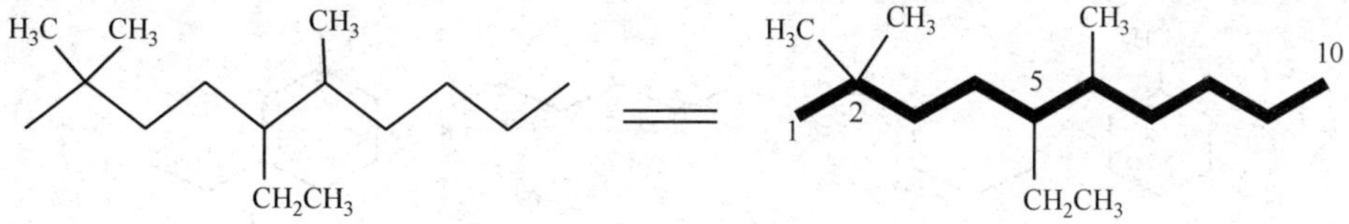

Fig.1.14　5-Ethyl-2,2,6-trimethyldecane

In some structures, it is difficult to decide which end of the chain to number from. For example, two different substituents might be placed at equal distances from either end of the chain. If that is the case, the group with alphabetical priority should be given the lowest numbering. For example, the structure in Fig.1.15a is 3-ethyl-5-methylheptane and not 5-ethyl-3-methylheptane.

However, there is another rule which might take precedence over the above rule. The structure (Fig.1.15c) has ethyl and methyl groups equally placed from each end of the chain, but there are two methyl groups to one ethyl group. Numbering should be chosen such that the smallest total is obtained. In this example, the structure is called 5-ethyl-3,3-dimethylheptane (Fig.1.15c) rather than 3-ethyl-5,5-dimethylheptane (Fig.1.15b) since 5+3+3 = 11 is less than 3+5+5 = 13.

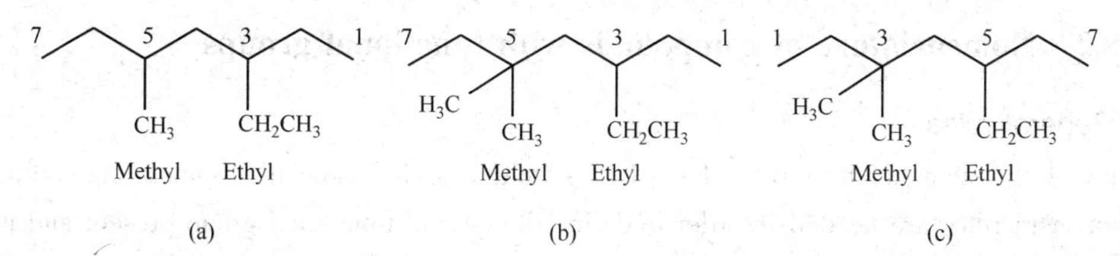

Fig.1.15　(a) 3-Ethyl-5-methylheptane; (b) incorrect numbering; (c) 5-ethyl-3,3-dimethylheptane

4. Cycloalkanes

Cycloalkanes are simply named by identifying the number of carbons in the ring and prefixing the alkane name with cyclo (Fig.1.16).

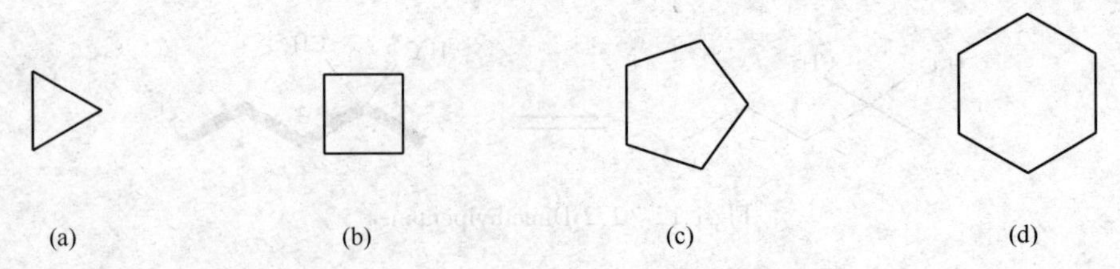

Fig.1.16 (a) Cyclopropane; (b) cyclobutane; (c) cyclopentane; (d) cyclohexane

5. Branched cyclohexanes

Cycloalkanes consisting of a cycloalkane moiety linked to an alkane moiety are usually named such that the cycloalkane is the parent system and the alkane moiety is considered to be an alkyl substituent. Therefore, the structure in Fig. 1.17a is methylcyclohexane and not cyclohexylmethane. Note that there is no need to number the cycloalkane ring when only one substituent is present.

If the alkane moiety contains more carbon atoms than the ring, the alkane moiety becomes the parent system and the cycloalkane group becomes the substituent. For example, the structure in Fig. 1. 17b is called 1-cyclohexyloctane and not octylcyclohexane. In this case, numbering is necessary to identify the position of the cycloalkane on the alkane chain.

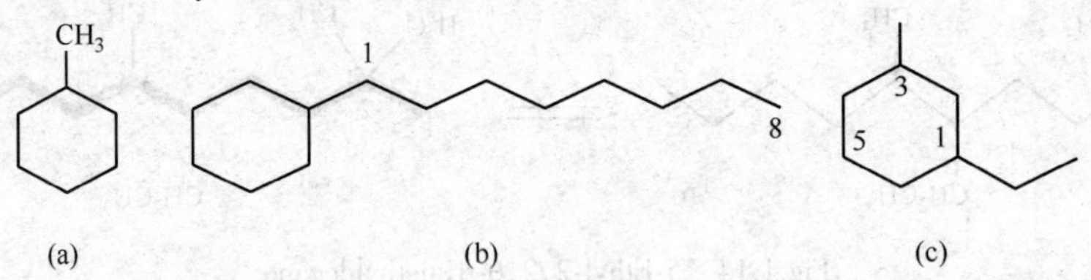

Fig.1.17 (a) Methylcyclohexane; (b) 1-cyclohexyloctane; (c) 1-ethyl-3-methylcyclohexane

6. Multi-branched cycloalkanes

Branched cycloalkanes having differentsubstituents are numbered such that the alkyl substituent having alphabetical priority is at position 1. The numbering of the rest of the ring is then carried out such that the substituent positions add up to a minimum. For example, the structure in Fig. 1. 17c is called 1-ethyl-3-methyl-cyclohexane rather than 1-methyl-3-ethylcy-clohexane or 1-ethyl-5-methylcyclo-hexane. The last name is incorrect since the total obtained by adding the substituent positions together is $5+1=6$ which is higher than the total obtained from the correct name (i.e. $1+3=4$).

1.8.2 Nomenclature of compounds with functional groups

1. General rules

Many of the nomenclature rules for alkanes hold true for molecules containing a functional group, but extra rules are needed in order to define the type of functional group present and its position within the molecule. The main rules are as follows:

① The main (or parent) chain must include the functional group, and so may not necessarily be the longest chain (Fig.1.18):

② The presence of some functional groups is indicated by replacing—ane for the parent alkane chain with the following suffixes:

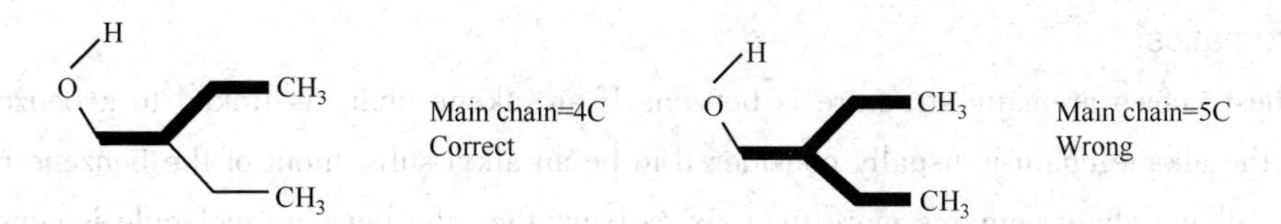

Fig.1.18 Identification of the main chain

functional group	suffix	functional group	suffix
alkene	-ene	alkyne	-yne
alcohol	-anol	aldehyde	-anal
ketone	-anone	carboxylic acid	-anoic acid
acid chloride	-anoyl chloride	amine	-ylamine.

The example in Fig.1.19 is a butanol.

③ Numbering must start from the end of the main chain nearest the functional group. Therefore, the numbering should place the alcohol (Fig.1.19) at position 1 and not position 4.

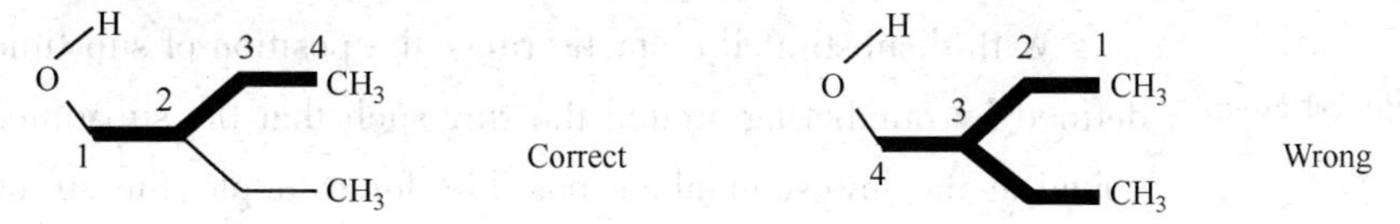

Fig.1.19 Numbering of the longest chain

④ The position of the functional group must be defined in the name. Therefore, the alcohol (Fig. 1.19) is a 1-butanol.

⑤ Other substituents are named and ordered in the same way as for alkanes. The alcohol (Fig. 1.19) has an ethyl group at position 2 and so the full name for the structure is 2-ethyl-1-butanol.

There are other rules designed for specific situations. For example, if the functional group is an equal distance from either end of the main chain, the numbering starts from the end of the chain nearest to any substituents. For example, the alcohol (Fig. 1.20) is 2-methyl-3-pentanol and not 4-methyl-3-pentanol.

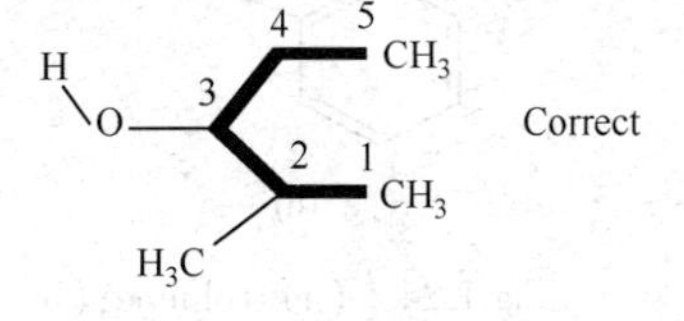

Fig.1.20 2-Methyl-3-pentanol

2. Alkenes and alkynes

Alkenes and alkynes have the suffixes—ene and—yne respectively (Fig. 1.21). With some alkenes it is necessary to define the stereochemistry of the double bond.

$$\overset{1}{H_3C}-\overset{2}{CH}=\overset{3}{CH}-\overset{4}{CH_3}$$

(a)

$$\overset{1}{H_3C}-\overset{2}{CH}=\overset{3}{C(CH_3)}-\overset{4}{C}\overset{5}{H_2CH_3}$$

(b)

$$\overset{1}{H_3C}-\overset{2}{C}\equiv\overset{3}{C}-\overset{4}{C}(CH_3)_2-\overset{5}{CH_3}$$

(c)

Fig.1.21 (a) 2-Butene; (b) 3-methyl-2-pentene; (c) 4,4-dimethyl-2-pentyne

3. Aromatics

The best known aromatic structure is benzene. If an alkane chain is linked to a benzene molecule, then the alkane chain is usually considered to be an alkyl substituent of the benzene ring. However, if the alkane chain contains more than six carbons, then the benzene molecule is considered to be a phenyl substituent of the alkane chain (Fig.1.22).

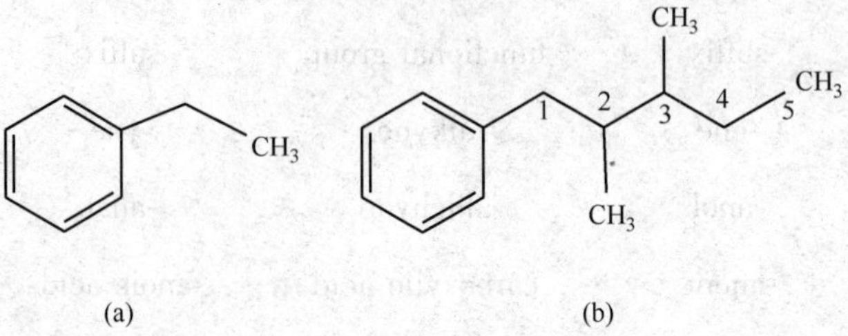

Fig.1.22 (a) Ethylbenzene; (b) 1-phenyl-2,3-dimethylpentane

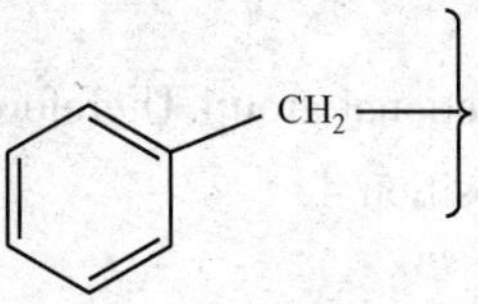

Fig.1.23 Benzyl group

Note that a benzyl group consists of an aromatic ring and a methylene group (Fig.1.23). Benzene is not the only parent name which can be used for aromatic compounds (Fig.1.24).

With disubstituted aromatic rings, the position of substituents must be defined by numbering around the ring such that the substituents are positioned at the lowest numbers possible, for example, the structure (Fig.1.25) is 1,3-dichlorobenzene and not 1,5-dichlorobenzene.

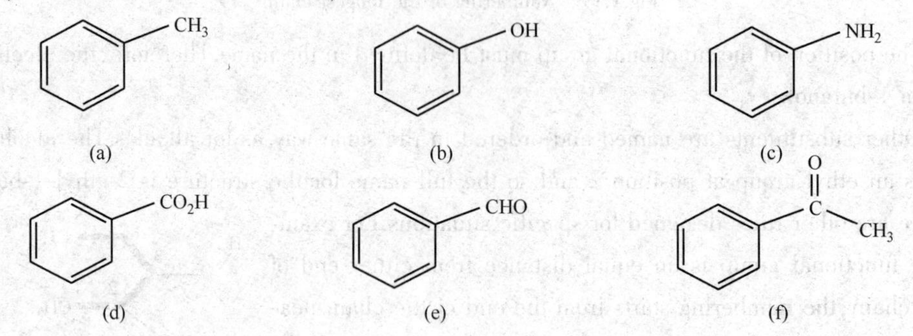

Fig.1.24 (a) Toluene; (b) phenol; (c) aniline; (d) benzoic; (e) benzaldehyde; (f) acetophenone

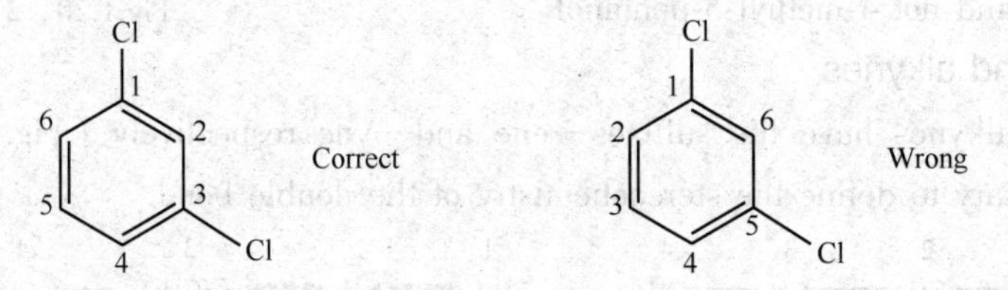

Fig.1.25 1,3-Dichlorobenzene

Alternatively, the terms ortho, meta, and para can be used. These terms define the relative position of one substituent to another (Fig.1.26). Thus, 1,3-dichlorobenzene can also be called meta-dichlorobenzene. This can be shortened to m-dichlorobenzene. The examples in Fig.1.27 illustrate how different parent names may be used. Notice that the substituent which defines the parent name is defined as position 1. For example, if the parent name is toluene, the methyl group must be at position 1.

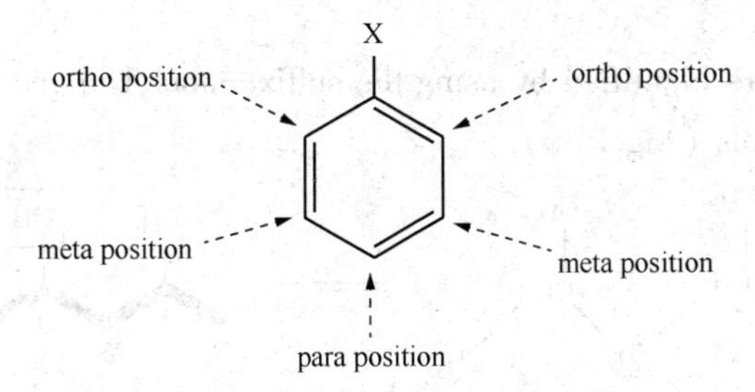

Fig.1.26 ortho, meta and para positions of an aromatic ring

When more than two substituents are present on the aromatic ring, the ortho, meta, para nomenclature is no longer valid and numbering has to be used (Fig.1.28).

Fig.1.27 (a); 2-Bromotoluene or O-bromotoluene; (b) 4-bromophenol or p-bromophenol; (c) 3-chloroaniline or m-chloroaniline

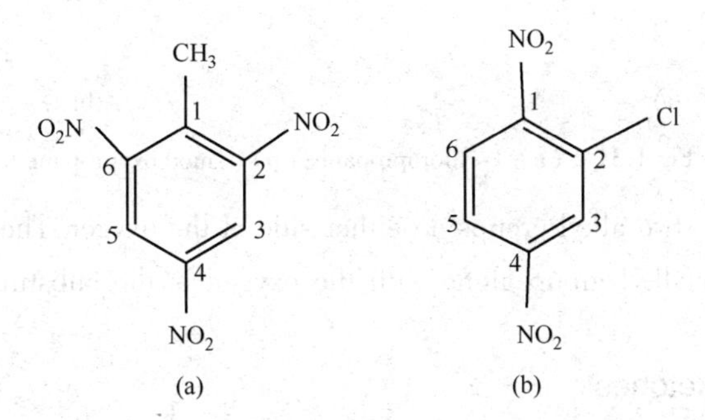

Fig.1.28 (a) 2,4,6-Trinitrotoluene; (b) 2-chloro-1,4-dinitrobenzene

Once again, the relevant substituent has to be placedat position 1 if the parent name is toluene, aniline, et cetera. If the parent name is benzene, the numbering is chosen such that the lowest possible numbers are used. In the example shown, any other numbering would result in the substituents having higher numbers (Fig.1.29).

Correct Substituent total =1+2+4=7

Wrong Substituent total =1+2+5=8

Wrong Substituent total =1+3+4=8

Fig.1.29 Possibile numbering systems of tri-substituted aromatic ring

4. Alcohols

Alcohols or alkanols are identified by using the suffix—anol. The general rules described earlier can be used to name alcohols (Fig.1.30).

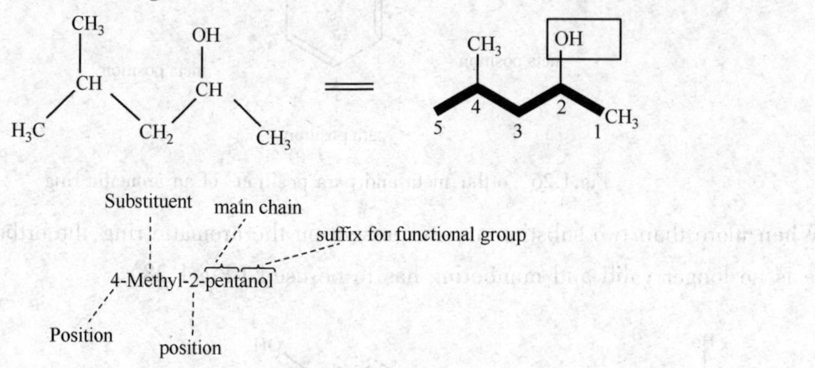

Fig.1.30 4-metyl-2-pentanol

5. Ethers and alkyl halides

The nomenclature for these compounds is slightly different from previous examples in that the functional group is considered to be a substituent of the main alkane chain. The functional group is numbered and named as a substituent (Fig.1.31).

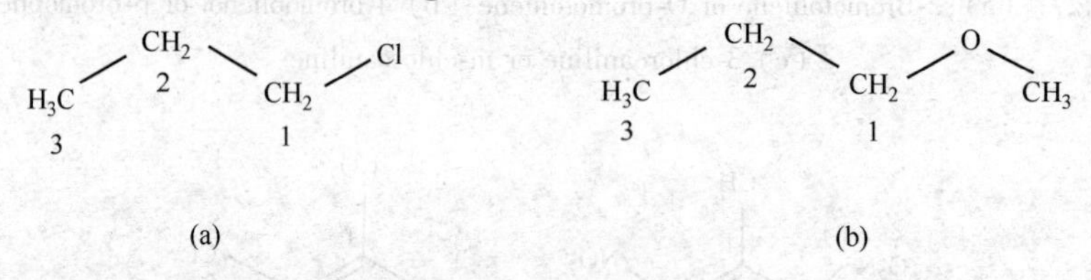

Fig.1.31 (a) 1-chloropropane; (b) 1-methoxypropane

Note that ethers have two alkyl groups on either side of the oxygen. The larger alkyl group is the parent alkane. The smaller alkyl group along with the oxygen is the substituent and is known as an alkoxy group.

6. Aldehydes and ketones

The suffix for an aldehyde (or alkanal) is—anal, while the suffix for a ketone (or alkanone) is-anone. The main chain must include the functional group and the numbering is such that the functional group is at the lowest number possible. If the functional group is in the center of the main chain, the numbering is chosen to ensure that other substituents have the lowest numbers possible (e.g. 2,2-dimethyl-3-pentanone and not 4,4-dimethyl-3-pentanone; Fig.1.32). 3-methyl-2-butanone can in fact be simplified to 3-methylbutanone. There is only one possible place for the ketone functional group in this molecule. If the carbonyl C=O group was at the end of the chain, it would be an aldehyde and not a ketone. Numbering is also not necessary in locating an aldehyde group since it can only be at the end of a chain (Fig.1.33).

7. Carboxylic acids and acid chlorides

Carboxylic acids and acid chlorides are identified by adding the suffix-anoic acid and-anoyl chloride respectively. Both these functional groups are always at the end of the main chain and do not need to be numbered (Fig.1.34).

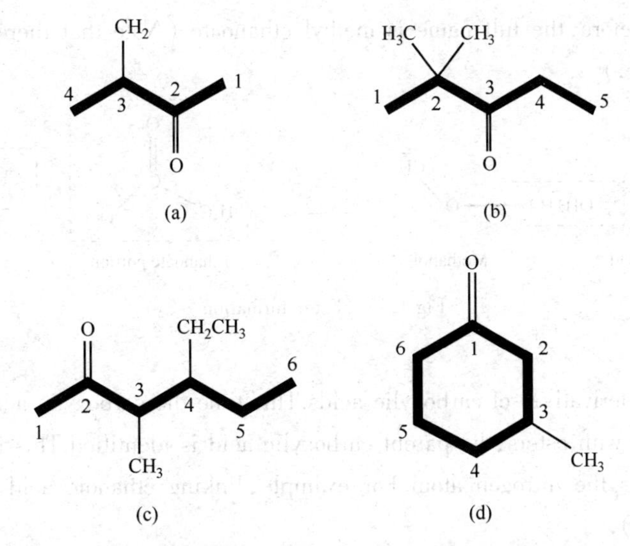

Fig.1.32 (a) 3-Methyl-2-butanone; (b) 2,2-dimethyl-3-pentanone; (c) 4-ethyl-3-methyl-2-hexanone; (d) 3-methylcyclohexanone

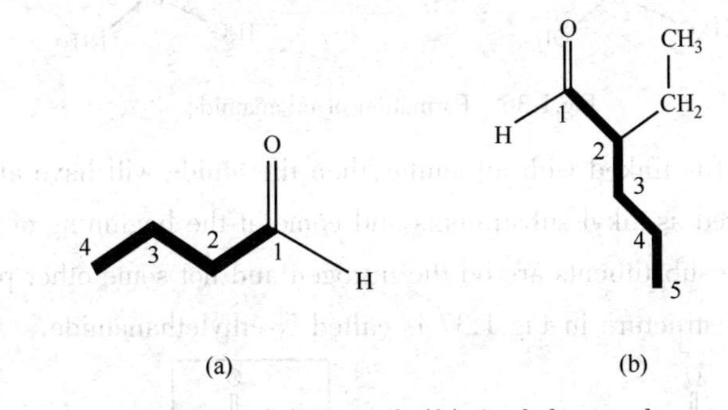

Fig.1.33 (a) Butanal; (b) 2-ethylpentanal

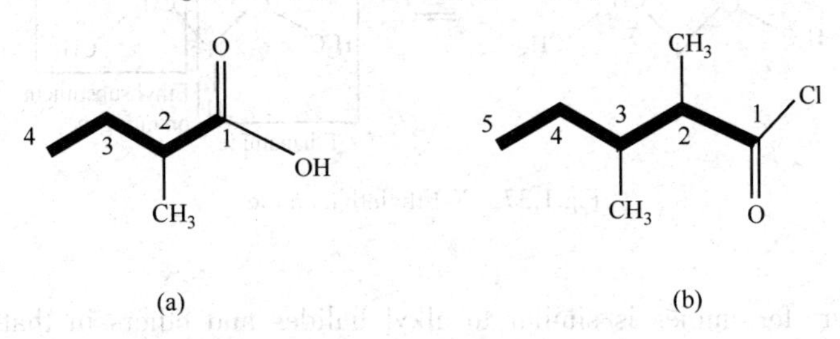

Fig.1.34 (a) 2-Methylbutanoic acid; (b) 2,3-dimethylpentanoyl chloride

8. Esters

To name an ester, the following procedure is carried out:

① identify the carboxylic acid (alkanoic acid) from which it was derived;

② change the name to an alkanoate rather than an alkanoic acid;

③ identify the alcohol from which the ester was derived and consider this as an alkyl substituent;

④ the name becomes an alkyl alkanoate.

For example, the ester (Fig.1.35) is derived from ethanoic acid and methanol. The ester would be an alkyl ethanoate since it is derived from ethanoic acid. The alkyl group comes from methanol and is amethyl group. Therefore, the full name is methyl ethanoate. (Note that there is a space between both parts of the name.)

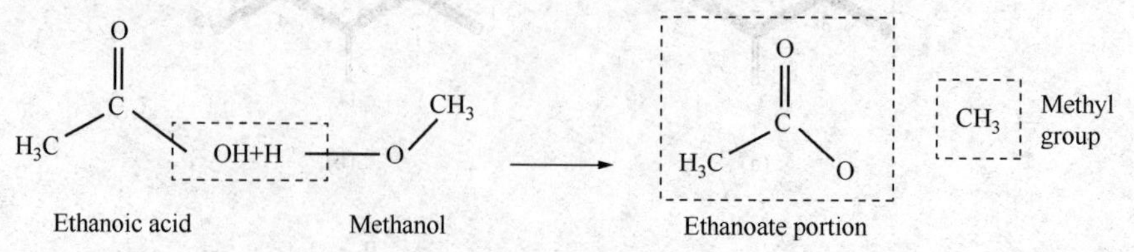

Fig. 1.35 Ester formation

9. Amides

Amides are also derivatives of carboxylic acids. This time the carboxylic acid is linked with ammonia or an amine. As with esters, the parent carboxylic acid is identified. This is then termed an alkanamide and includes the nitrogen atom. For example, linking ethanoic acid with ammonia gives ethanamide (Fig. 1.36).

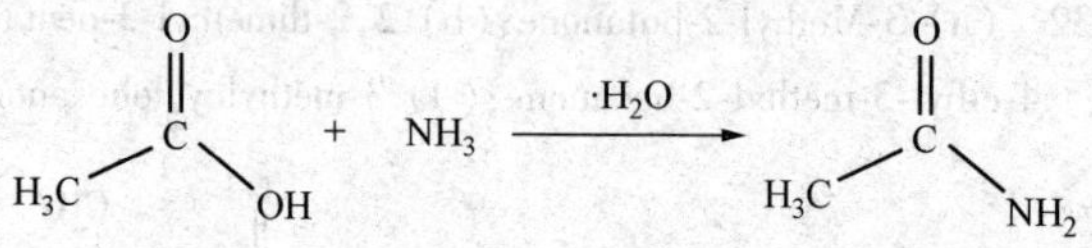

$$H_3C\text{-}C(=O)\text{-}OH + NH_3 \xrightarrow{-H_2O} H_3C\text{-}C(=O)\text{-}NH_2$$

Fig. 1.36 Formation of ethanamide

If the carboxylic acid is linked with an amine, then the amide will have alkyl groups on the nitrogen. These are considered as alkyl substituents and come at the beginning of the name. The symbol N is used to show that the substituents are on the nitrogen and not some other part of the alkanamide skeleton. For example, the structure in Fig. 1.37 is called N-ethylethanamide.

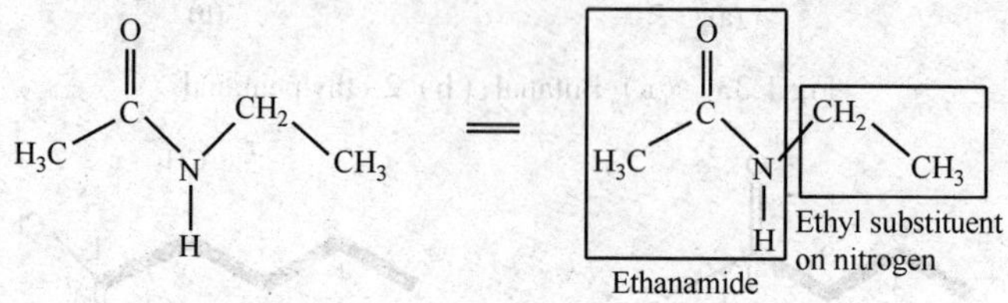

Ethanamide
Ethyl substituent on nitrogen

Fig. 1.37 N-Ethylethanamide

10. Amines

The nomenclature for amines is similar to alkyl halides and ethers in that the main part (or root) of the name is an alkane and the amino group is considered to be a substituent (Fig. 1.38). Simple amines are sometimes named by placing the suffix—ylamine after the main part of the name (Fig. 1.39).

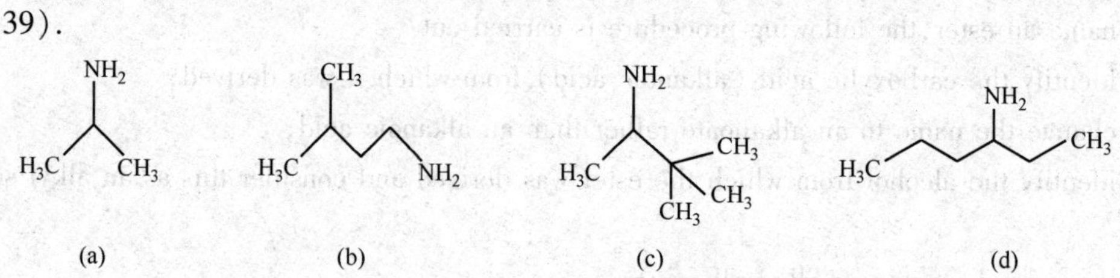

(a) (b) (c) (d)

Fig. 1.38 (a) 2-Aminopropane; (b) 1-amino-3-methylbutane; (c) 2-amino-3,3-dimethylbutane; (d) 3-aminohexane

H_3C —— NH_2 (a)　　H_3C—CH_2—NH_2 (b)

Fig.1.39 (a) Methylamine; (b) ethylamine

Amines having more than one alkyl group attached are named by identifying the longest carbon chain attached to the nitrogen. In the example (Fig.1.40), that is an ethane chain and so this molecule is an aminoethane (N, N-dimethylaminoethane).

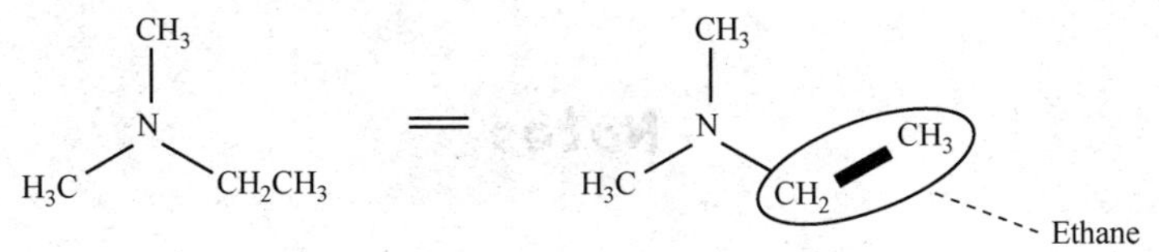

Fig.1.40 N, N-Dimethylaminoethane

Some simple secondary and tertiary amines have common names (Fig.1.41).

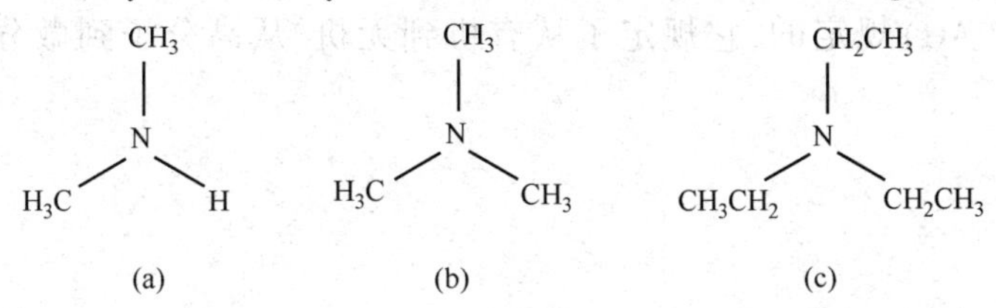

Fig.1.41 (a) Dimethylamine; (b) trimethylamine; (c) trithylamine

11. Thiols and thioethers

Thiols are named by adding the suffix thiol to the name of the parent alkane (Fig. 1.42a). Thioethers are named in the same way as ethers using the prefix alkylthio, for example, 1-(methylthio)propane (Fig.1.42c). Simple thioethers can be named by identifying the thioether as a sulfide and prefixing this term with the alkyl substituents, for example, dimethyl sulfide (Fig.1.42b).

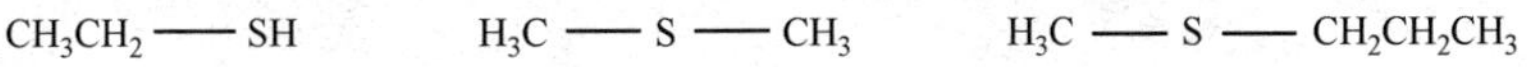

CH_3CH_2 —— SH (a)　　H_3C —— S —— CH_3 (b)　　H_3C —— S —— $CH_2CH_2CH_3$ (c)

Fig.42 (a) Ethanethiol; (b) dimethylsulfide; (c) 1-(methylthio) propane

From *organic chemistry* by *Graham Patrick*

New Words and Expressions

nomenclature [nəu'menklətʃə] 系统命名法；命名；术语；专门名称

et cetera [ˌet 'setərə] 〈法〉及其他，等等；如此等等；以此类推

ortho ['ɔːθəʊ] 正的；邻位的

stereochemistry [stɪərɪə'kemɪstrɪ] 立体化学

trinitrotoluene [traɪ'naɪtrəʊ'tɒljʊiːn] 三硝基甲苯，TNT

alkoxy ['ɔːlkɒksɪ] 烷氧基

alkanoate [ɔːlkæˈnəʊt] 链烷酸酯(或盐)
benzyl group 苯甲基
aldehyde [ˈældɪhaɪd] 醛,乙醛
ethanamide [ɪˈθænəmaɪd] 乙酰胺
aminoethane [æmɪnəˈʊeθən] 氨基乙烷
thiol [ˈθaɪəʊl] 硫醇;巯基
thioether [θaɪəʊˈiːθə] 硫醚

Notes

IUPAC,是国际纯粹和应用化学联合会(International Union of Pure and Applied Chemistry)的缩写。IUPAC 命名法是一种有系统命名有机化合物的方法。该命名法是由国际纯粹与应用化学联合会(IUPAC)规定的,它规定了从有机到无机、从高分子到微分子及各方面化学术语。

Unit 2 Fine Chemistry

2.1 Introduction of Fine Chemicals

Fine chemicals are complex, single, pure chemical substances. They are produced in limited volumes (<1000 tons/year) and at relatively high prices (> $10/kg) according to exacting specifications, mainly by traditional organic synthesis in multipurpose plants. *Biotechnical* processes are gaining ground. The global production value is about $85 billion. Fine chemicals are used as starting materials for specialty chemicals, particularly pharmaceuticals, biopharmaceuticals and agrochemicals. Custom manufacturing for the life science industry plays a big role; however, a significant portion of the fine chemicals total production volume is manufactured in house by large users. The industry is fragmented and extends from small, privately owned companies to divisions of big, diversified chemical enterprises. The term "fine chemicals" is used in distinction to "heavy chemicals", which are produced and handled in large lots and are often in a crude state.

Since their inception in the late 1970s, fine chemicals have become an important part of the chemical universe. The total production value of $85 billion is split about 60/40 among in-house production by the main consumers, the life science industry, on the one hand, and the fine chemicals industry on the other hand. The latter pursues both a "supply push" strategy, whereby standard products are developed in-house and offered ubiquitously, and a "demand pull" strategy, whereby products or services determined by the customer are provided exclusively on a "one customer/one supplier" basis. The products are mainly used as building blocks for proprietary products. The hardware of the top tier fine chemical companies has become almost identical. The design, lay-out and equipment of the plants and laboratories has become practically the same all over the world. Most chemical reactions performed go back to the days of the dyestuff industry. Numerous regulations determine the way labs and plants have to be operated, thereby contributing to the uniformity.

The roots of both the term "fine chemicals" and the emergence of the fine chemical industry as a distinct entity date back to the late 1970s, when the overwhelming success of the histamine H_2 receptor antagonists Tagamet (cimetidine) and Zantac (ranitidine hydrochloride) created a strong demand for advanced organic chemicals used in their manufacturing processes. As the in-house production capacities of the originators, the pharmaceutical companies Smith, Kline & French and Glaxo, could not keep pace with the rapidly increasing requirements, both companies (now merged as GlaxoSmithKline) outsourced part of the manufacturing to chemical companies experienced in producing relatively sophisticated organic molecules. Lonza, Switzerland, which already had supplied an early intermediate, methyl acetoacetate, during drug development, soon became the main supplier of more and more advanced precursors. The signature of a first, simple supply contract is generally acknowledged as the historical document marking the beginning of the fine chemical industry.

The beginning supply contract between Smith Kline French and Lonza for cimetidine precursors. In the subsequent years, the business developed favorably and Lonza was the first fine chemical company entering in a strategic partnership with SKF. In a similar way, Fine Organics, UK became the supplier of the thioethyl-*N'*-methyl-2-nitro-1,1-ethenediamine moiety of ranitidine, the second H_2 receptor antagonist, marketed as Zantac by Glaxo. Other pharmaceutical and agrochemical companies gradually followed suit and also started outsourcing the procurement of fine chemicals. An example in case is F.I.S., Italy, which partnered with Roche, Switzerland for custom manufacturing precursors of the benzodiazepine class of tranquilizers, such as Librium (chlordiazepoxide HCl) and Valium (diazepam).

The growing complexity and potency of new pharmaceuticals and agrochemicals requiring production in multipurpose, instead of dedicated plants and, more recently, the advent of biopharmaceuticals had a major impact on the demand for fine chemicals and the evolution of the fine chemical industry as a distinct entity. For many years, however, the life science industry continued considering captive production of the active ingredients of their drugs and agrochemicals as a core competency. Outsourcing was recurred to only in exceptional cases, such as capacity shortfalls, processes requiring hazardous chemistry or new products, where uncertainties subsisted about the chance of a successful launch.

The underlying principle for definition of the term "fine chemicals" is a three-tier segmentation of the universe of chemicals into commodities, fine chemicals, and specialty chemicals. Fine chemicals account for the smallest part, about 4% of the total $2500 billion turnover of the global chemical industry.

Commodities are large-volume, low-price homogeneous, and standardized chemicals produced in dedicated plants and used for a large variety of applications. Prices, typically less than $1/kg, are cyclic and are fully transparent. Petrochemicals, basic chemicals, heavy organic and inorganic chemicals, (large-volume) monomers, commodity fibers, and plastics are all part of commodities. Typical examples of single products are ethylene, propylene, acrylonitrile, caprolactam, methanol, toluene, o-xylene, phthalic anhydride, poly (vinyl chloride) soda, and sulfuric acid.

Fine chemicals are complex, single, pure chemical substances. They are produced in limited quantities (up to 1000 metric tons per year) in multipurpose plants by multistep batch chemical or biotechnological processes. They are based on exacting specifications, are used for further processing within the chemical industry, and are sold for more than $10/kg.

Fine chemicals are "*high-value chemicals purchased for their molecular qualities rather than for their functional performance, usually to make drugs.*"

The category is further subdivided on the basis of either the added value (building blocks, advanced intermediates, or active ingredients) or the type of business transaction (standard or exclusive products). As the term indicates, exclusive products are made exclusively by one manufacturer for one customer, which typically uses them for the manufacture of a patented specialty chemical, primarily a drug or agrochemical. Typical examples of single products are β-lactates, imidazoles, pyrazoles, triazoles, tetrazoles, pyridine, pyrimidines, and other N-heterocyclic compounds. A third way of differentiation is the regulatory status, which governs the manufacture. Active pharmaceutical ingredients and advanced intermediates thereof have to be produced under current Good Manufacturing Practice (cGMP) regulations. They are established by the (US) Food and Drug Administration (FDA) in order to guarantee the highest possible safety of the drugs made thereof. All advanced intermediates and APIs destined for drugs and other specialty chemicals destined for human consumption on the US market have to be produced according to cGMP rules, regardless of the location of the plant. The regulations apply to all manufacturing processes, such as chemical synthesis, bio-technology, extraction, and recovery from natural sources. All in all, the majority of fine chemicals have to be manufactured according to the cGMP regime (Table 2.1).

Table 2.1 Definition of Fine Chemicals (as opposed to Commodities and Specialties)

Commodities	Fine chemicals	Specialities
Single pure chemical substances···	Single pure chemical substances···	Mixture
Produced in dedicated plants	Produced in multipurpose plants	Formulated
High volume/low price	Low vol. (<1000 mtpa) High price(> $ 10/kg)	Undifferentiated
Many applications	Few applications	Undifferentiated
Sold on specifications	Sold on specifications "what they are"	Sold on performance "what they can do"

A precise distinction between commodities and fine chemicals is not feasible. In very broad terms, commodities are made by chemical engineersand fine chemicals by chemists. Both commodities and fine chemicals are identified according to specifications. Both are sold within the chemical industry, and customers know how to use them better than do suppliers. In terms of volume, the dividing line comes at about 1000 tons/year; in terms of unit sales prices, this is set at about $ 10/kg. Both numbers are somewhat arbitrary and controversial. Many large chemical companies include larger-volume/lower-unit-price products, so they can claim to have a large fine chemicals business (which is more appealing than commodities!). The threshold numbers also cut sometimes right into otherwise consistent product groups. This is, for instance, the case for active pharmaceutical ingredients, amino acids, and vitamins. In all three cases the two largest-volume products, namely, acetyl salicylic acid and paracetamol; L-lysine and D, L-methionine, and ascorbic acid and niacin, respectively, are produced in quantities exceeding 10,000 tons/year, and sold at prices below the $ 10 /kg level.

Specialty chemicals are formulations of chemicals containing one or more fine chemicals as active ingredients. They are identified according to performance properties. Customers are trades outside the chemical industry and the public. Specialty chemicals are usually sold under brand names. Suppliers have to provide product information. Subcategories are adhesives, agrochemicals, biocides, catalysts, dyestuffs and pigments, enzymes, electronic chemicals, flavors and fragrances, food and feed additives, pharmaceuticals, and specialty polymers.

The distinction between fine and specialty chemicals is net. The former are sold on the basis of "what they are"; the latter, on "what they can do." In the life science industry, the active ingredients of drugs are fine chemicals, the formulated drugs specialties.

Electronic chemicals provide another illustrative example of the difference between fine and specialty chemicals: Merck KGaA produces a range of individual fine chemicals as active substances for liquid crystals in a modern multipurpose plant in Darmstadt, Germany. An example is (*trans*, *trans*)-4-(difluoromethoxy)-3,5-difluorophenyl-4'-propyl-1,1'-bicyclohexyl. Merck ships the active ingredients to its secondary plants in Japan, South Korea, and Taiwan, where they are compounded into liquid crystal formulations. These specialties have to comply with stringent use-related specifications (electrical and color properties, etc.) of the Asian producers of consumer electronics such as cellular phones, DVD players, and flat-screen TV sets.

"Commoditized" specialty chemicals contain commodities as active ingredients and are interchangeable. Thus, ethylene glycol "99%" is a commodity. If it is diluted with water, enhanced with a colorant, and sold as "super antifreeze" in a retail shop, it becomes a commoditized specialty.

Note: Sometimes fine chemicals are considered as a subcategory of specialty chemicals. On the basis of the definitions given above this classification should be avoided.

From *Fine Chemicals: The Industry and the Business* by *Peter Pollak*

New Words and Expressions

pharmaceuticals [fɑrmə'sjʊtiklz] 药物(pharmaceutical 的复数)
methyl acetoacetate[æsətəu'æsiteit] 乙酰乙酸盐
cimetidine [sə'mɛtədin] 甲氰咪胍,甲腈咪胍(抗消化性溃疡药)
ranitidine [ræ'naitidin] 雷尼替丁;甲胺呋硫(一种抗溃疡药)
potency['potnsi] 效能;力量;潜力;权势
outsourcing ['aʊt sɔrsɪŋ] 外包;外购;外部采办
homogeneous [homə'dʒinɪəs] 均匀的;[数] 齐次的;同种的
acrylonitrile[ækrəlo'naɪtrəl] [有化] 丙烯腈;氰乙烯
caprolactam[kæprəu'læktəm] [有化] 己内酰胺
pyridine ['piri din] [有化] 吡啶;氮苯(等于 pyridina)
arbitrary['ɑrbətrɛri] [数] 任意的;武断的;专制的
salicylic [sælə'silik] 水杨酸的;得自水杨酸的
paracetamol[pærə'sitəmɑl] [药] 扑热息痛
glycol ['glaikɒl] 乙二醇;甘醇;二羟基醇

Notes

(1) **Fine chemicals** 精细化学品,是指那些具有特定的应用功能,技术密集,商品性强,产品附加值较高的化工产品。欧美一些国家把产量小、按不同化学结构进行生产和销售的化学物质,称为精细化学品。

(2) **Specialty chemicals** 专用化学品,是一种特定的化学品,各个国家都有不同的定义。

欧美一些国家把产量小,经过改性或复配加工,具有多功能或专用功能,既按其规格说明书,又根据其使用效果进行小批量生产和小包装销售的化学品称为专用化学品。我国指水处理化学品、造纸化学品、皮革化学品、油脂化学品、油田化学品、生物工程化学品、日化产品专用化学品、化学陶瓷纤维等特种纤维及高功能化工产品,以及其他各种用途的专项化学用品的制造。把产量小、经过加工配制、具有专门功能或最终使用性能的产品,称为专用化学品。中国、日本等则把 fine chemicals 和 specialty chemicals 统称为精细化学品。

(3) Heavy chemicals 大宗化学品,指生产过程中化工技术要求高、产量大、应用范围广泛的大宗化学品,例如石油化工中的合成树脂、合成橡胶及合成纤维三大合成材料,无机化工产品中的三酸两碱、合成氨等。

(4) GlaxoSmithKline 葛兰素史克公司,由葛兰素威康(Glaxò Wellcome)和史克必成(SmithKline Beecham)强强联合,于 2000 年 12 月成立,是世界领先的制药业巨擘,总部设在英国。2013 年 7 月爆出药品行业的行贿受贿事件,促进我国进一步解决药品行业不正当竞争,导致药品行业价格不断上涨的问题。

2.2 Dyes and Pigments

Natural dyes are dyes or colorants derived from plants, invertebrates, or minerals. The majority of natural dyes are vegetable dyes from plant sources-roots, berries, bark, leaves, and wood—and other organic sources such as fungi and lichens.

Archaeologists have found evidence of textile dyeing dating back to the Neolithic period. In China, dyeing with plants, barks and insects has been traced back more than 5,000 years. Many natural dyes require the use of chemicals called mordants to bind the dye to the textile fibers; tannin from oak galls, salt, natural alum, vinegar, and ammonia from stale urine were used by early dyers. Many mordants, and some dyes themselves, produce strong odors, and large-scale dyeworks were often isolated in their own districts.

The first human-made (synthetic) organic dye, mauveine, was discovered by William Henry Perkin in 1856. Many thousands of syntheticdyes have since been prepared. Synthetic dyes quickly replaced the traditional natural dyes. They cost less, they offered a vast range of new colors, and they imparted better properties upon the dyed materials. Unlike dyes, pigments do not dissolve; they are applied as fine solid particles mixed with a liquid. A distinction is usually made between a pigment, which is insoluble in the vehicle (resulting in a suspension), and a dye, which either is itself a liquid or is soluble in its vehicle (resulting in a solution). The term biological pigment is used for all colored substances independent of their solubility. A colorant can be both a pigment and a dye depending on the vehicle it is used in. In some cases, a pigment can be manufactured from a dye by precipitating a soluble dye with a metallic salt. The resulting pigment is called a lake pigment.

In general, the same ones are used in oil-, and water-based paints, printing inks, and plastics. They may be inorganic compounds (usually brighter and longer-lasting) or organic compounds. Natural organic pigments have been used for centuries, but today most are synthetic or inorganic. The primary white pigment is titanium dioxide. Carbon black is the most usual black pigment. Iron oxides give browns, ranging from yellowish through orange to dark brown. Chromium compounds yield chrome yellows, oranges, and greens; cadmium compounds brilliant yellows, oranges, and reds. The most common blues, Prussian blue and ultramarine, are also inorganic. Organic pigments, usually synthesized from aromatic hydrocarbons, include the nitrogen-containing azo pigments (red, orange, and yellow; see azo dyes) and the copper phthalocyanines (brilliant, strong blues and greens). Chlorophyll, carotene, rhodopsin, and melanin are pigments produced by plants and animals for specialized purposes.

Pigments are used for coloring paint, ink, plastic, fabric, cosmetics, food and other materials. Most pigments used in manufacturing and the visual arts are dry colorants, usually ground into a fine powder. This powder is added to a vehicle (or binder), a relatively neutral or colorless material that suspends the pigment and gives the paint its adhesion.

Dyes and pigments can be classified according to their chemical structure or their mode of application. The most important commercial products are the azo, anthraquinone, sulphur, indigoid, triphenylmethane and phthalocyanine dyes.

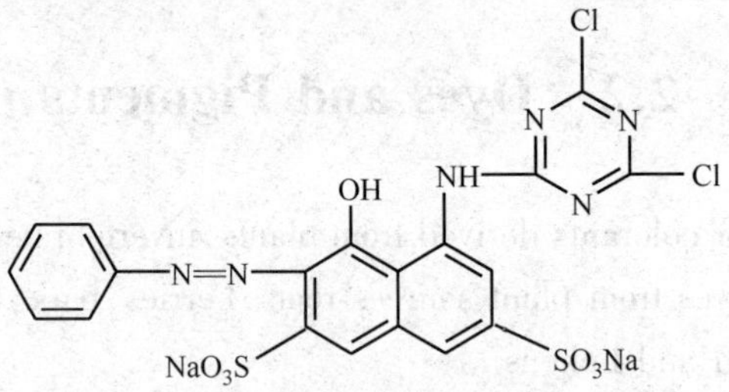

Reactive Brilliant Red X-3B

1. Reactive dyes

Reactive dyes utilize a chromophore attached to a substituent that is capable of directly reacting with the fibre substrate. The covalent bonds that attach reactive dye to natural fibers make them among the most permanent of dyes. "Cold" reactive dyes, such as Procion MX, Cibacron F, and Drimarene K, are very easy to use because the dye can be applied at room temperature. Reactive dyes are by far the best choice for dyeing cotton and other cellulose fibers at home or in the art studio.

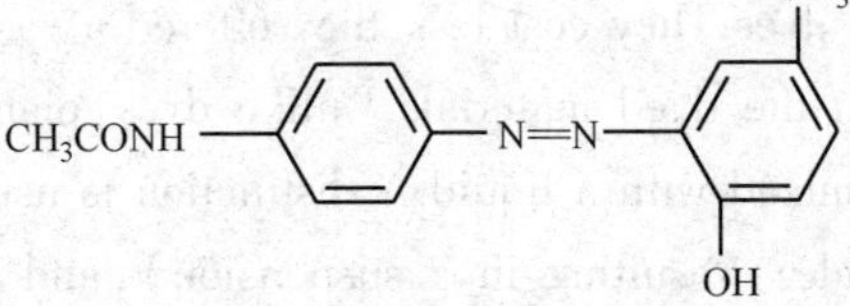

Disperse Yellow 3, 11855

2. Disperse dyes

Disperse dyes were originally developed for the dyeing of cellulose acetate, and are substantially water insoluble. The dyes are finely ground in the presence of a dispersing agent and then sold as a paste, or spray-dried and sold as a powder. Their main use is to dye polyester but they can also be used to dye nylon, cellulose triacetate, and acrylic fibres. In some cases, a dyeing temperature of 130℃ is required, and a pressurised dyebath is used. The very fine particle size gives a large surface area that aids dissolution to allow uptake by the fibre. The dyeing rate can be significantly influenced by the choice of dispersing agent used during the grinding.

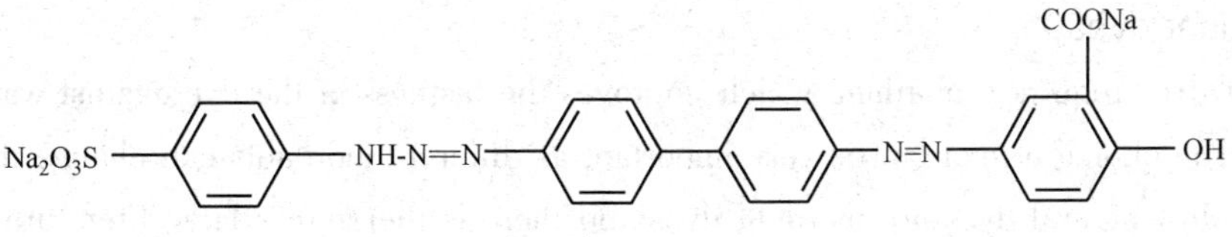

Direct Yellow GR

3. Direct dyes

Direct or substantive dyeing is normally carried out in a neutral or slightly alkaline dyebath, at or near boiling point, with the addition of either sodium chloride (NaCl) or sodium sulfate (Na_2SO_4). Direct dyes are used on cotton, paper, leather, wool, silk and nylon. They are also used as pH indicators and as biological stains.

4. Vat dyes

Vat dyes are essentially insoluble in water and incapable of dyeing fibres directly. However, reduction in alkaline liquor produces the water soluble alkali metal salt of the dye, which, in this leuco form, has an affinity for the textile fibre. Subsequent oxidation reforms the original insoluble dye. The color of denim is due to indigo, the original vat dye.

5. Sulfur dyes

Sulfur dyes are two part "developed" dyes used to dye cotton with dark colors. The initial bath imparts a yellow or pale chartreuse color, This is aftertreated with a sulfur compound in place to produce the dark black we are familiar with in socks for instance. Sulfur Black 1 is the largest selling dye by volume.

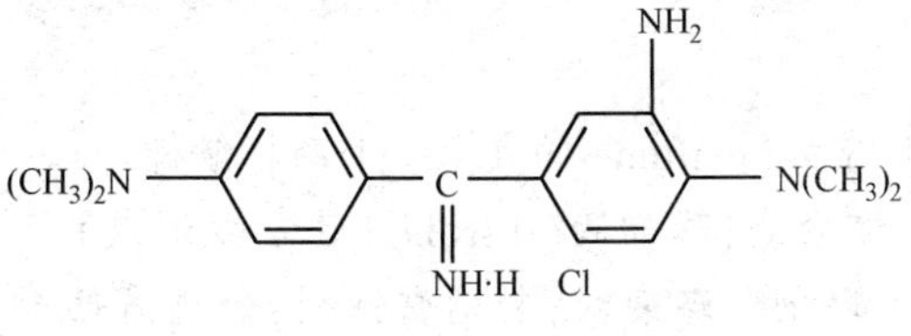

Basic Light Yellow O

6. Basic dyes

Basic dyes are water-soluble cationic dyes that are mainly applied to acrylic fibers, but find some use for wool and silk. Usually acetic acid is added to the dyebath to help the uptake of the dye onto the fiber. Basic dyes are also used in the coloration of paper.

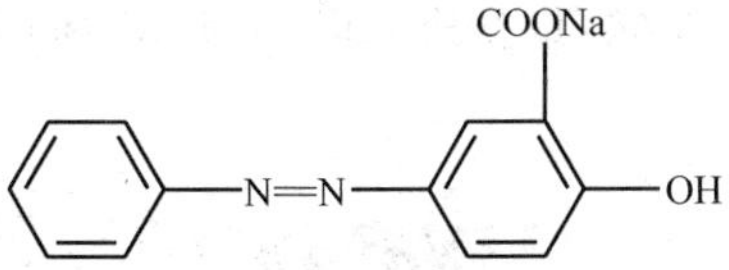

Acid Mordant Yellow 3G

7. Acid dyes

Acid dyes are water-soluble anionic dyes that are applied to fibers such as silk, wool, nylon and modified acrylic fibers using neutral to acid dyebaths. Attachment to the fiber is attributed, at least partly, to salt formation between anionic groups in the dyes and cationic groups in the fiber. Acid dyes are not substantive to cellulosic fibers. Most synthetic food colors fall in this category.

8. Mordant dyes

Mordant dyes require a mordant, which improves the fastness of the dye against water, light and perspiration. The choice of mordant is very important as different mordants can change the final color significantly. Most natural dyes are mordant dyes and there is therefore a large literature base describing dyeing techniques. The most important mordant dyes are the synthetic mordant dyes, or chrome dyes, used for wool; these comprise some 30% of dyes used for wool, and are especially useful for black and navy shades. The mordant, potassium dichromate, is applied as an after-treatment. It is important to note that many mordants, particularly those in the heavy metal category, can be hazardous to health and extreme care must be taken in using them.

9. Azo dyes

Azo dyeing is a technique in which an insoluble azoic dye is produced directly onto or within the fibre. This is achieved by treating a fibre with both diazoic and coupling components. With suitable adjustment of dyebath conditions the two components react to produce the required insoluble azo dye. This technique of dyeing is unique, in that the final color is controlled by the choice of the diazoic and coupling components.

From *Best Available Techniques Reference Document for the Manufacture of Organic Fine Chemicals* by *European commission.*

New Words and Expressions

perfumes [pə'fjuːms] 香水(perfume 的复数);[轻] 香料
distillates ['distiləts] 馏分油;蒸馏油(distillate 的复数形式)
mordants ['mɔrdnts] [助剂] 媒染剂;金属腐蚀剂;金属箔粘着剂
switch [switʃ] 开关;转换;鞭子
toxins ['tɔksin] [毒物] 毒素,毒质;毒素类(toxin 的复数)
mauveine [məʊ'veɪn] [染料] 苯胺紫
phthalocyanines [fθələsaɪə'naɪnz] 酞菁
indigo ['indigo] 靛蓝,靛蓝染料;靛蓝色;槐蓝属植物
heralded ['hɛrəld] 先驱;传令官;报信者;通报;预示…的来临

Notes

(1) **Reactive dyes**,活性染料含有能与纤维分子中的羟基、氨基等发生反应的基团,在染色时能与纤维形成共价键,所以特别耐光,广泛用于纤维素染色的一类染料,即生成"染料-纤维"化合物。

(2) **Disperse dyes**,分散染料不溶于水,在水中呈分散微粒状态,它们被表面活性剂分散

成稳定的悬乳液,在高温下染色,染料迁移到纤维上,在纤维上形成固体溶液,并被次价键力所固着,因不溶于水,所以耐洗度较高。

(3) Direct dyes,直接染料的分子较长,对纤维有较高的直接性,因此它在用于纤维染色时,不需要媒染剂处理,可直接上接。含有大量羟基能够形成氢键,主要用于棉、麻、人造纤维的染色。

(4) Acid dyes,酸性染料一般都含有磺酸基、羧酸基和羟基等可溶性基团,可溶解于水中,这类染料大都在酸性染浴中染色。

(5) Basic dyes,碱性染料分子中含有碱性基团如氨基或取代氨基,能与蛋白质纤维上的羧基成盐而染色-盐基染料。碱性染料溶于水能电离生成有色离子,所以碱性染料也属于阳离子染料,但应用性能与阳离子染料不同。

2.3 Food Additives

The pursuit of happiness through the enjoymentof food is a centuries old human endeavor.Taste, texture, freshness and eye appeal are major contributors to such enjoyment, made possible in our modern lifestyle through the use of highly specialized ingredients known as food additives.

Food additives are substances added to food to preserve flavor or enhance its taste and appearance.Some additives have been used for centuries; for example, preserving food by pickling (with vinegar), salting, as with bacon, preserving sweets or using sulfur dioxide as in some wines.With the advent of processed foods in the second half of the 20th century, many more additives have been introduced, of both natural and artificial origin.

Food additives afford us the convenience and enjoyment of a wide variety of appetizing, nutritious, fresh, and palatable foods.Their quantities in food are small, yet their impact is great.Without additives, we would be unfortunately lacking in the abundant and varied foods that we enjoy today.

2.3.1 Purposes of using food additives

Direct food additives serve four major purposes in our foods:

① To provide nutrition—to improve or maintain the nutritional quality of food.For example, the addition of iodine to salt has contributed to the virtual elimination of simple goiter.The addition of Vitamin D to milk and other dairy products has accomplished the same thing with respect to rickets.Niacin in bread, cornmeal and cereals has helped eliminate pellagra, a disease characterized by central nervous system and skin disorders.

Other nutritional food additives (such as thiamine and iron) are used for further fortification in the diet and as a result, diseases due to nutritional deficiencies, common in lesser developed countries, are now very rare in the United States.

② To maintain product quality and freshness—fresh foods do not stay that way for long periods of time; they rapidly deteriorate, turn rancid and spoil.Food additives delay significantly this deterioration and prevent spoilage caused by growth of microorganisms, bacteria and yeast and also by oxidation (oxygen in air coming into contact with the foods).For example, if you were to cut slices of fresh fruits such as apples, bananas or pears, they would rapidly turn brown as a result of this oxidation process. However, placing these slices in juice from lemons, limes or oranges can stop this process. Food processors do the same thing by using ascorbic acid—the principal active ingredient in citrus juice—when packaging fruit slices. Propionates, which naturally occur in cheese, are used similarly in bakery goods to prevent the growth of molds.

③ To aid in the processing and preparation of foods-additives impart and/or maintain certain desirable qualities associated with various foods. For example, we expect salad dressings to stay mixed once they have been shaken.

Emulsifiers such as lecithin from soybeans maintain mixture and improve texture in dressings and other foods. They are used in ice cream where smoothness is desired, in breads to increase volume and impart fine grain quality, and in cake mixes to achieve better consistency.

Pectin, derived from citrus peels and used in jellies and preserves when thickening is desired, belongs in the category of stabilizers and thickeners. Leaveners used to make breads, biscuits and rolls rise, include yeast, baking powder and baking soda. Humectants, like sorbitol that naturally occurs in apples, are used when moisture retention is necessary, such as in the packaging of shredded coconut.

④ To make foods appealing—the majority of food additives are most often used for this purpose. Unless foods look appetizing and appeal to our senses, they will most likely go uneaten and valuable nutrients will be lost. Food additives such as flavoring agents and enhancers, coloring agents and sweeteners are included by food processors because we demand foods that look and taste good.

There are many other food additives than those listed here. However, most of them fall into the four functional classes described.

2.3.2 Common food additives

The following is our basic layout of the summary several common food additives.

1. Sodium benzoate, benzoic acid

Preservative: Fruit juice, carbonated drinks, pickles. Manufacturers have used sodium benzoate (and its close relative benzoic acid) for a century to prevent the growth of microorganisms in acidic foods. The substances occur naturally in many plants and animals. They appear to be safe for most people, though they cause hives, asthma, or other allergic reactions in sensitive individuals.

Anotherproblem occurs when sodium benzoate is used in beverages that also contain ascorbic acid (vitamin C). The two substances, in an acidic solution, can react together to form small amounts of benzene, a chemical that causes leukemia and other cancers. Though the amounts of benzene that form are small, leading to only a very small risk of cancer, there is no need for consumers to experience any risk. In the early 1990s the FDA had urged companies not to use benzoate in products that also contain ascorbic acid, but in the 2000s companies were still using that combination. A lawsuit filed in 2006 by private attorneys ultimately forced Coca-Cola, PepsiCo, and other soft-drink makers in the U.S. to reformulate affected beverages, typically fruit-flavored products.

2. Silicon dioxide, silica, calcium silicate

Anti-caking agent: salt, soups, coffee creamer, and other dry, powdery foods. This chemical is just sand. Silicon dioxide occurs naturally in foods, especially foods derived from plants. Breathed in, silica dust can cause lung disease, even cancer. For use in food, it is finely ground and added to salt and other foods to help powders flow more easily. Aluminium calcium silicate and tricalcium silicate are also used in foods, and calcium silicate is used in some dietary supplements.

3. Potassium sorbate

A preservative that is the potassium salt of sorbic acid, prevents growth of mold: Cheese, syrup, jelly, cake, wine, dry fruits. Sorbic acid occurs naturally in many plants. These additives are safe. It is a white crystalline powder which is very soluble in water, with a solubility of 139 g in 100 mL at 20℃. This solubility allows for solutions of high concentration which can be used for dipping and spraying. It is effective up to pH 6.5. It has approximately 74 percent of the activity of sorbic acid, therefore requiring higher concentrations to obtain comparable results as sorbic acid. It is effective against yeasts and is used in cheese, bread, beverages, margarine, and dry sausage. Typical usage levels are 0.025 to 0.10 percent.

4. Sodium Bicarbonate

A leavening agent with a pH of approximately 8.5 in a 1 percent solution at 25℃. It functions with food grade phosphates (acidic leavening compound) to release carbon dioxide which expands during the baking process to provide the baked good with increased volume and tender eating qualities. It is also used in dry-mix beverages to obtain carbonation, which results when water is added to the mix containing the sodium bicarbonate and an acid. It is a component of baking powder. It is also termed baking soda, bicarbonate of soda, sodium acid carbonate, and sodium hydrogen carbonate.

5. Artificial sweeteners

Artificial sweetener: Baked goods, chewing gum, gelatin desserts, diet soda, Sunette.

They areused in beverages, candies, chewing gum, yogurts, and many other products to provide sweetness without the calories. The question is: are they safe? Controversies have swirled around most of the additives. Sucralose, rebiana, and neotame appear to be safe, but acesulfame-potassium, aspartame, and saccharin may pose a slight risk of cancer (read about each ingredient elsewhere on this page). But research on all of them is relatively limited, especially considering how widely they are used, and surprises might occur. For instance, a 2010 study found that artificially sweetened drinks probably caused preterm deliveries; the researchers suspected that aspartame was the culprit (the study needs to be confirmed by other scientists). Synthetic "high-potency" sweeteners were the rule until about 2009 when rebiana, which is purified out of stevia leaves, became marketed widely in the United States. Rebiana, which has "taste challenges," allowed companies to claim for the first time "all natural" on their artificially sweetened (with a natural ingredient, that is) products.

Companies advertise their artificially sweetened foods as being almost magical weight-loss potions. The fact is, though, that losing weight is real hard, and people need to make a real concerted effort to eat fewer calories and exercise more. Artificial sweeteners can make the struggle a little more pleasant.

6. Artificial colorings

Most artificially colored foods are colored with synthetic chemicals that do not occur in nature. Because colorings are used almost solely in foods of low nutritional value (candy, soda pop, gelatin desserts, etc.), you should simply avoid all artificially colored foods. In addition to problems mentioned below, colorings cause hyperactivity in some sensitive children. The use of coloring usually indicates that fruit or other natural ingredient has not been used. Some foods are artificially colored with natural substances, such as beta-carotene and carmine. Please refer elsewhere in Chemical Cuisine for information about them.

(7) Antioxidants

Oxidation reactions happen when chemicals in the food are exposed to oxygen in the air. In natural conditions, animal and plant tissues contain their own antioxidants but infoods, these natural systems break down and oxidation is bound to follow.

Oxidation of food is a destructive process, causing loss of nutritional value and changes in chemical composition. Oxidation of fats and oils leads to rancidity and, in fruits such as apples, it can result in the formation of compounds which discolour the fruit. Antioxidants are added to food to slow the rate of oxidation and, if used properly, they can extend the shelf life of the food in which they have been used.

Fats and oils, or foods containing them, are the most likely to have problems with oxidation. Fats react with oxygen and even if a food has a very low fat content it may still need the addition of an antioxidant. They are commonly used in: vegetable oil, snacks (extruded), animal fat, meat, fish, poultry, margarine, dairy products, mayonnaise/salad dressing, baked products, potato products (instant mashed potato).

New Words and Expressions

palatable [ˈpælətəbl] 美味的，可口的；愉快的
pickles [ˈpɪkl] 咸菜（pickle 的复数形式）；腌渍物；各式腌菜
lecithin [ˈlɛsiθin] [生化] 卵磷脂；蛋黄素
benzoate [ˈbenzəʊeɪt] 苯酸盐；安息香酸盐
sorbate [ˈsɔːbeit] 吸着物；山梨酸酯
beverages [ˈbɛvəridʒ] 饮料；酒水；饮料类（beverage 的复数形式）
discolour [dɪsˈkʌlə] 污染；使褪色

Notes

(1) Artificial colorings，人工合成色素是指用人工化学合成方法所制得的有机色素，主要是以煤焦油中分离出来的苯胺染料为原料制成的。按结构，可分类偶氮类、氧蒽类和二苯甲烷类等。按溶解性，可分为脂溶性着色剂和水溶性着色剂。按来源，国家列入卫生使用标准的人工色素为以下 8 种：胭脂红，苋菜红，日落黄，赤藓红，柠檬黄，新红，靛蓝，亮蓝。

(2) Rebiana，甜菊糖，是一种从菊科草本植物甜叶菊（或称甜菊叶）中精提的新型天然甜味剂，而南美洲使用甜叶菊作为药草和代糖已经有几百年历史。它具有高甜度、低热能的特

点,其甜度是蔗糖的200~300倍,热值仅为蔗糖的1/300。国际甜味剂行业的资料显示,甜菊糖已在亚洲、南美洲和美国广泛应用于食品生产中。中国是全球主要甜菊糖生产国之一。

2.4 Equipment and Unit Operation

2.4.1 Reactors

The main equipment in multipurpose plants is the stirred tank reactor, which fulfils the flexibility requirements arising from the varied physical states of the materials being used (e.g.dry powders, wet solids, pastes, liquids, emulsions, gases).

The vessels are required to withstand a range of process conditions (e.g.temperature, pressure, corrosion) and are thus usually made of stainless steel, rubber—or glass-lined steel, enamel coated, or other special materials.The mechanical design of the agitator baffles and cooling systems is constrained by the need to attach and maintain the rubber or glass lining.

Other characteristics: used for both batch and continuousmode, as well as in cascades; sized up to 60 m^3 (fermentation reactors up to about 1000 m^3); usually dished bottom (reactions may be carried out under pressure); equipped with one or more stirrers to ensure the requested mixing degree, heat-exchange performance; jackets or half pipe coils are often fitted around the vessel to provide heat transfer; wall baffles are installed inside to prevent the gross rotation ("swirl") of the contents with the stirrer.

2.4.2 Equipment and operations for product work-up

1.Drying

Many different dryer types are available and are in use.They include among others: fluidised bed dryers, vacuum dryers, spray dryers, band/belt dryers.

2.Liquid-solid separation

Liquid-solid separation is used, e.g.for the separation of a precipitated product, catalyst, solid impurities.The spectrum of the available and used equipment is wide and includes decanters, decanter centrifuges, sieves, sand filters, rotary drum filters, band filters, plate filters, nutsche filters, membrane systems, centrifuges.

3.Distillation

Distillation is carried out to separate or purify volatile components from less volatile components.A distillation unit always consists of a means of heating the feed, the column or the vapour line (with many alternatives of packing to achieve specific results) and a heat—exchanger to condense the vapours.Distillation is still the most widely used method of separation in the manufacture of fine chemicals and is often the first choice in view of low costs, wealth of experience, and proven performance.

Batch columns are widely used in industrial practice for making fine chemicals.Very often these are packed columns and structured packings of different configurations are now widely used.These packings offer very low values of height equivalent to a theoretical stage; values of 0.15 m to 0.20 m are common, and if desired even a value of 0.10 m can be realized. Structured packings also exhibit low pressure drop, which is a great advantage in distillation at reduced pressure commonly employed in the fine chemicals industry. Absolute pressures at the top of the column of 5 mmHg are common and at times even have values as low as 1 mmHg.

We may encounter problems in the purification of substances with a high normal boiling point. If purification only requires a small number of theoretical stages, Short Path Distillation (SPD), in which pressures can be as low as 0.001 bar, can prove useful. Many vitamins and pharmaceuticals can be processed without deterioration of quality.

Strategies for optimum reflux ratio are covered in standard texts on the subject. In the recent past the subject of batch distillation has attracted attention, and different column configurations and operating policies have been suggested. The column configuration is concerned with the arrangement of the column as to where the product is withdrawn. The operating policy is concerned with how to operate the column, e.g. at constant reflux ratio or constant distillate composition. Inverted, middle-vessel, and multi-vessel configurations have been suggested.

In the inverted configuration the feed is charged to the reflux drum and then continuously added to the top of the column. The inverted column seems to be better suited for cases in which the products are to be recovered at high purity from a feed low in light components, whereas the regular column appears to be best for cases in which the products are to be recovered at high purity from a feed rich in light components.

4. Liquid-liquid extraction

Liquid-liquid extraction or solvent extraction is a separation process which is based on the different distribution of the components to be separated between two liquid phases.

Liquid-liquid extraction is primarily applied where direct separation methods such as distillation and crystallisation cannot be used or are too costly. Liquid-liquid extraction is also used when the components to be separated are heat sensitive (e.g. antibiotics) or relatively non-volatile. Extraction apparatus can be classified into countercurrent columns, centrifugal extractors, and mixer-settlers. In a simple case, even a stirred tank may be applicable. All industrial equipment designs use the principle of dispersing one of the two liquids into the other in order to enlarge the contact area for mass transfer.

2.4.3 Cooling

Cooling can be carried out directly or indirectly (Table 2.2). Direct cooling is also used as a reaction stopper in emergency situations.

Table 2.2 Direct and indirect cooling

	Operation	Description	Environmental
Direct	Water injection	Direct cooling is carried out by injection of water, usually to cool down vapour phases	Waste water streams loaded with vapour contaminants
	Addition of ice or water	Addition of ice or water is carried out to adjust temperature of processes (e.g. to enable temperature jumps or shocks)	Increased volume of waste water streams

续表

	Operation	Description	Environmental
Indirect	Surface heat-exchange	Indirect cooling is provided by surface heat-exchangers, where the cooling medium (e.g. water, brines) is pumped in a separate circuit	Cooling waters and spent brines

2.4.4 Cleaning

Due to the frequent product changes, well established cleaning procedures are required to avoid cross-contamination, e.g. for the production of intermediates and APIs.

The cleaning of equipment, such as reactors, centrifuges and sieves is carried out using water, sodium hydroxide, hydrochloric acid, acetone, specific solvents and steam, depending on the equipment or substances to be cleaned. The cleaning process is finished with water to rinse or with an organic solvent (water free rinsing) where the drying of the equipment is important.

The cleaning process can be carried out in different ways:

① With hoses: Cleaning is carried out with pressurised water from a hose to reduce water consumption.

② Cleaning-in-place units (CIP): Different systems of cleaning have been established to limit emissions and to improve efficiency, such as the CIP system. Cleaning-in-place allows equipment to be cleaned directly inside with water scatterers under pressure and allows cleaning liquids to be recovered (where the operator is not concerned about cross-contamination). CIP also enables the operator to carry out the cleaning process without the need to take the equipment apart or for workers to enter the vessels.

2.4.5 Energy supply

Two main sources of energy are consumed on the typical site: steam, electricity.

Generally, only steam is produced on-site and electricity is supplied by an external source. Cogeneration by self-production of electricity and steam is advantageous on large sites. Energy is normally provided by boilers equipped with turbines fitted with burners for natural gas and fuel oil, with gas being the main fuel (about 95%). Spent solvents are often used as a fuel together with gas. Steam and electricity can also be provided by on-site combined cycle power plants, thermal oxidisers or incinerators.

2.4.6 Vacuum systems

Many processes in organic chemistry are operated under vacuum. A number of criteria influence the selection of a vacuum pump, such as the required pressure difference, volume flows, temperature, etc. The choice of the pump type is relevant also from the environmental point of view. Table 2.3 gives an overview to some pump types and the environmental issues.

Table 2.3 Some pump types and their main environmental issues

Pump type	Medium	Main environmental issues
Liquid ring vacuum pump	Water	Water ring pumps cause relatively large amounts of waste waters streams. If VOCs are present, these contaminate the waste water stream. Especially halogenated hydrocarbons can be a problem
	Solvent	Contamination with the pumped substance, typically connected to a recovery system
Dry vacuum pump	No medium, no lubrication	No contamination of any medium
Dry vacuum pump	No medium, with lubrication	The lubrication oil has to be collected and disposed of

2.4.7 Solvent recovery

Solvent recovery is carried out on-site or off-site. Factors that influence whether a solvent is recovered are:

① purity requirements for internal re-use in the process (e.g. cGMP requirements)

② purity requirements for commercial re-use

③ complexity of the purification process to reach the required purity, e.g. if mixtures form azeotropes

④ gap between the boiling points in the case of solvent mixtures

⑤ purchase costs for the fresh solvent compared to the work-up costs

⑥ amount of waste streams created

⑦ safety requirements

The waste streams can be re-used or recovered, but many need to be considered for a disposal route.

From *Best Available Techniques Reference Document for the Manufacture of Organic Fine Chemicals by European commission.*

New Words and Expressions

vessel [ˈves(ə)l] 血管;船舶;容器

enamel [iˈnæm(ə)l] 搪瓷;珐琅;瓷釉;指甲油

fluidise [ˈfluːidaiz] 使…液化(等于 fluidize)

crystallisation [ˌkristəliˈzeiʃən] 结晶;结晶作用

nutsch 吸滤器

membrane [ˈmembrein] 膜;薄膜;羊皮纸

recuperator [riˈkjupəˌretɚ] 恢复者;同流换热器

azeotrope [ˈeiziətrəup] [化学] 恒沸物;共沸混合物

Notes

(1) Bar,巴,气压单位,“巴”源自希腊语:βάρος(baros),意思是重量。巴和毫巴的概念由英国气象学家 Napier Shaw 先生于 1909 年发明,于 1929 年为国际所接受。气压通常以毫巴作单位,如标准大气压被定义为 1013.25 毫巴(百帕),等于 1.01325 巴。长期以来,世界各地的气象学者即使用毫巴作为测量大气压的众多单位之一。这导致国际单位帕花了一段时间才为人们所广泛采用。毫巴现在仍然被广泛使用,虽然各国官方都渐渐过渡到其实数字是一样的国际单位:hPa(百帕)。

(2) VOCs,volatile organic chemcials,挥发性有机化合物,1989 年 WHO 定义 VOCs 是一组沸点从 50℃至 260℃、室温下饱和蒸气压超过 133.322Pa 的易挥发性化合物。其主要成分为烃类、氧烃类、含卤烃类、氮烃及硫烃类、低沸点的多环芳烃类等。大气污染是直接影响人类以及整个生态环境的因素之一,挥发性有机化合物(VOCs)就是一类较严重的大气污染物,人类也越来越重视对它的治理。

2.5 Processes in Fine Chemical Industry

Fine chemicals are not always produced on a small scale. Production may be measured in thousands of tonnes per year for certain foodadditives and drug intermediates. Although no single production process applies, most fine chemical manufacture is carried out in batch processes with synthesis being followed by separation and purification steps; typically, organic synthesis processes are the most complex. Plant is not dedicated to a single process; it is not unusual for a single plant to be capable of producing 100 separate products.

2.5.1 Raw materials and their delivery to site

The range and diversity of products categorised as fine chemicals means that an equally diverse range of raw materials is required as precursors. They are usually delivered to site by road, in containers ranging from 25 to 40 kg kegs through 200 litre drums to 1000 litre demountable tanks. Dry raw materials may be supplied in bags or sacks. Other principal deliveries to site will include carrier solvents, acids bases and industrial gases.

Carrier solvents are used to transfer material mixes through the reaction steps. They may include a wide range of organic solvents including toluene, methanol, dichloromethane and acetone. Acids and bases are used as buffers in various reaction steps as well as in support services such as wet scrubber systems (acidic or alkali scrubber liquors), wastewater treatment (pH adjustment), and deionised water production (regeneration of ion-exchange beds). Carrier solvents, acids and bases are usually supplied in bulk and delivered by road tanker.

The following industrial gases are used: purge gases (such as nitrogen) and, in certain circumstances, the rare gases (argon, xenon, krypton etc). Hot nitrogen may also be used in product drying.

Gases such as chlorine, hydrogen chloride, hydrogen etc, may also be used in reaction steps. Gases may be supplied in pressurised cylinders, demountable tanks or may be delivered in bulk by road tanker. The latter is most likely for bulk liquefied gases such as oxygen or nitrogen.

2.5.2 Raw materials storage and transfer

Drummed raw materials are usually stored inside purpose-built chemical storage warehouses, equipped with fire suppression equipment, secondary containment to receive spills and, in the event of emergency and fire practices, fire-fighting water. Certain storage methods may have been used in the past that are unacceptable today; these include storage in unprotected outdoor compounds with no spill protection or secondary containment.

Drummed materials are typically moved into and out of storage by fork-lift truck, using dedicated drum clamps. It is not uncommon (and in the past it would have been normal) for drums to be loosely stacked on pallets in transit to and from storage areas—a practice which may have led frequently to accidental spillage. Materials required for one batch process are usually transferred in a single operation. Inside the plant, drums are mostly transported on clamped drum trolleys.

Bulk liquids (solvents, acids and bases) are transferred directly to above-or below-ground bulk storage tanks. According to recommendations for good practice, these should incorporate secondary contained tankage and tanker off-loading stances, the provision of tank overfill protection systems and interlock systems, and adequate standard operating procedures to ensure the safe and secure off-loading of tankers. Less satisfactory procedures may have been common in the past.

Bulk liquids and gases are generally piped into the plant. Pipework is supported on overhead pipe racks or is at ground level within lined pipe trenches. Underground pipework may be present in older facilities and could be acause of leakage.

In the plant itself, raw materials are frequently charged to reactors through ports. Transfer is typically by barrel pump for liquids; dry materials are generally manually charged.

2.5.3 Outline of production processes

Process operations typically involve a batch reactor into which various raw materials, including catalysts, may be added along with a carrier solvent (eg. toluene, methanol, acetone). The reaction may be carried out under pressure, under vacuum or at elevated temperatures. The reaction stage normally finishes with a product separation step involving filtration (eg. under vacuum or pressure) and solvent extraction and/or distillation to remove the product from the mother liquors. For inorganic chemicals, one reaction stage is often sufficient, and there may be little or no wastage. Complex high-value organic chemicals, dyes or vitamins may require six or more stages, in which case the loss of yield at each step becomes very important The raw material consumed can far outweigh the final product. The product is then dried either by vacuum techniques or by flow of heated nitrogen through the slurry. They undergo further purification, depending solid or liquid is usually stored and then product is removed and stored or may on the required specification. The final packaged into containers prior to distribution to customers.

(1) Specialist gases

Specialist gases of high purity include oxygen, nitrogen, argon, neon, krypton and xenon all of which can be manufactured as by-products from cryogenic air separation plants where the air is liquefied and distilled. Filtered air is compressed in a centrifugal compressor and is cooled to near its dew point, in order to dry it, in a heat exchanger. The resulting dry air is passed through an adsorber to remove traces of hydrocarbons and carbon dioxide, and from there into the bottom of the separation column. The column is operated under constant reflux with nitrogen and oxygen removal at the top and bottom respectively. Other gases are removed at differing column levels. Simpler pressure-absorption plants are used where only oxygen or nitrogen are needed. The gases are stored prior to shipment in bulk or pressurised containers.

(2) Food additives

Food additives are generally those chemicals that are combined with foods to effect certain modifications; they include preservatives, colourants, flavourings and stabilisers. Common additives are propionic and benzoic acids (preservatives), acetic acid and sodium citrate (buffers), saccharin (sweetener), ascorbic acid (Vitamin C) and other vitamins, natural thickeners and spices and essential oils (flavourings). Food additives are typically organic compounds. Monosodium glutamate is a common additive used to enhance natural flavours. Its manufacture involves the fermentation of sugar and ammonia in a reactor, followed by centrifugation, evaporation and hydrolysis, normally in sodium hydroxide solution. The liquor is then neutralised and acidifies prior to filtration. The pre-purification process involves crystallisation and colour removal by adding activated carbon. Purification is achieved by centrifugal separation, further crystallisations and product drying.

(3) Dyestuffs and pigments

Dyestuffs and pigments are synthesised from a relatively small number of primary raw materials, principally cyclic aromatic compounds including benzene, toluene, xylenes, naphthalene and aniline; many aliphatic organics are also used. Inorganic feedstocks include sulphuric acid and oleum for sulphonation, chlorine and bromine for alogenation, nitric acid for nitration, sodium nitrite for diazotisation as well as hydrochloric acid, chlorosulphonic acid and ammonia. Batch processing remains the norm in the dye production industry with reaction steps being carried out in traditional cast iron, mild steel, glass-lined or stainless steel reactors. Typically, dye intermediates go to subsequent process steps in the liquid state, although vacuum dryers, spray dryers and rotary dryers may be used where appropriate.

2.5.4 Transfer of the finished product

The final product from the fine chemicals manufacturing industry comprises high purity chemical reagents of high value which form the raw material inputs to subsequent process industries. Products are typically packaged in 200 litre drums or smaller containers and stored in warehouses under carefully controlled conditions. Final products are usually dispatched by lorry in mixed loads to customers.

2.5.5 Wastes

Wastes arising from fine chemicals manufacture will be wastewaters, liquid effluents, solid wastes and spent solvents. Waste waters are typically discharged via on-site treatment facilities to the sewer as trade effluent. On-site treatment may comprise simple primary settlement and pH adjustment or more sophisticated treatment such as anaerobic digestion or wet air oxidation.

Liquid effluents which cannot be discharged to wastewater treatment systems will typically be tinkered to appropriate off-site treatment or disposal facilities, which may include co-disposal landfills, hazardous waste treatment plants (for neutralization, precipitation etc) or high temperature incineration facilities. In the past, spillages were disposed to soakaways and drains.

Solid wastes will typically comprise purification residues such as spent activated carbon and will probably be disposed of off-site to landfill or high temperature incinerator as appropriate. Spent solvents may be collected and recovered on site or sent for off-site recovery. There may be on-site landfill sites or lagoons at older sites.

From *Chemical works: fine chemicals manufacturing works by DOE (Department of the Environment) Industry Profile.*

New Words and Expressions

distillation [disti'leiʃən] 精馏,蒸馏,净化;蒸馏法;精华,蒸馏物
filtration [fil'treiʃən] 过滤;筛选
crystallization [kristəlai'zeiʃən] 结晶化;具体化
extraction [ik'strækʃən] 取出;抽出;拔出;抽出物;出身
deionize [di'aiə,naiz] 除去离子;消电离
oleum ['oliəm] [无化] 发烟硫酸
diazotisation [dai,æzətai'zeiʃən] 重氮化
demountable [di'maʊntəbl] 可拆卸的;可分离的
cryogenic [kraiə'dʒenik] 冷冻的;低温学的;低温实验法的
benzoic [ben'zəʊɪk] 安息香的
citrate ['sɪtreɪt] 柠檬酸盐
saccharin ['sækərɪn] [有化] 糖精;邻磺酰苯甲酰亚胺
monosodium [mɒnəʊ'səʊdiəm] 味精;谷氨酸钠
warehouse ['weəhaʊs] 仓库,货栈
effluent ['eflʊənt] 污水;流出物;废气
soakaway ['səʊkəweɪ] 渗水坑
off-site n.工地外;厂区外
lagoon [lə'guːn] [地理][水文] 泻湖;环礁湖;咸水湖

Notes

(1) **Deionized water**,去离子水,是指除去了呈离子形式杂质后的纯水。国际标准化组织

ISO/TC 147 规定的“去离子”定义为:“去离子水完全或不完全地去除离子物质,主要指采用离子交换树脂处理方法。”现在的工艺主要采用 RO 反渗透的方法制取。应用离子交换树脂去除水中的阴离子和阳离子,但水中仍然存在可溶性的有机物,可以污染离子交换柱从而降低其功效,去离子水存放后也容易引起细菌的繁殖。在半导体行业中,去离子水被称为“超纯水”或是“18 兆欧水”。

(2) **Fire suppression equipment**,灭火设备,消防泡沫、二氧化碳灭火器、干粉灭火器、消防用水、消防沙、石棉被、低压蒸汽等是精细化工厂必备的灭火设备,以应对可能发生的火灾事故。

2.6 Synthesis of Vitamin K1:Quinones

2.6.1 1,2-Naphthoquinone

In a 125-mL Erlenmeyer flask dissolve 3.9 g of the dye Orange Ⅱ in 50 mL of water and warm the solution to 40~50℃.Add 4.5 g of sodium hydrosulfite dihydrate and swirl until the red color is discharged and a cream-colored or pink precipitate of l-amino-2-naphthol separates.To coagulate the product,heat the mixture nearly to boiling until it begins to froth,then cool in an ice bath,collect the product on a suction filter,and wash the residue with water.Prepare a solution of 1 mL of concentrated hydrochloric acid,20 mL of water,and an estimated 50 mg of tin(Ⅱ) chloride (antioxidant); transfer the precipitate of aminonaphthol to this solution and wash in material adhering to the funnel. Swirl,warm gently,and when all but a little fluffy material has dissolved,filter the solution by suction through a thin layer of decolorizing charcoal.Transfer the filtered solution to a clean flask,add 4 mL of concentrated hydrochloric acid,heat over a hot plate until the precipitated aminonaphthol hydrochloride has been brought into solution,and then cool thoroughly in an ice bath.Collect the crystalline,colorless hydrochloride and wash it with a mixture of 1 mL of concentrated hydrochloric acid and 4 mL of water.Leave the air-sensitive crystalline product in the funnel while preparing a solution for its oxidation.Dissolve 5.5 g of iron(Ⅲ) chloride crystals ($FeCl_3 \cdot 6H_2O$) in 2 mL of concentrated hydrochloric acid and 10 mL of water by heating,cool to room temperature,and filter by suction.

Wash the crystalline aminonaphthol hydrochloride into a beaker,stir,add more water,and warm to about 35℃ until the salt is dissolved.Filter the solution quickly by suction from a trace of residue and stir in the iron(Ⅲ) chloride solution.1,2-Naphthoquinone separates at once as a voluminous precipitate and is collected on a suction filter and washed thoroughly to remove all traces of acid.The yield from pure,salt-free Orange Ⅱ is usually about 75%.

1,2-Naphthoquinone,highly sensitive and reactive,does not have a well-defined melting point but decomposes at about 145~147℃.Suspend a sample in hot water and add concentrated hydrochloric acid.Dissolve a small sample in cold methanol and add a drop of aniline;the red product is 4-anilino-l,2-naphthoquinone.

2.6.2 2-Methyl-1,4-naphthoquinone

In the hood,clamp a separatory funnel in place to deliver into a 600-mL beaker,which can be cooled in an ice bath when required. The oxidizing solution to be placed in the funnel is prepared by dissolving 50 g of chromium(Ⅵ) oxide (CrO_3, chromic anhydride) in 35 mL of water and diluting the dark red solution with 35 mL of acetic acid. In the beaker prepare a mixture of 14.2 g of 2-methylnaphthalene and 150 mL of acetic acid, and without cooling run in small portions of the oxidizing solution. Stir with a thermometer until the temperature rises to 60℃. At this point ice cooling will be required to prevent a further rise in temperature. By alternate addition of reagent and cooling, the temperature is maintained close to 60℃ throughout the addition, which can be completed in about 10 min. When the temperature begins to drop spontaneously the solution is heated gently on the steam bath (85~90℃) for 1 h to complete the oxidation.

Dilute the dark green solution with water nearly to the top of the beaker, stir well for a few minutes to coagulate the yellow quinone, collect the product on a Büchner funnel, and wash it thoroughly with water to remove chromium(Ⅲ) acetate. The crude material can be crystallized from methanol (40mL) while still moist, and gives 6.5~7.3 g of satisfactory 2-methyl-1,4-naphthoquinone, mp 105~106℃. The product is to be saved for the preparation of 6, 9, and 11. This quinone must be kept away from light, which converts it into a pale yellow, sparingly soluble polymer.

2.6.3 2-Methyl-1,4-naphthohydroquinone

In an Erlenmeyer flask dissolve 2 g of 2-methyl-l,4-naphthoquinone in 35 mL of ether by warming on a steam bath, pour the solution into a separatory funnel, and shake with a fresh solution of 4 g of sodium hydrosulfite in 30 mL of water. After passing through a brown phase (quinhydrone) the solution should become colorless or pale yellow in a few minutes; if not, add more hydrosulfite solution. After removing the aqueous layer, shake the ethereal solution with 25 mL of saturated sodium chloride solution and 1~2 mL of saturated hydrosulfite solution to remove the bulk of the water. Filter the ethereal layer by gravity through a filter paper moistened with ether and about one-third filled with anhydrous sodium sulfate. Evaporate the filtrate on the steam bath until nearly all the solvent has been removed, cool, and add petroleum ether. The hydroquinone separating as a white or grayish powder is collected, washed with petroleum ether, and dried; the yield is about 1.9 g (the substance has no sharp mp).

2.6.4 Vitamin K_1(2-Methyl-3-phytyl-1Anaphthoquinone)

Place 1.5 g of phytol 2 and 10 mL of dioxane (see margin note) in a 50mL Erlenmeyer flask and warm to 50℃ on a hot plate. Prepare a solution of 1.5 g of 2-methyl-1,4-naphth-ohydroquinone and 1.5 mL of boron trifluoride etherate in 10 mL of dioxane, and add this in portions with a Pasteur pipette in the course of 15 min with constant swirling and while maintaining a temperature of 50℃ (do not overheat). Continue in the same way for 20 min longer. Cool to 25℃, wash the solution into a separatory funnel with 40 mL of ether, and wash the orange-colored ethereal solution with two 40mL portions of water to remove boron trifluoride and dioxane. Extract the unchanged hydroquinone with a freshly prepared solution of 2 g of sodium hydrosulfite in 40 mL of 2% aqueous sodium hydroxide and 10 mL of a saturated sodium chloride solution (which helps break in the resulting emulsion). Shake vigorously for a few minutes, during which time any red color should disappear and the alkaline layer should acquire a bright yellow color. After releasing the pressure through the stopcock, allow the layers to separate, keeping the funnel stoppered as a precaution against oxidation. Draw off the yellow liquor and repeat the extraction a second and a third time, or until the alkaline layer remains practically colorless. Separate the faintly colored ethereal solution, dry it over anhydrous sodium sulfate, filter into a tared flask, and evaporate the filtrate on the steam bath—eventually with evacuation at the aspirator. The total oil, which becomes waxy on cooling, amounts to 1.7~1.9 g.

Add 10 mL of petroleum ether (bp 35~60℃) and boil and manipulate with a spatula until the brown mass has changed to a white paste. Wash the paste into small centrifuge tubes with 10~20 mL of fresh petroleum ether, make up the volume of paired tubes to the same point, cool well in ice, and centrifuge. Decant the brown supernatant liquor into the original tared flask, fill the tubes with fresh solvent, and stir the white sludge to an even suspension. Then cool, centrifuge, and decant as before. Evaporation of the decanted liquor and washings gives 1.1~1.3 g of residual oil, which can be discarded. Dissolve the portions of washed white sludge of vitamin Kj hydroquinone in a total of 10~15 mL of absolute ether, and add a little decolorizing charcoal for clarification, if the solution is pink or dark. Add 1 g of silver oxide and 1 g of anhydrous sodium sulfate. Shake for 20 min, filter into a tared flask, and evaporate the clear yellow solution on the steam bath, removing traces of solvent at the water pump. Undue exposure to light should be avoided when the material is in the quinone form. The residue is a light yellow, rather mobile oil consisting of pure vitamin Kj; the yield is about 0.6~0.9 g. A sample for preservation is transferred with a Pasteur pipette to a small specimen vial wrapped in metal foil or black paper to exclude light.

To observe a characteristic color reaction, transfer a small bit of vitamin on the end of a stirring rod to a test tube, stir with 0.1 mL of alcohol, and add 0.1 mL of 10% alcoholic potassium hydroxide solution; the end pigment responsible for the red color is phthiocol.

2.6.5 Phthiocol (2-Methyl-3-hydroxy-1,4-naphthoquinone)

5 $\xrightarrow{HOONa}$ 10 $\xrightarrow{H_2SO_4}$ 11

10: Oxide, MW 188.17, m.p. 93.5~94.5℃

11: Phthiocol, MW 188.17, m.p. 172~173℃

Dissolve 1 g of 2-methyl-1,4-naphthoquinone in 10 mL of alcohol by heating, and let the solution stand while the second reagent is prepared by dissolving 0.2 g of anhydrous sodium carbonate in 5 mL of water and adding (cold) 1 mL of 30% hydrogen peroxide solution. Cool the quinine solution under the tap until crystallization begins, add the peroxide solution all at once, and cool the mixture. The yellow color of the quinone should be discharged immediately. Add about 100 mL of water, cool in ice, and collect the colorless, crystalline epoxide; yield 0.97g, m.p.93.5 ~ 94.5℃ (pure: 95.5 ~ 96.5℃).

To 1 g of 10 in a 25-mL Erlenmeyer flask add 5 mL of concentrated sulfuric acid; stir if necessary to produce a homogeneous deep red solution, and after 10 min cool this in ice and slowly add 20 mL of water. The precipitated phthiocol can be collected, washed, and crystallized by dissolving in methanol (25 mL), adding a few drops of hydrochloric acid to give a pure yellow color, treating with decolorizing charcoal, concentrating the filtered solution, and diluting to the saturation point. Alternatively, the yellow suspension is washed into a separatory funnel and the product extracted with a mixture of 25 mL each of toluene and ether. The organic layer is dried over anhydrous sodium sulfate and evaporated to a volume of about 10 mL for crystallization. The total yield of pure phthiocol, mp 172 ~ 173℃, is about 0.84 ~ 0.88 g.

From *organic experiments* by Louis F.Fieser

New Words and Expressions

coagulate [kəʊˈægjʊleɪt] 使…凝结 *vi.* 凝结
quinone [ˈkwɪnəʊn; kwɪˈnəʊn] 醌(等于 chinone)
aminonaphthol 氨基萘酚
sodium thiosulfate [无化] 硫代硫酸钠
Erlenmeyer flask 锥形烧瓶;爱伦美氏烧瓶
separatory funnel [化工] 分液漏斗
quinhydrone[有化] 醌氢醌
petroleum ether [油气] 石油醚;石油精
dioxane [daɪˈɒkseɪn] 二氧六环;二氧己环
Pasteur pipette 巴斯德吸管
boron trifluoride 三氟化硼
sodium hydrosulfite 保险粉;[无化] 连二硫酸钠
spatula [ˈspætjʊlə] (调和、涂抹用)抹刀,小铲;[医] 压舌板
supernatant 浮在表面的;上层的
specimen vial 样品瓶
sodium carbonate 碳酸钠

Questions

1. To what general class of compounds does phytol belong?

2. Write a mechanism for the reaction of 2-methyl-1, 4-naphthoquinone epoxide with sulfuric acid to form phthiocol.

Unit 3 Polymer Chemistry

3.1 Introduction of Polymers

A polymer is a large moleculeor macromolecule composed of repeating structural units connected by covalent chemical bonds. Well known examples of polymers include plastics, rybbers and fibres. A simple example is polypropylene whose repeating unit structure is shown below.

$$\text{—}CH_2\text{—}\underset{\displaystyle |}{\overset{CH_3}{}}\!\!CH\text{—}\left[CH_2\text{—}\overset{CH_3}{\overset{|}{CH}}\right]_n\text{—}CH_2\text{—}\overset{CH_3}{\overset{|}{CH}}\text{—}$$

While "polymer" in popular usage suggests "plastic", the term actually refers to a large class of natural and synthetic materials with a variety of properties and purposes. Natural polymer materials such as shellac and amber have been in use for centuries. Biopolymers such as proteins and nucleic acids play crucial roles in biological processes. A variety of other natural polymers exist, such as cellulose, which is the main constituent of wood and paper. Some common synthetic polymers are Bakelite, neoprene, nylon, PVC (polyvinylchloride), polystyrene, polyacryonitrile and PVB (polyvinyl butyral).

3.1.1 Type of polymers

On the basis of origin or source, the polymers are classified into two types: (a) Natural Polymers, (b) Synthetic Polymers. Naturalpolymers are isolated from natural materials, mostly plants and animal sources. Starch and cellulose are very common examples of polysaccharides (Fig.3.1). Synthetic polymers are used extensively in industry. In the intervening century, synthetic polymer materials such as Nylon, polyethylene, Teflon, and silicone have formed the basis for a burgeoning polymer industry. These years have also shown significant developments in rational polymer synthesis. Most commercially important polymers today are entirely synthetic and produced in high volume, on appropriately scaled organic synthetic techniques. Synthetic polymers today find application in nearly every industry and area of life. Polymers are widely used as adhesives and lubricants, as well as structural components for products ranging from children's toys to aircraft. They have been employed in a variety of biomedical applications ranging from implantable devices to controlled drug delivery. Polymers such as poly (methyl methacrylate) find application as photoresist materials used in semiconductor manufacturing and low-k dielectrics for use in high-performance microprocessors. Recently polymers have also been employed in the development of flexible polymer-based substrates for electronic displays.

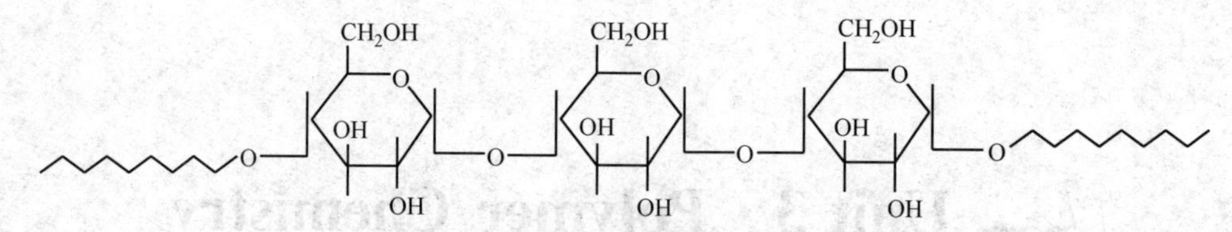

Fig.3.1 Structure of Polysaccharides

Based on their structure, the polymers are classified as: (a) Linear Polymers, (b) Branched Chain Polymers, (c) Cross-linked Polymers or Network Polymers. Linear polymers are the polymers where monomeric units are linked together to form long straight chains. The polymeric chains are stacked over one another to give a well packed structure. As a result of close packing, such polymers have high densities, high tensile strength and high melting points. Common examples of these type polymers are polyethylene, polyester and nylon etc. Branched chain polymers are the polymers where the monomeric units are linked to constitute long chains, which are also called main-chain. There are side chains of different lengths which constitute branches. Branched chain polymers are irregularly packed and thus, they have low density, lower tensile strength and lower melting points as compared to linear polymers. Amylopectin and glycogen are common examples of such type. Cross-linked polymers or network polymers are the polymers where the monomeric units are linked together to constitute a three dimensional network. The links involved are called cross links. Cross-linked polymers are hard, rigid and brittle because of their network structure. Cross-linked polymers are bakelite, formaldehyde resin, melamine, etc.

On the basis of the mode of synthesis, the polymers are classified as: (a) Addition Polymers, (b) Condensation Polymers. When the monomer units are repeatedly added to form long chains without the elimination of any by-product molecules, the product formed is called addition polymer and the process involved is called addition polymerisation. The monomer units are unsaturated compounds and are usually of alkenes. The molecular formula and hence the molecular mass of the addition polymer is an integral multiple of that of the monomer units. A few examples of addition polymerisation are showed Fig.3.2 Condensation polymers are the polymers where the monomers react together with the elimination of a simple molecule like H_2O, NH_3 or ROH, etc. The reaction is called condensation and the product formed is called condensation polymer. As the process involves the elimination of by-product molecules, the molecular mass of the polymer is not the integral multiple of the monomer units. For example, nylon-66 is a condensation polymer of hexamethylene diamine and adipic acid (Fig.3.3).

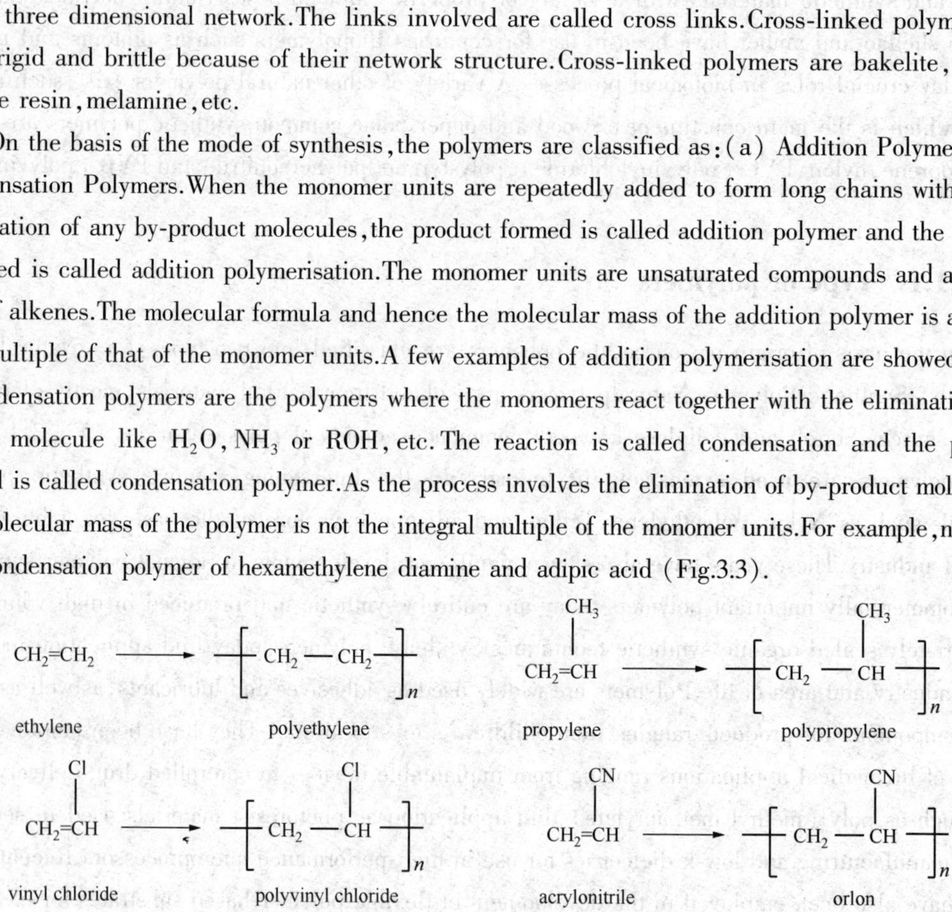

$$\underset{\text{ethylene}}{CH_2{=}CH_2} \longrightarrow \underset{\text{polyethylene}}{\left[CH_2 - CH_2 \right]_n}$$

$$\underset{\text{propylene}}{CH_2{=}CH(CH_3)} \longrightarrow \underset{\text{polypropylene}}{\left[CH_2 - CH(CH_3) \right]_n}$$

$$\underset{\text{vinyl chloride}}{CH_2{=}CH(Cl)} \longrightarrow \underset{\text{polyvinyl chloride}}{\left[CH_2 - CH(Cl) \right]_n}$$

$$\underset{\text{acrylonitrile}}{CH_2{=}CH(CN)} \longrightarrow \underset{\text{orlon}}{\left[CH_2 - CH(CN) \right]_n}$$

Fig.3.2 Examples of addition polymerisation

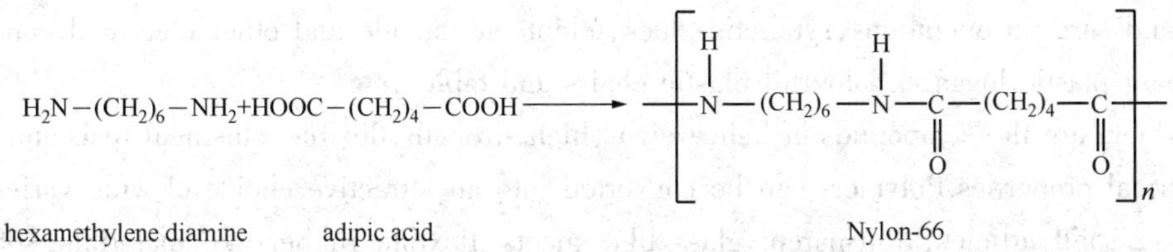

$$H_2N-(CH_2)_6-NH_2+HOOC-(CH_2)_4-COOH \longrightarrow \left[-\overset{H}{\overset{|}{N}}-(CH_2)_6-\overset{H}{\overset{|}{N}}-\underset{O}{\underset{\|}{C}}-(CH_2)_4-\underset{O}{\underset{\|}{C}}- \right]_n$$

hexamethylene diamine　　adipic acid　　Nylon-66

Fig.3.3　Example of condensation polymerisation

On the basis of the magnitude of intermolecular forces, the polymers have been classified into the following four categories: (a) Elastomers, (b) Fibers, (c) Thermoplastics, (d) Thermosetting Polymers. Elastomers are the polymers in which the polymer chains are held up by weakest attractive forces. They are amorphous polymers having high degree of elasticity. The weak forces permit the polymer to be stretched out about ten times their normal length but they return to their original position when the stretching forces is withdrawn. In fact, these polymers consist of randomly coiled molecular chains having few cross links. When the stress is applied, these randomly cross chains straighten out and the polymer gets stretched. As soon as the stretching force is released, the polymer regain the original shape because weak forces do not allow the polymer to remain in the stretched form.

The best known elastomer is rubber, whether synthetic or natural. The elasticity of such polymers can be further modified by introducing few cross-links between the chains. For example, natural rubber, a gummy material, has a poor elasticity. Fibers are the polymers which have quite strong interparticle forces such as Hydrogen-bonds. They have high modulus and high tensile strength. These are thread-like polymers and can be woven into fabrics. Silk, terylene, nylon, etc., are some common examples of such types of polymers. Thermoplastie are the polymers in which the interparticle forces of attraction are in between those of elastomers and fibers. The polymers can be easily moulded into desired shapes by heating and subsequent cooling to room temperature. There is no cross-linking between the polymer chains. In fact, thermoplastic polymers soften on heating and becomes fluids, but on cooling they become hard. They are capable of undergoing such reversible changes on heating and cooling repeatedly. A few examples of thermoplastics are polyethylene, polystyrene, PVC, etc. Thermosetting polymers are the polymers which become hard and infusible on heating. They are normally made from semi-fluid substances with low molecular masses, by heating in mould. Heating results in excessive cross-linking between the chains forming three dimensional network of bonds as a consequence of which a non-fusible and insoluble hard material is produced. Bakelite is a common example of thermosetting polymer. A thermoplastic material can be remelted without any change, while a thermosetting material undergoes a permanent change upon melting and thereafter sets to a solid which can not be remelted.

3.1.2　Importance of polymers

Polymers are the chief products of modern chemical industry which form the backbone of present society. They have become so much a part of our daily life that it appears almost impossible that we could ever do without them. The materials made of polymers find multifarious uses and applications in all walks of our society. Common examples of these include plastic dishes, cups, non-stick pans, kitchen utensils, plastic pipes and fittings, plastic bags, rain coats, automobile tyres, seat covers, TV, radio, computer, transistor, cabinets, synthetic fibres, flooring materials, materials for biomedical and surgical operations, synthetic glues, telephone, mobile and other electrical components, light elegant plastic luggage, colourful plastic chairs and tables, etc.

Polymers are the compounds of light weight, high strength, flexible, chemical resistant with special electrical properties. Polymers can be converted into an attractive choice of wide variety of colours, strong solid articles, transparent glass like sheets, flexible rubber-like materials, soft foams, smooth and fine fibres, jelly-like food materials etc. Polymers can be used to seal joints, bear loads, fill cavities, jerk resistant in between glasswares, and bond objects. Today the polymers are enriching the quality of human life.

From *polymer chemistry by Alka L.Gupta*

New Words and Expressions

shellac[ʃə'læk] 虫胶,虫漆
bakelite['beɪkəlaɪt] 酚醛树脂
neoprene['niːəpriːn] 氯丁橡胶
polyvinylchloride [pɒliː'vaɪnɪlklɒraɪd] 聚氯乙烯
polyacryonitrile [ˌpɒlɪ'ækrələʊ'naɪtrɪl] 聚丙烯腈
polyvinyl butyral[ˌpɔli'vainil 'bjuːtiræl] 聚乙烯醇缩丁醛
polysaccharides['pɒlɪsækærɪdiːz] 多糖
burgeoning ['bɜːdʒənɪŋ] 迅速成长的,迅速发展的
photoresist materials 光致抗蚀剂材料
monomeric units 单体单元
amylopectin [ˌæmələʊ'pektɪn] 支链淀粉,胶淀粉
formaldehyde resin 甲醛树脂
addition polymerisation 加成聚合

Notes

(1) **Teflon**,是杜邦公司使用在其一系列氟聚合物产品上的注册商标。大约70年前,化学家Roy J.Plunkett博士在杜邦位于美国新泽西州的Jackson实验室中发明了聚四氟乙烯树脂,杜邦公司以"Teflon"作为该产品的商标名称。

(2) **Thermoplastics** 热塑性塑料指具有加热软化、冷却硬化特性的塑料。加热时变软以至流动,冷却变硬,这种过程是可逆的,可以反复进行。聚乙烯、聚丙烯、聚氯乙烯、聚苯乙烯、聚甲醛,聚碳酸酯,聚酰胺、丙烯酸类塑料、其它聚烯侵及其共聚物、聚讽、聚苯醚,氯化聚醚等都是热塑性塑料。热塑性塑料中树脂分子链都是线型或带支链的结构,分子链之间无化学键产生,加热时软化流动。

(3) **Thermosetting polymer** 热固性聚合物,又称体形聚合物,加热条件下发生了交联反应,形成了网状或体型结构,再加热时不能熔融塑化,也不溶于溶剂。

3.2 Repeat Units and Degree of Polymerization

A polymer is made up of many small molecules which have combined to form a single large molecule. The individual small molecules which constitute the repeating units in a polymer are known as *monomers* (means, "single parts"). The process by which the monomer molecules are linked to form a big polymer molecule is called "*polymerisation*". For example, polyethylene is a polymer which is obtained by the polymerisation of ethylene. The ethylene molecules are referred to as monomer units (Fig.3.4).

$$\underset{\text{ethylene}}{CH_2{=}CH_2} \longrightarrow \underset{\text{polyethylene}}{\left[CH_2 - CH_2 \right]_n}$$

Fig.3.4 Polymerisation of ethylene

Similarly, butadiene is a gaseous compound, with a molecular weight of 54. It combines about 4000 times and forms a polymer known as polybutadiene, with about 2,00,000 molecular weight (Fig.3.5).

$$\underset{(4000\text{times})}{\text{nButadiene}} \longrightarrow \underset{(\text{Synthetic rubber})}{\text{Polybutadiene}}$$

Fig.3.5 Preparationof synthetic rubber

Polymers are divided into two broad categories depending upon the nature of the repeating units. These are: (1) Homopolymers, (2) Copolymers (mixed polymers).

The polymer formed from one kind of monomers is called homopolymers. For example, polyethylene is an example of homopolymer. The polymer formed from more than one kind of monomer units is called co-polymer or mixed polymer. For example, Buna-S rubber which is formed from 1,3-butadiene ($CH_2{=}CH{-}CH{=}CH_2$) and styrene ($C_6HCH{=}CH_2$) is an example of co-polymer (Fig.3.6).

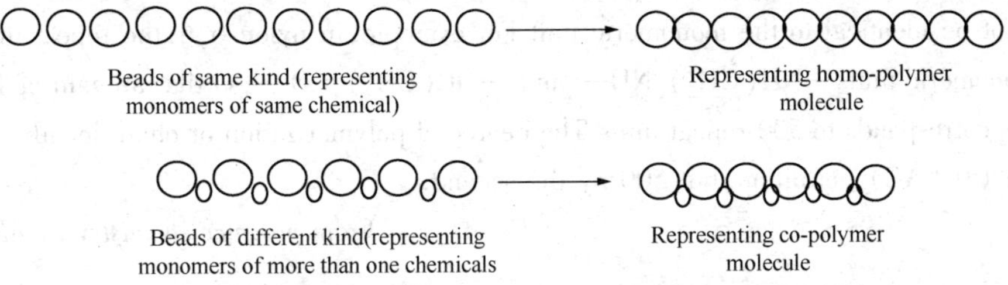

Fig.3.6 Preparation of homopolymer and copolymer

As it is stated earlier that polymerisation is possible with molecules of same or of different monomeric compounds. When molecules just add and form the polymer, the process is called addition polymerisation. In this case the monomer units retain their structural identity when it gets transformed into a polymer. For example, the molecule of ethylene monomer can undergo addition polymerisation and form polyethylene, in which the structural identity of ethylene is retained.

When the two monomers (of the same or different molecules) link with each other by the elimination of a small molecule, such as water or methyl alcohol, as a by-product to form a polymer, the process is called condensation polymerisation. The condensation takes place between two reactive functional groups, like the carboxyl group (—COOH) of an acid and the hydroxy group (—OH) of an alcohol. It is, therefore, observed that in addition polymerisation the molecular weight of the polymer is almost equal to that of all the molecules which combine to form the polymer, while in "condensation polymerisation" the molecular weight of the polymer is lesser than the weight of the simple molecules eliminated during the condensation process (Fig.3.7).

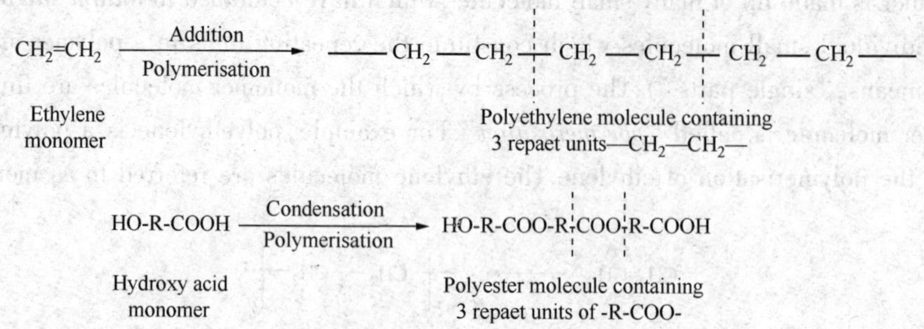

CH_2=CH_2 —Addition Polymerisation→ —CH_2—CH_2—CH_2—CH_2—CH_2—CH_2—

Ethylene monomer

Polyethylene molecule containing 3 repaet units—CH_2—CH_2—

HO-R-COOH —Condensation Polymerisation→ HO-R-COO-R-COO-R-COOH

Hydroxy acid monomer

Polyester molecule containing 3 repaet units of -R-COO-

Fig.3.7 Repeat units of polyethylene and polyester

Polymer bulk properties may be strongly dependent on the size of the polymer chain. Like any molecule, a polymer molecule's size may be described in terms of molecular weight or mass. In polymers, however, the molecular mass may be expressed in terms of degree of polymerization, essentially the number of monomer units which comprise the polymer.

For a homopolymer, there is only one type of monomeric unit and the number-average degree of polymerization is given below.

$$DP_n = X_n = \frac{M_n}{M_0}$$

where M_n is the number-average molecular weight and M_0 is the molecular weight of the monomer unit. For most industrial purposes, degrees of polymerization in the thousands or tens of thousands are desired.

Somepeople, however, define DP as the number of repeat units, where for copolymers the repeat unit may not be identical to the monomeric unit. For example, in nylon-6,6, the repeat unit contains the two monomeric units—$NH(CH_2)_6NH$— and —$OC(CH_2)_4CO$—, so that a chain of 1000 monomeric units corresponds to 500 repeat units. The degree of polymerization or chain length is then 1000 by the first (IUPAC) definition, but 500 by the second.

From *polymer chemistry by Alka L. Gupta*

New Words and Expressions

degree of polymerization 聚合度

polybutadiene [pɒlɪbjuːtə'daɪiːn] 聚丁二烯

homopolymers [həʊmə'pɒlɪmə] 均聚物

Buna-S rubber 丁苯橡胶

methyl alcohol 甲醇

reactive functional groups 反应官能团

number-average molecular weight 数均分子量
monomeric unit 单体单元
step-growth polymerization 逐步聚合
monomer conversion 单体转化率

Notes

(1) Carothers' equation 卡罗瑟斯方程是美国化学家 Wallace Hume Carothers(华莱士·休姆·卡罗瑟斯)在研究逐步聚合理论过程中推导出的。Wallace Hume Carothers 是尼龙和氯丁橡胶的发明者。

(2) Buna 是丁二烯的意思,BUNA 它起源于二战期间,首先由德国合成它用钠做催化剂,故名丁钠橡胶。BUNA 是国外丁钠橡胶的一个牌号,现俄罗斯用的比较多它有两种结构一种是 BUNA-S(相当国内的顺丁胶)BUNA-N(相当国内的丁晴胶)。

3.3 Free-radical Chain Polymerization

This is perhaps the most well-known method of polymerization, and as the name implies, involves the continuous addition of monomer units to a growing freeradical chain. The general mechanism of this process in relation to the polymerization of a vinyl monomer is shown in Fig.3.8. As Fig. 3.8 shows, initiation is a two-stage process in which, first a free radical is formed, and second this radical adds on to a monomer unit. The second stage is essentially the same all the related processes. However, the first step can be achieved in a variety of ways, and the type of initiator depends on the nature of the polymerization experiment.

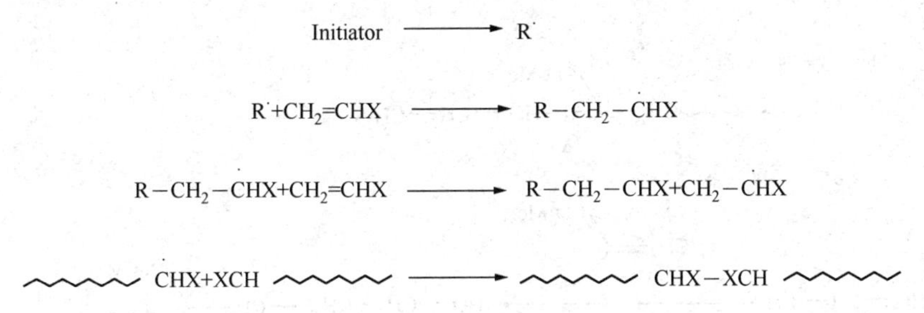

Fig.3.8 Mechanism of free-radical polymerization

In a laboratory, 2, 2′-azo-bisisobutyronitrile (AIBN), in a sealed tube is usually the initiator of choice for this and other free-radical processes presumably because of the convenient timescale of its decomposition (Fig.3.9) of about 18 h at 60℃. More commonly used in an industrial setting are peroxides and, in an aqueous or mixed environment, inorganic initiators such as persulfate and other redox systems. Electromagnetic radiation, usually visible or ultraviolet light but occasionally higher energy radiations such as X-and γ-rays are also of some importance, photoinitiators often being used to cure preformed polymer chains by the polymerization of pendant polymerizable side-groups. In some examples of free-radical initiators, polymerization apparently occurs in the absence of any added initiator, where polymerization is induced by adventitious free-radical production. The propagation step is, of course, the core of the process, but as in all chaingrowth processes, it is the number of monomer units that are added for each initiator molecule that determines the molecular weight of the final material. In the case of free-radical polymerization this is controlled by considering the processes involved in terminating the growing chain, and often these involve radical radical combinations and high molecular weights are favoured by keeping the concentration of free radicals low. In the ideal case, that is, where there is no chain transfer, the number average degree of polymerization is related to the initiator concentration by eqn (1).

$$X_n = \frac{[Monomer]}{\sqrt{Initiator}} \tag{1}$$

$$CH_3-\underset{CH_3}{\overset{CN}{C}}-N{=}N-\underset{CH_3}{\overset{CN}{C}}-CH_3 \longrightarrow 2CH_3-\underset{CH_3}{\overset{CN}{\dot{C}}}+N_2$$

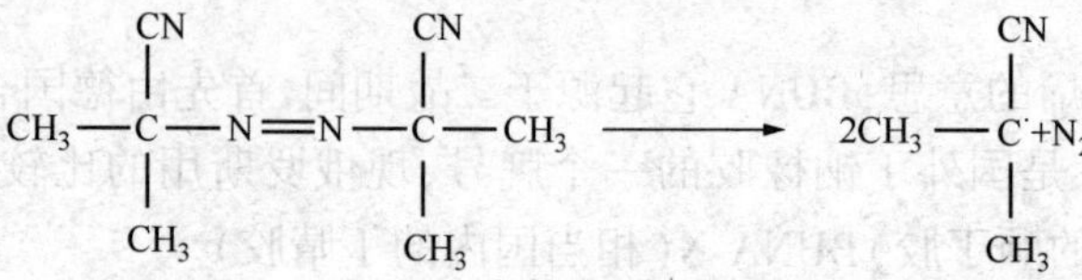

Fig.3.9 Thermal decomposition of the initiator AIBN

It is often found that the molecular weight of the material is rather higher than is convenient, and in these processes, chain-transfer agents may be used. Common chain-transfer agents in use include thiols and halogenoalkanes, and in some cases the solvent may be used to control molecular weight [e.g. both toluene and tetrahydrofuran (THF) may act in this way]. The use of functionalized chain-transfer agents such as 2-mercaptoethanol can lead to monofunctional polymers, as shown in Fig.3.10, and these can be subsequently reacted to form, for example, a block copolymer.

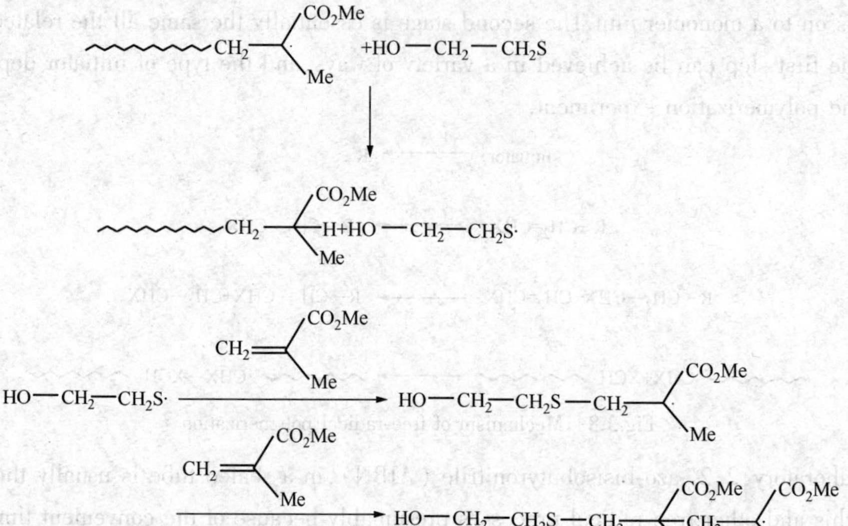

$$\sim\sim CH_2-\dot{C}(CO_2Me)(Me) + HO-CH_2-CH_2S$$

$$\downarrow$$

$$\sim\sim CH_2-C(CO_2Me)(Me)-H + HO-CH_2-CH_2S\cdot$$

$$HO-CH_2-CH_2S\cdot \xrightarrow{CH_2=C(CO_2Me)(Me)} HO-CH_2-CH_2S-CH_2-\dot{C}(CO_2Me)(Me)$$

$$\xrightarrow{CH_2=C(CO_2Me)(Me)} HO-CH_2-CH_2S-CH_2-C(CO_2Me)(Me)-CH_2-\dot{C}(CO_2Me)(Me)$$

Fig.3.10 Chain-transfer process with 2-mercaptoethanol to give a hydroxyl-terminated polymer

3.3.1 Homopolymerization

Although the presence of water is generally not an issue in free-radical chain polymerization (indeed water may be a suitable medium for polymerization as in Protocols) unlike, for example, chain-growth polymerization initiated by anionic species, it is always advisable to use solvents of the highest purity and this will generally include some element of predrying.

For example the polymerization of styrene is shown Fig.3.11. In this reactive process, solvents should be distilled, particularly as a number of suitable solvents for polymerization reactions contain stabilizers which usually serve to mop up free radicals and therefore inhibit the polymerization process. It is usually advisable to predry solvents often with calcium hydrideand then distil them from another drying agent. $LiAlH_4$ is not nowadays a favoured reagent in this latter capacity. Ethereal solvents are not usually suitable for free-radical processes (unless a restricted molecular weight is required). However, where they are used, they are generally dried on a mixture of sodium and benzophenone under a nitrogen atmosphere, and the deep blue colour of the sodium benzophenone ketyl indicating that the solvent is dry. This ketyl also acts to remove oxygen. The demand for dry THF in many laboratories is such that a semi-permanent solvent-still is required. It is important that the users remember the possibility and danger of peroxide build-up in such systems and regularly test for these. This is done by adding 1 mL of the solvent to 1 mL of a 10% (w/v) solution of sodium iodide in acetic acid, and a yellow colour indicates peroxides present, but in low concentrations; a brown colour high concentrations. In any case, a still should be dismantled regularly and cleaned. Solvent-stills are a clear fire risk and should not be used without appropriate precautions and should be carefully monitored during use.

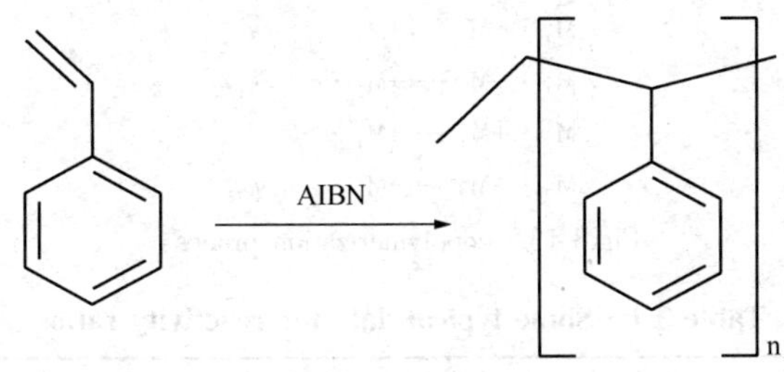

Fig.3.11 Chain-transfer process with 2-mercaptoethanol to give a hydroxyl-terminated polymer

3.3.2 Copolymerization

The introduction of a second monomer into a free-radical polymerization is a useful tool to modify the properties of the resultant polymer. For example, preparation of a poly(styrene-acrylic acid) copolymer by free-radical polymerization is shown Fig.3.12. Such an approach may offer advantages over the blending of the two polymers since the latter procedure does not guarantee a miscible material due to the poor entropy of mixing of large molecules. One simple application of copolymerization might be the introduction of chemical reactive units to allow the incorporation of other units following polymerization. Two examples might be the incorporation of chromophores containing nitro compounds to generate a liquid crystalline compound with specific optical properties or the introduction of hydroxyl or other units to provide site for subsequent cross-linking. This approach can be useful in providing materials that display a permanent memory of their orientation at the time of cross-linking.

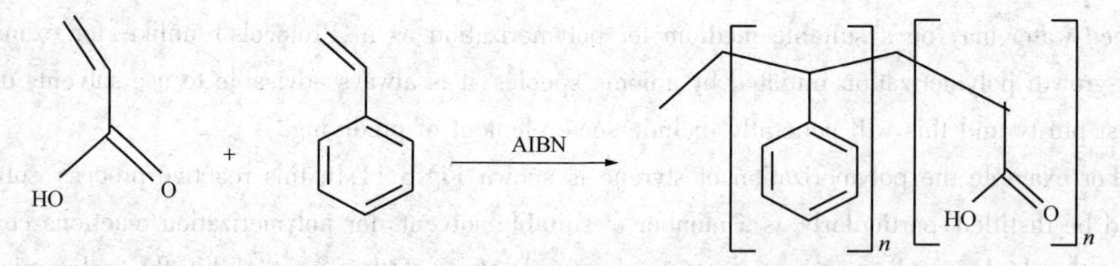

Fig.3.12 A copolymer formed from styrene and acrylic acid

For free-radical copolymers the incorporation of a second monomer is not straightfor—ward. The composition of the final copolymer is determined by the kinetics in a way first described by Dorstal but later elaborated by Alfrey, Mayo and Walling, and others. The kinetic model assumes that the kinetics depends on the end group of the radical chain and the new monomer in a way commonly described for monomers M_1 and M_2 as shown in Fig.3.13.

The monomer reactivity ratios $r_1(=k_{11}/k_{12})$ and $r_2(=k_{22}/k_{21})$ (Table 3.1) reflect the relative rate constants for a given radical adding to its precursor monomer and to the alternative. If the monomers are very similar for example two slightly different acrylates then the values of r_1 and r_2 are close to equal and unity. In such a case, the composition of the polymer is equal to the composition of the feedstock at all stages of the polymerization. If on the other hand, the values are both small as in the case of maleic anhydride and styrene then each monomer is reluctant to react with itself, and the result is an alternating copolymer.

$$M_1\cdot + M_1 \longrightarrow M_1\cdot \qquad k_{11}$$
$$M_1\cdot + M_2 \longrightarrow M_2\cdot \qquad k_{12}$$
$$M_2\cdot + M_1 \longrightarrow M_1\cdot \qquad k_{21}$$
$$M_2\cdot + M_2 \longrightarrow M_2\cdot \qquad k_{22}$$

Fig.3.13 copolymerization process

Table 3.1 Some typical data for reactivity ratios

M_1	M_2	r_1	r_2	Conditions
Styrene	Acrylonitrile	0.4	0.04	60℃
Styrene	Maleic anhydride	0.04	0.01	60℃
Styrene	Methyl methacrylate	0.52	0.46	60℃
Methyl methacrylate	Styrene	0.46	0.42	60℃
Acrylonitrile	Methyl methacrylate	0.15	1.22	80℃

From *polymer chemistry by Fred J.Davis*

New Words and Expressions

freeradical chain 自由基链

electromagnetic radiation 电磁辐射

photoinitiators 光引发剂

propagation[ˌprɒpəˈgeɪʃn] 增长,传播,传输
adventitious [ˌædvenˈtɪʃəs] 外来的,偶然的
thiols [θiˈːəʊlz] 硫醇
halogenoalkanes 卤代烃
2-mercaptoethanol 2-巯基硫醇
benzophenone [ˈbenzəfɪnən] 苯甲酮
peroxides[pəˈrɔkˌsaɪdz] 过氧化物,过氧化氢
chromophores 发色团,生色团
maleic anhydride 马来酸酐,顺丁烯二酸酐
Acrylonitrile [ækrələʊˈnaɪtrɪl] 丙烯腈

Notes

(1) Free-radical chain polymerization 自由基聚合是以自由基为活性链的活性中心的链式聚合。按反应体系的物理状态自由基聚合的实施方法有本体聚合、溶液聚合、悬浮聚合、乳液聚合和超临界二氧化碳聚合五种聚合方法。

(2) Reactivity ratio 竞聚率是指在链式共聚合时,以某一单体的结构单元为末端的活性链分别与该单体及参与共聚的另一种单体的加成反应的速度常数之比。

3.4 Glass Transition Temperature

Almost all amorphous polymers and many crystalline polymers possess a temperature boundary. Above this temperature the substance remains soft, rubbery and flexible, and below this temperature it becomes hard, glassy and brittle. The temperature, below which a polymer is hard and above which it is soft is called the glass transition temperature. The hard, glassy, brittle state is known as the *glassy state* and the soft, rubbery, flexible state is the *rubbery* or *viscoelastic state*. The glass transition temperature is denoted by T_g. T_f is another term for temperature. When a polymer is heated further, it forms a viscous liquid and starts flowing, this state is known as viscous-fluid state and the temperature is termed as flow temperature (T_f). The different states with change of temperature are as:

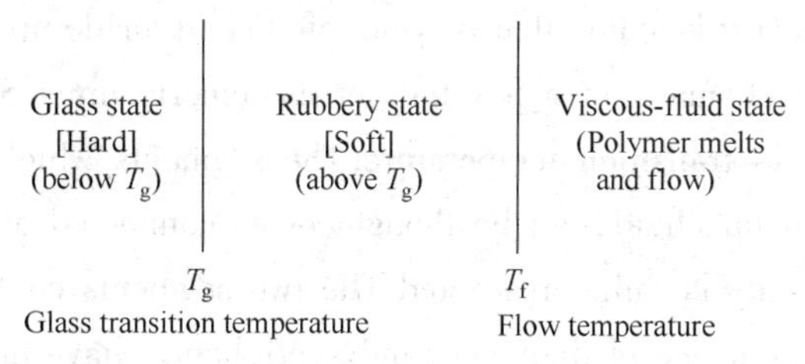

An ordinary natural rubber ball if cooled below -70℃ becomes so hard and brittle that it will break into several pieces like a glass ball falling on a hard surface. This happens because there is a temperature boundary for amorphous and many crystalline polymers. The transition from the rubber to the glass-like state is an important feature of polymer behaviour, marking as it does a region where dramatic changes in the physical properties, such as hardness and elasticity, are observed. The changes are completely reversible and the transition from a rubber to a glass is a function of molecular motion, not polymer structure. In the rubber-like state or in the melt the chains are in relatively rapid motion, but as the temperature is lowered the movement becomes progressively slower until the available thermal energy is insufficient to overcome the rotational energy barrier in the chain. At this temperature, T_g, the chains become locked in whichever, the conformation they possessed when T_g was reached. Below T_g, the polymer is in the frozen liquid (glassy) with a completely random structure. It is quite obvious that T_g is an important characteristic property of any polymer as it has an important bearing on the potential application of a polymer.

3.4.1 Experimental Demonstration of T_g

The glass transition temperature is not specific to long chain polymers. Any substance, which can be cooled to a sufficient degree below its melting point (T_m) without crystallizing, will form a glass. The phenomenon can be conveniently demonstrated using glucose penta-acetate (CPA). A crystalline sample of CPA is melted, then chilled rapidly in ice-water to form a brittle amorphous mass. By working the hard material between one's fingers, the transition from glass to rubber will be felt when the sample is warmed up. A little perseverence, with further rubbing and pulling, will result in the recrystallization of the rubbery phase, which then crumbles to a powder.

3.4.2 Effect of Molecular Weight on T_g

The molecular weight of a polymer affects on glass transition temperature (T_g), but up to a limited value, *i.e.*, 20,000. Above this value, the effect of the molecular weight does not account. The mathematical relationships between T_g and molecular weight of a polymer have been made as:

$$T_g = T_g^{\infty} - \frac{K}{M_n} \quad (1)$$

$$\frac{1}{T_g} = T_g^{\infty} - \frac{K}{M_n} \quad (2)$$

where, T_g^{∞} = glass transition temperature at infinite molecular weight

K and A = arbitrary constants

To understand this effect it is stated that polymer chains are made up of several monomeric units comprise many segments, each made of a few tens of monomeric units. Since the mobility of chain segments influences the glass transition temperature. The segments which are away from the chain ends, although part of the main chain, can be thought of as connected at the two ends to the main chain and, hence their mobility is rather restricted. The two segments containing the chain ends, are connected to the main chain at one of their ends only and, hence, have more freedom for motion.

For a definite weight of the polymer, a low molecular sample will have more chain end segments than a high molecular weight polymer sample. The larger the number of chain end segments, the larger will be the effective segmental motion. Thus, the T_g value will be lower for low molecular weight polymers.

3.4.3 Effect of Chemical Structure on T_g

The chain repeat units, intermolecular forces, chain stiffness, symmetry, stiff bonds, hindrance to free rotation along the polymer chain, and bulky side groups effects glass transition temperature. A comparison of T_g among various chemicals gives an idea about the T_g effect on chemical structure, as follow(Table 3.2):

Table 3.2 T_g of polymer with chemical structure

Polymers	T_g(℃)
Polybutadiene	-85
Styrene-butadiene copolymer(25 : 75)	-55
Polystyrene	+100
Poly (a-methyl styrene)	+150
Polyacenaphthalene	+285

3.4.4 Effect of Chain Topology on T_g

The glass transition temperature(T_g) of random copolymers lies between their corresponding homopolymers; T_g for the copolymer often being a weighted average given by:

$$\alpha_1 w_1 (T_g - T_{g1}) + \alpha_2 w_2 (T_g - T_{g2}) = 0 \tag{3}$$

where, T_{g1} and T_{g2} = homopolymers

w_1 and w_2 = Weight fractions of monomers 1 and 2 in the copolymer

α_1 and α_2 = depend on monomer type.

From the above linear relationships, there are numerous deviations, both positive are negative. The common depression of T_m by copolymerisation occurs, since the changes at T_g do not require fitting a structure into a crystal lattice, and therefore, structure irregularity does not affect T_g as it does T_m. The above considerations for random copolymers lead to low ratios of T_m to T_g. Sometimes the block and graft copolymers consist of long homogeneous chain segments to show the properties of both homopolymers rather than intermediate values. Thus a block copolymer in which one homopolymer has high, and the other low, softening and brittleness temperatures may contain both a high softening point and a low brittleness temperature. Hence, it may have T_m/T_g values higher than the other polymers.

3.4.5 Effect of Chain Branching and Cross-linking on T_g

The effect of chain branching and cross-linking on glass transition temperature can be explained in terms of free volume. If the concentration of chain ends in a branched polymer is high, it increases the free volume, hence causes lowering in glass transition temperature, whereas cross-linking of polymer decreases free volume and increases T_g. Roughly, the above change in cross-linked polymer may be accounted for in terms of the average molecular weight of the segment by the following equation:

$$T_g = T_g^{\infty} - \frac{k}{M_n} \tag{4}$$

where, k = Arbitrary constant

The glass transition temperature is an important property for a polymer. It provides following information: (a) It is used to know whether a polymer molecule is flexible or rigid and brittle. (b) T_g is used to measure the type of response of the polymer whether it exhibits mechanical stress. (c) Glass transition temperature gives an idea about the polymeric material whether it will behave like plastic or rubber. (d) T_g value along with T_m value provides an indication of the temperature region at which a polymer transforms from solid rigid state to a soft viscous state. (e) T_g informed about a right processing temperature. (f) Various processing techniques such as moulding, calendering and extrusion can be used to convert polymeric materials into finished products by knowing their temperature regions.

Polymers above their T_g will be soft and flexible. They possess a delayed elastic response; *i.e.*, viscoelasticity. Polymers below their T_m will be hard and brittle, and will contain dimensional mobility.

From *polymer chemistry by Alka L. Gupta*

New Words and Expressions

glass transition temperature 玻璃化温度
amorphous polymers 非晶态聚合物
crystalline polymers 结晶聚合物
temperature boundary 温度边界,边界层温度
viscoelastic state 黏弹性态
glassy state 玻璃态
rotational energy barrier 转动能垒
conformation[ˌkɒnfɔːˈmeɪʃn] 构造
random structure 随机结构,随机构造
glucose penta-acetate 葡萄糖五乙酸酯
recrystallization 重结晶
segment[ˈsegmənt] 部分,片段,短节
effective segment 有效链节
crystal lattice 晶格
intermediate values 介值
free volume 自由体积
calendering 压延

Notes

Glass transition temperature 玻璃化温度是指高聚物由高弹态转变为玻璃态的温度,指无定型聚合物(包括结晶型聚合物中的非结晶部分)由玻璃态向高弹态或者由后者向前者的转

变温度，是无定型聚合物大分子链段自由运动的最低温度，通常用 T_g 表示。没有很固定的数值，往往随着测定的方法和条件而改变。高聚物的一种重要的工艺指标。在此温度以上，高聚物表现出弹性；在此温度以下，高聚物表现出脆性，在用作塑料、橡胶、合成纤维等时必须加以考虑。

3.5 Polyethylene

Polyethylene is a commonly used polymer. It was first produced in England in 1933 by polymerising the ethylene monomer by the Imperial Chemical Industries, Ltd. (ICI). It is the simplest hydrocarbon polymer and has the following structure:

$$\underset{\text{ethylene}}{CH_2{=}CH_2} \longrightarrow \underset{\text{polyethylene}}{\left[CH_2{-}CH_2 \right]_n}$$

For the production of a polyethylene, ethylene monomer can be prepared either by the dehydration of ethanol or by the hydrogenation of acetylene. Sometimes, it is produced from petroleum products by the cracking process. In cracking process, saturated hydrocarbons such as ethane or propane undergo catalytic degradation and dehydrogenation, producing ethylene. At room temperature, ethylene is a gas and its boiling point is −104℃. There are various varieties of polyethylene, such as:

(a) Low-density (Branched) polyethylene.

(b) High-density (Linear) polyethylene

(c) High and ultra-high molecular weight polyethylene

(d) Cross-linked polyethylene

3.5.1 Low-density (Branched) Polyethylene

Branched polyethylene was the first commercial ethylene polymer, commonly designated as low-density or high pressure material. Low-density polyethylene consists of molecules with branches and is produced by the high pressure polymerisation of ethylene, and using oxygen as the initiator. The reaction occurs at 1400 atm of pressure and in the temperature range of 180~250℃. Besides of oxygen, other initiators are peroxides, hydroperoxides, and azo compounds.

1. Structure of Low-density Polyethylene

Low density polyethylene melts at 110~125℃ and is a partially 50%~60% crystalline solid; with density in the range 0.91~0.94. While practically no solvent dissolves it at room temperature. It is soluble in many solvents at temperature above 100℃. Some of the useful solvents for polyethylene at high temperatures are carbon tetrachloride, toluene, dec aline, trichloro ethylene and xylene. The dissolved polymer precipitates out as the solution cools down to room temperature.

As the low-density polyethylene consists of molecules with branches, this branching occurs during the process of polymerisation either by intermolecular or intramolecular chain transfer reactions as follows:

$$\sim\sim\sim CH_2-\dot{C}H_2 + \sim\sim\sim CH_2-CH_2-CH_2-CH_2-R \longrightarrow$$

$$\sim\sim\sim CH_2-CH_3 + \sim\sim\sim CH_2-\dot{C}H-CH_2-CH_2-R$$

Intennolecular chain transfer

$$\sim\sim\sim CH_2-CH_2-CH_2-CH_2-CH_2-\dot{C}H_2 \longrightarrow$$

$$\sim\sim\sim CH_2-CH_2-CH_2-\dot{C}H-CH_2-CH_3$$

Intramolecular chain transfer

2.Properties of Low-density Polyethylene

The physical properties of low-density polyethylene are functions of three structural variables such as, molecular weight, molecular weight distribution or long-chain branching, and short chain branching. Short-chain branching has a predominant effect on the degree of crystallinity and therefore on the density of polyethylene. Actually these properties are influenced by total chain branching, but the number of long-chain branch points per molecule in polyethylene is too less than the number of short chain branch points that the former can be neglected. Hence, the properties which dependon crystallinity, such as hardness, stiffness, chemical resistance, tear strength, softening temperature and yield point, increase with increasing density or decreasing amount of short-chain branching in the polymer, while toughness, permeability to liquids and gases, and flex life decrease under the same conditions. Polyethylene has good toughness and pliability over a wide temperature range. Its denSity falls rapidly above room temperature; and resulting much dimensional changes which cause difficulty in some fabrication methods. Low crystalline melting point, *i.e.*, about 115℃, limits the temperature range of good mechanical properties.

3.Application of Low-density Polyethylene

For many years, almost two-thirds of the low and medium density branched polyethylene produced has gone into film and sheeting uses. Over three-fourths of the polyethylene film produced goes into packing applications, including pouches, bags, and wrapping for produce, textile products, merchandise, frozen and perishable foods, and many other products. Other film uses include drapes and tablecloths. It is also used in agriculture such as, canal, tank and pond liners, green houses, ground covers etc. and in construction as moisture barriers and utility covering material.

Polyethylene is used as extrusion coating for packaging materials. Low-density polye—thylene is inert to chemicals and resistance to breakage, these properties find used in the manufacture of milk-type cartons for wide variety of foods and drinks. The non-polar nature of the polymer makes it ideal for providing insulations to electric cables. Low-density polyethylene is also used in toys, various housewares, containers, lids, domestic water line connections, and several other purposes.

3.5.2 High-density (linear) Polyethylene

High-density linear polyethylene can be prepared by various methods such as, coordination polymerization of ethylene, polymerisation of ethylene with supported metal oxide catalysts, and radical polymerization of ethylene at extremely high pressures. In coordination polymerisation of ethylene, a catalyst is used. The catalyst prepared as a colloidal dispersion by reacting an aluminium alkyl and $TiC1_4$ in heptane solvent. At relatively low pressure, ethylene is added to the reaction vessel at 50-75℃. After cooling the vessel, polymer is obtained as a powder or granules which is filtered and centrifuged off, washed and dried. At the end of the reaction, the catalyst is destroyed by using alcohol or water.

In another method, supported metal-oxide catalysts such as chromium, or molybdenum oxides, supported over alumina-silica bases is used. The ethylene is fed with a paraffin or cycloparaffin diluent, at 60~200℃ and around 500 psi pressure. The polymer is recovered either by cooling or by solvent evaporation.

High-density linear polyethylenes are highly crystalline polymers containing less than one side chain per 200 carbon atoms in the main chain. Its density is in the range of 0.95-0.97 and melting point is above 127℃.

Linear polyethylenes are hard, stiffed and have higher crystallinity. They have higher melting point, greater tensile strength, good chemical resistance, low temperature brittleness, and low permeability to gases.

High density linear polyethylene is used in the manufacture of toys, containers, films, sheets, wire and cable insulation, extrusion coating, pipe, rotational molding, injection molding, crates, tubs, pails, caps, closures and several other household articles. Whenever high tensile strength and toughness are required, high-density linear polyethylene finds better use.

3.5.3 High and Ultra-high Molecular weight Polyethylene

The average molecular weight of high-density linear polyethylene is 100,000-200,000 whereas, high molecular-weight polyethylene has *Mw* between 300,000 and 500,000. This polymer is widely used commercially. It has long-term stress retention, and improved environmental stress crack resistance, impact and tensile strength. High molecular weight polymer is used in the manufacturing of pipe, films, and large blow-molded containers.

Ultra-high molecular weight poly-ethylene consists of *Mw* between 300,000 and 600,000. This polymer does not flow and melt in the normal plastic manner. It has exceptional impact resistance and abrasion quality. It is used in the high-wear resistance applications such as gaskets, conveyor-belt parts, valve seats etc.

From *polymer chemistry by Alka L. Gupta*

New Words and Expressions

Hydrogenation [ˌhaidrədʒiˈneiʃən] 加氢，氢化作用
Acetylene [əˈsɛtɪˌliːn] 乙炔

Low density polyethylene 低密度聚乙烯
Intermolecular 分子间
predominant effect 主导效应
permeability [pɜˈmɪəˈbɪləti] 渗透性
containers [kənˈtenɚ] 容器，集装箱
granule [ˈgrænjʊl] 颗粒
alumina-silica 硅酸铝
cycloparaffin [ˌsaɪkləʊˈpærəfɪn] 环烷烃
retention [riˈtenʃən] 保留，滞留

Notes

(1) **Low-density polyethylene (LLDPE)** 低密度聚乙烯是指主链中平均每1000个碳原子带有约20~30个乙基、丁基或其他支链，密度通常为0.910~0.925g/cm³ 的聚乙烯。新一代LLDPE将其密度扩大至塑性体(0.890~0.915g/cm³)和弹性体(<0.890g/cm³)。但美国塑料工业协会(SPI)和美国塑料工业委员会(APC)只将LLDPE的范围扩大至塑性体，不包括弹性体。20世纪80年代，Union Carbide和Dow Chemical公司将其早期销售的塑性体和弹性体称之为非常低密度的聚乙烯(VLDPE)和超低密度聚乙烯(ULDPE)树脂。

(2) **Imperial Chemical Industries** 帝国化学工业公司，是英国最大的化工产品生产企业，世界最大化工垄断集团之一。总部设在伦敦。公司是1926年由勃仑纳·蒙特公司、诺贝尔工业公司、联合制碱公司和英国染料公司合并而成的。2007年已被荷兰阿克苏诺贝尔公司收购。

(3) **High-density polyethylene (HDPE)** 高密度聚乙烯，是指主链中平均每1000个碳原子仅含几个支链，密度通常为0.946~0.976g/cm³ 的聚乙烯。HDPE是一种结晶度高、非极性的热塑性树脂。

3.6 Hydrophobically Associating Acrylamide Polymer

Water-soluble polymers are used in the oilfield in polymer flooding, in drilling and completion fluids, in acid stimulation treatments, as drag reducing agents and in profile modification. These water-soluble polymers are usually acrylamide copolymers, partially hydrolyzed polyacrylamide (HPAM), or biopolymers such as xanthan or guar gum. Water-soluble hydrophobically associating polymers (AP) are water-soluble polymers that contain a small number (less than one mole percent) of hydrophobic groups attached directly to the polymer backbone (Fig.3.13). In aqueous solutions, these hydrophobic groups can associate to minimize their exposure to the solvent, similar to the formation of micelles by a surfactant above its critical micelle concentration (cmc) (Fig.3.14).

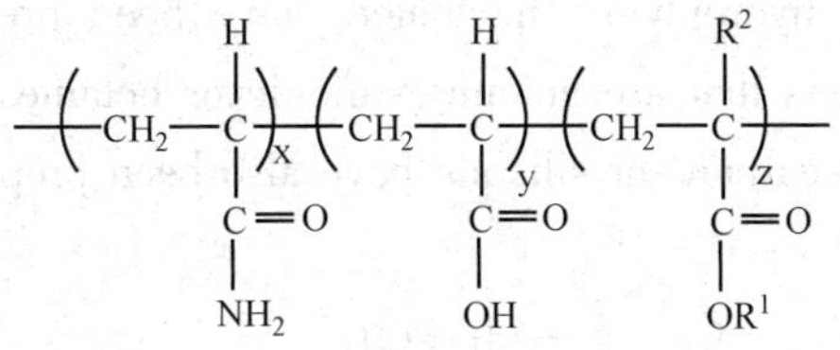

$$\left(CH_2-\underset{\substack{|\\C=O\\|\\NH_2}}{\overset{\substack{H\\|}}{C}}\right)_x\left(CH_2-\underset{\substack{|\\C=O\\|\\OH}}{\overset{\substack{H\\|}}{C}}\right)_y\left(CH_2-\underset{\substack{|\\C=O\\|\\OR^1}}{\overset{\substack{R^2\\|}}{C}}\right)_z$$

R^1=hydrophobic group (8-20)

R^2=H or CH_3

Fig.3.13 Associating Acrylamide Copolymer

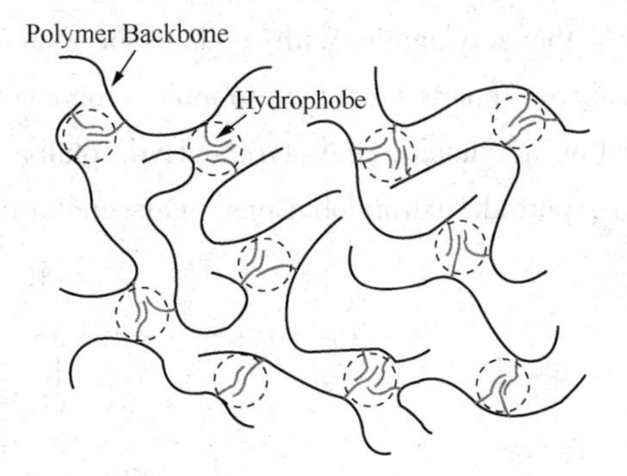

Fig.3.14 Two-Dimensional Representation of Interactions of Hydrophobes in an Associating Polymer

Polymer association can result in solution properties not available with conventional water-soluble polymers. The potential for using hydrophobically associating polymers in the oilfield was extensively reviewed in 1998. In that work, Taylor and Nasr-El-Din examined the synthesis, characterization, stability and rheology of acrylamide-based associating polymers for improved oil recovery. A number of significant developments have occurred since that time. More associating polymers have become commercially available, and new applications for their use in the oilfield have been reported.

3.6.1 Synthesis

Associating acrylamide polymers have most commonly been prepared by copolymerizing acrylamide with a hydrophobic monomer and other monomers such as acrylic acid. Fig.3.15 shows a hydrophobicaiiy associating acrylamide/acrylic acid/dodecyl methacrylate copolymer. The hydrophobic monomer used to prepare the copolymer was dodecyl methacrylate. Carboxylic acid groups have been incorporated into associating polyacrylamides by base hydrolysis after polymerization, but copolymerization is more common.

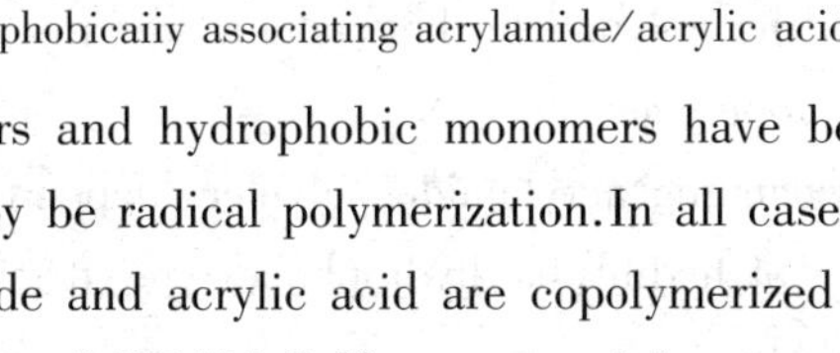

$$\left(CH_2-\underset{\substack{|\\C=O\\|\\NH_2}}{\overset{\substack{H\\|}}{C}}\right)_x\left(CH_2-\underset{\substack{|\\C=O\\|\\OH}}{\overset{\substack{H\\|}}{C}}\right)_y\left(CH_2-\underset{\substack{|\\C=O\\|\\O(CH_2)_{11}CH_3}}{\overset{\substack{CH_3\\|}}{C}}\right)_z$$

x:30~100 y:0~70 z:0.01~1

Fig.3.15 Structure of hydrophobicaiiy associating acrylamide/acrylic acid/dodecyl methacrylate copolymer

Many different monomers and hydrophobic monomers have been used to prepare acrylamide-based associating polymers by be radical polymerization. In all cases, acrylamide is the major monomer. In many cases, acrylamide and acrylic acid are copolymerized with a hydrophobic monomer to produce associating analogues of HPAM. Sulfonate-containing monomers including vinyl sulfonate, 4-vinyl benzene sulfonate and 2-acrylamide-2-methyl-1-propanesulfonic acid (AMPS, Fig. 3. 16) have been used to replace acrylic acid and improve salt sensitivity. N-vinyl pyrrolidone (NVP) has been used in large proportions to make the resulting polymer more resistant to base-catalyzed hydrolysis of the acrylamide. With acylates or methacrylates, n-alkyl esters and polyethoxy n-alkyl esters have been reported as hydrophobic monomers. Other hydrophobic monomers have been prepared based on acrylamide and styrene. Hydrophobic monomers that are anionic, cationic or betaines have been reported. Hydrophobic monomers containing fluorocarbons or silicone have also been prepared.

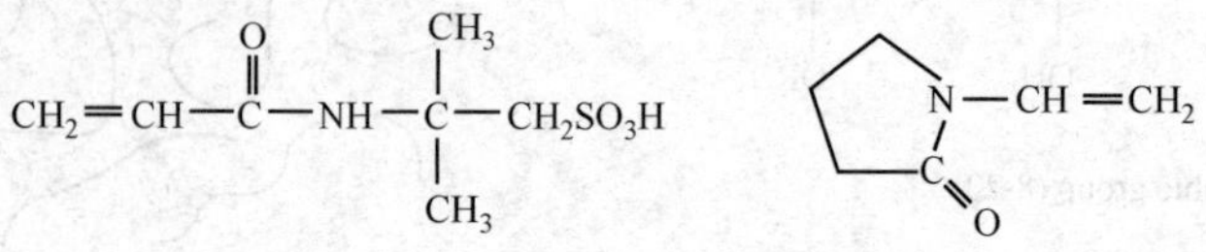

$CH_2{=}CH{-}C(=O){-}NH{-}C(CH_3)_2{-}CH_2SO_3H$

AMPS

$N{-}CH{=}CH_2$ (N-vinylpyrrolidone ring, $C{=}O$)

NVP

Fig.3.16 Structure of monomers

Associating acrylamide polymers have been prepared by incorporating hydrophobic groups into the polymer after the polymerization process. The advantage of this approach is that commercially available polymers can be used as starting material. A disadvantage is that reactions involving viscous polymer solutions are not easily carried out because of problems associated with mixing and reaction homogeneity.

Most of the preparations of associatingacrylamide polymers have used a micellar polymerization technique, in which a surfactant such as sodium dodecyl sulfate (SDS) is used in an aqueous solution to solubilize the hydrophobic monomer. Micelles of SDS may then contain molecules of the hydrophobic monomer. Although other surfactants can be used, SDS is readily available in a pure form. Impurities such as alcohols or heavy metal cations could interfere with the polymerization, resulting in polymers of reduced molecular weight.

3.6.2 Properties

Associating acrylamide polymers have good hydrolytic stability, polymer solutions at a concentration of 2000 ppm in brine (3 mass% NaCl + 0.3 mass% calcium chloride) were heated at 93℃ for 100 days. The increase in the degree of hydrolysis was much lower when AMPS was incorporated into the polymer. For instance, at 93℃, 78% of the amide groups in polyacrylamide (PAM) were hydrolyzed after 100 days. For an associating polymer containing 40 mole percent AMPS, only 36% of the amide groups were hydrolyzed during this time period. When the associating polymer contained 52 mole percent NVP, only 5 percent of the amide groups were hydrolyzed under the same conditions. Associating polymers containing NVP were hydrolyzed more slowly than commercial PAM at 40℃. Surprisingly, associating groups in the polymer further reduced the rate of amide group hydrolysis, as compared to an acrylamide/NVP copolymer.

Rheological properties of associating acrylamide polymers depend on several factors, including the total molecular weight, degree of hydrolysis, hydrophobe type, degree of incorporation of hydrophobe, and distribution of hydrophobe .

Solubility of associating polymers is an important consideration in oilfield applicaitions, since a lack of solubility can lead to formation damage. The solubility of water-soluble associating acrylamide polymers decreases as the hydrophobe content increases. As molecular weight of the polymer increases, or hydrophobe chain length increases, the amount of hydrophobe required to make the polymer insoluble decreases. Obviously, this will limit the maximum hydrophobe content that can be introduced into an associating polymer.When fluorocarbon-containing hydrophobic groups are used, much lower concentrations of the hydrophobic group are required to make the resulting associating polymer insoluble.One way to increase the solubility of associating polymers in water is to introduce ionic character on the polymer backbone.

3.6.3 Application

Water-soluble associatingacrylamide polymers are used in many oilfield operations including drilling, polymer-augmented water fiooding, chemical fiooding and profiie modification.The role of the polymer in most enhanced oil recovery (EOR) field applications is to increse the viscosity of the mobile phase: This increase in viscosity can improve sweep efficiency during enhanced oil recovery processes.In drilling fluids, the solution rheology is very important.Shear thinning fluids are desired that can suspend cuttings at low shear rates, but offer little resistance to flow at high shear rates.

1. Flooding

In the most EOR field applications, one of the most important effects of water-soluble associating acrylamide polymers is to increase the viscosity of the water solution.They can remarkably enlarge the sweep volume and enhance the sweep effective, thus highly increase the produce of petroleum. In Daqing Oil Field China, the produce of petroleum is 12% higher when polymer flooding is used instead of water flooding, The oil increment for every ton of polymer injected is 120 tons and more than 7 million tons have been produced by polymer flooding in 1998.

The partially hydrolytized polyacrylamide (HPAM) is among these polymers and is widely used in polymer flooding and combinational chemical flooding in China.The electrostatic resistance among the core groups makes the molecular chains of HPAM extend and this extension can effectively enlarge the fluid mechanical volume of the polymer and then increase the viscosity.However, in salting solution, the existence of sal ions weakens the electrostatic resistance and makes the chains coil up, which results in an obvious decrease of the fluid mechanical volume and the viscosity.Furthermore, HPAM is easy to be attacked by thermal oxidation degradation when the temperature of oil reservoir is above 60°C and completely lose its viscosity.As the consequences, a high concentration of the polymer is required in actual applications and HPAM is not suitable for oil reservoir with high temperature and high saltness. All these disadvantages seriously retard the propagation of combinational chemical flooding in China.

The synthesized hydrophobically associating polymer has a structure of hydrophilic molecular main chains with a few hydrophobic groups. In the polymer water solution, the hydrophobic groups gather together as pushed by water.It results in physical association both among and within polymer molecular chains.This association can increase the fluid mechanical volume, therefore it has a strong effect on increasing the viscosity of the water solution.Adding small molecular electrolytes into water solution can then increase the polarizability of the water solution and then reinforce the hydrophobic association.The salt-resistance thus improved.Furthermore, the hydrophobic effect can be regarded as a process of entropy increase, which means that as temperature increases, the hydrophobic effect also increase.The hydrophobic association is favored when temperature increase in certain conditions. It means that water-soluble hydrophobically associating polymers should have higher temperature resistance than ordinary water-soluble polymers.

2. Acid Diversion

Associative polymers have been used to achieve fluid diversion during an acid stimulation treatment. Polymers based on dimethylaminoethyl methacrylate were investigated, with hydrophobic groups of C10, C16 or C18. This associating polymer produces a very low viscosity in the acidizing fluid, but reacts with the formation surface to reduce aqueous phase permeability. It is claimed that permanent water permeability reduction occurs in sandstone formations. Coreflood testing in sandstone and carbonate cores showed that the C16 modified polymer demonstrated the highest level of permeability reduction.

Use of associating polymers for simultaneous acid diversion and water control in carbonate reservoirs was reported by Al-Taq et al. Experimental studies included coreflood experiments on reservoir cores at downhole conditions. Extensive lab testing showed that the associative polymers had no significant effect on the relative permeability to oil, butrelative permeability to water was significantly reduced. The associative polymer was used during a matrix acid treatment of a damaged well and included stages of associative polymer and 20 wt% HCl with additives.

From *Water-Soluble Hydrophobically Associating Polymers for Improved Oil Recovery*:
A Literature Review by K.C, Taylor

New Words and Expressions

hydrophobically associating polymers 疏水缔合型聚合物
hydrophobic groups 疏水基团
dodecyl methacrylate 甲基丙烯酸月桂酯
propanesulfonic acid 丙磺酸
N-vinylpyrrolidone *N*-乙烯基吡咯烷酮
polyethoxy 聚醚
betaines [ˈbiːtəiːn] 甜菜碱
polymer-augmented water flooding 聚合物加强注水驱油
hydrolyzed polyacrylamide 水解聚丙烯酰胺
hydrophobic effect 疏水效应
dimethylaminoethyl methacrylate 甲基丙烯酸二甲胺乙酯

Notes

(1) **Hydrophobically associating polymers** 疏水缔合聚合物是指在聚合物亲水性大分子链上带有少量疏水基团的水溶性聚合物。其溶液具有独特的性能,在水溶液中,此类聚合物的疏水基团由于疏水作用而发生聚集,使大分子链产生分子内和分子间缔合。当聚合物浓度高于某一临界浓度(CAC)后,大分子链通过疏水缔合作用聚集,形成以分子间缔合为主的超分子结构—动态物理交联网络,流体力学体积增加,溶液粘度大幅度升高,小分子电解质的加入和升高温度均可增加溶剂的极性,使疏水缔合作用增强。

(2) **Enhanced oil recovery (EOR)** 提高原油采收率是在80年代提出的,它的前身是三次采油,是一种用来提高油田原油采收率的技术,通过气体注入、化学注入、超声波刺激、微生物注入或热回收等方法来实现。

3.7 Some examples of dendrimer synthesis

3.7.1 Introduction

Dendrimers are highly branched macromolecules with unique structural properties. They may be thought of as core-shell type macromolecules wherein they amplify their mass and terminal groups as a function of growth stages. These growth stages are referred to as generations (i.e. G = 0,1,2,...). They possess three key architectural features: (i) a core region; (ii) interior shell zones containing cascading tiers of branch cells (generations) with radial connectivity to the initiator core; and (iii) an exterior or surface region of terminal moieties attached to the outermost generation. With this architecture, a careful choice of building blocks and functional groups can provide control over shape, dimensions, density, polarity, reactivity, and solubility.

One of the earlier dendrimers made, using a divergent strategy, is the Starburst© poly(amidoamine) (PAMAM) dendrimer family. This method involved assembling repeat units to introduce branch cells around the initiator core through successive chemical reactions at the periphery of the growing macromolecule. The first step of PAMAM synthesis involves Michael addition of four moles of methyl acrylate to the nucleophilic ethylenediamine core. This leads to an electrophilic carbomethoxy surface, which is then allowed to react with an excess of 1,2-diaminoethane to give a nucleophilic surface at generation zero. Reiteration of these two steps now involves addition of 8 mol of methyl acrylate to give G=0.5 (electrophilic, carbomethoxy surface). This is followed by amidation to return to a nucleophilic surface at G = 1.0. As a result of this reiterative branch cell assembly, it is apparent that these constructions follow systematic dendritic branching rules, with radial symmetry giving a well-defined three-dimensional geometry to the final dendritic product.

In general, the placement of reactive functionalities on the exterior surface of the dendrimers allows introduction of a wide variety of terminal moieties. In alternate synthetic approaches, spacer groups have been deliberately introduced to relieve the steric hindrance in order to facilitate construction of the next generation. This may provide the possibility of enhancing interior cargo spaces for "guest-host" type chemistry.

3.7.2 Excess reagent method

Synthesis of star-branched ester-terminated precursor: (core: 1,2-diaminoethane; G = 0); [dendri-PAMAM(CO_2Me)$_4$]

(1) Prepare a solution of methyl acrylate (35 g,37 mL,0.407 mol) in methanol (10 mL) and transfer it to the two-necked round-bottomed flask in an ice-bath.

(2) Prepare a solution of 1,2-diaminoethane (5 g,5.5 mL,0.083 mol) in methanol (10 mL) and transfer it to the addition funnel. Add the solution slowly over a period of 2 h, and monitor the rate of addition periodically to ensure that approximately 1.25 mL of this solution is added every 10 min. The final mixture must be stirred for 30 min at 0℃ and then allowed to warm to room temperature followed by stirring for a further 24 h.

(3) Remove the excess solvent under reduced pressure at 40℃, re-dissolve in (20 mL) methanol and evacuate as before, followed by drying the resulting colourless oil under vacuum (10^{-1} mmHg, 40℃) overnight.

(4) Obtain NMR (^1H, ^{13}C), mass spectra, and size exclusion chromatography (SEC) of this product to assure product identity and quality for use in the next growth step.

Synthesis of PAMAM star-branched amine-terminated precursor: (core: 1,2-diaminoethane; G = 0); [dendri-PAMAM(NH_2)$_4$]

Efficient branching amplification requires reactions with a very high degree of selectivity to minimize any structural defects. The tetradirectional amine terminated (G=0) star-branched compound core (1,2-diaminoethane) is a key intermediate in the synthesis of highly pure dendrimeric macromolecules. This generation zero PAMAM intermediate is made by the controlled addition of the ester terminated precursor to a 100-fold excess of 1,2-diaminoethane. Formation of the amide linkage is highly exothermic and it is absolutely essential to keep the reaction below 40℃. Control of the reaction is obtained by adding the ester terminated dendrimer at −5℃. Dendrimer amidation versus bridging amidation are kinetically similar. To prevent both intradendrimeric cyclization as well as interdendrimer bridgings, a large excess of 1,2-diaminoethane is used. The excess can be removed to an undetectable level by azeotropic techniques.

(5) Prepare a solution of ethylenediamine (37.56 g,43 mL,0.625 mol) in methanol (50 mL) and transfer it to a two-necked round-bottomed flask in an ice-bath.

(6) Prepare a solution of PAMAM (G = 0) (5 g,0.0125 mol) in methanol (20 mL) and transfer it to the addition funnel. Add the solution slowly over a period of 10 min and keep the temperature below 0℃. Stir the final mixture for 96 h at room temperature.

(7) When ester groups are no longer detectable by NMR spectroscopy, remove the solvents under reduced pressure maintaining the temperature no higher than 40℃. Remove the excess 1,2-diaminoethane by using an azeotropic mixture of toluene and methanol. The remaining toluene can be removed by azeotropic distillation using methanol. Finally, remove the remaining methanol under vacuum (10^{-1} mmHg, 40℃, 48 h).

(8) Dry the resulting colourless oil under vacuum (10^{-1} mmHg, 40℃) overnight.

(9) Obtain NMR (^1H, ^{13}C), mass spectra, and SEC as this product will be used in the next step.

Synthesis of PAMAM dendrimer ester terminated: (core: 1,2-diaminoethane; G = 1.0); [dendri-PAMAM(CO_2Me)$_8$]

(10) Prepare a solution of methyl acrylate (12.9 g, 13.5 mL, 0.15 mol) in methanol (20 mL) and transfer it to a two-necked round-bottomed flask in an ice-bath.

(11) Prepare a solution of PAMAM (G = 0, amine terminated) (8 g, 0.015 mol) in methanol (20 mL) and transfer it to the addition funnel. Add the solution slowly over a period of1 h and keep the temperature below 0℃. Stir the final mixture for 24 h at room temperature.

(12) Remove the excess solvent under reduced pressure at 50℃ and dry the resulting colourless oil under vacuum (10^{-1} mmHg, 40℃) overnight.

(13) Obtain NMR (^1H, ^{13}C), mass spectra, and SEC as this product will be used in the next step.

Synthesis of PAMAM dendrimer amine terminated: (core: 1,2-diaminoethane; G = 1.0); [dendri-PAMAM(NH_2)$_8$]

(14) Prepare a solution of ethylenediamine (60 g, 65 mL, 0.994 mol) in methanol (100 mL) and transfer it to a two-necked round-bottomed flask in an ice-bath.

(15) Prepare a solution of PAMAM (G = 1.0, ester terminated) (5 g, 0.004 mol) in methanol (20 mL) and transfer it to the addition funnel. Add the solution slowly over a period of 10 min and keep the temperature below 0℃. Stir the final mixture for 96 h at room temperature.

(16) When ester groups are no longer detectable by NMR spectroscopy, remove the solvents under reduced pressure maintaining the temperatureno higher than 40℃. Remove the excess 1,2-diaminoethane by using an azeotropic mixture of toluene and methanol. The remaining toluene can be removed by azeotropic distillation using methanol. Finally, remove the remaining methanol under vacuum (10^{-1} mmHg, 40℃, 48 h).

(17) Dry the resulting colourless oil under vacuum (10^{-1} mmHg, 40℃) overnight.

(18) Obtain NMR (^1H, ^{13}C), mass spectra, and SEC.

3.7.3 Protection - deprotection method

A more compact Starburst® dendrimer can be designed which eliminates the need for excess reagent, by using "protect-deprotect" schemes. In this synthetic approach, the reactive branch cell reagent contains multiple functionalities that are masked in a cyclic structure. For example, a bicyclic ortho ester structure may be used to mask three hydroxyl groups of pentaerythritol which leaves one unprotected hydroxyl group for coupling. An important requirement for this synthesis is the efficient formation of ether linkages. The synthetic amplication is initiated from a tetrabromide PE-Br_4 core. Branch cell reiteration is a four-step process involving: (a) nucleophilic displacement of bromide ions by alkoxide functionality; (b) mild acid hydrolysis of the bicyclic orthoester group to deprotected three hydroxyl groups/orthoester moiety; (c) tosylation of the hydroxyl groups; and finally (d) bromide ion displacement of the tosylate groups to continue the sequence to the next generation level.

Synthesis of the reactive branch cell Reagent: 1-methyl-4-hydroxymethyl-2,6,7-trioxabi-

cyclo-[2,2,2]-octane (MHTBO)

(1) To a round-bottomed flask (500 mL) equipped with a Dean-Stark trap fitted with a reflux condenser and a magnetic stirrer bar, add pentaerythritol (27.2 g, 200 mmol), triethyl orthoacetate (32.44g, 36.6mL, 200mmol), PPTS (1g, 4mmol), and 200mL of dioctylphthalate.

(2) Attach the condenser to the dual manifold using a gas-inlet adapter, and place the system under a nitrogen atmosphere. Heat the reaction mixture on an oil-bath, with stirring at 140℃ for 2~3 h under nitrogen until quantitative recovery of ethanol (32 mL theoretical) is obtained. Replace the nitrogen line with an aspirator vacuum (~25 mmHg) to remove the residual ethanol.

(3) Replace the trap with a large Dewar condenser containing ice-water and evacuate the mixture at < 0.1 mmHg.

(4) Raise the bath temperature to 160℃ to remove any residue product.

(5) Dissolve the crude product in 250 mL of refluxing toluene, filter hot, and allow to cool to room temperature for 3 h. Leave this mixture in the freezer (−10℃) for 18 h. Filter the mixture in a Buchner funnel containing fast flow filter paper.

(6) Allow the white solid to dry in air for 15 min and then vacuum dry at 25℃ overnight.

(7) Obtain NMR (^1H, ^{13}C), and mass spectra, as this product will be used in the next step.

Synthesis of poly(ether) dendrimers from a pentaerythritol core PE(MBO)$_4$

(8) To a two-necked 500 mL round-bottomed flask equipped with a mechanical stirrer, a powder addition funnel, c and a rubber septum, add sodium hydride (7.3 g, 304 mmol, 12 g of a 60% dispersion in mineral oil) and 100 mL of hexane. Stir the mixture for 5 min then allow the reactants to settle to give a clear mixture and decant into a beaker of containing methanol. Repeat this procedure three times d and then remove the mechanical stirrer and connect the system to the dual manifold using a gas inlet. Evacuate the greyish-white slurry at high vacuum to a constant weight of sodium hydride (6.5 g, 270 mmol). Then, with the flask under an atmosphere of nitrogen, add anhydrous DMF (150 mL) via a cannula.

(9) Add MHTBO (39.2 g, 245 mmol) to the addition funnel and place under an inert atmosphere using the manifold system. Then add the MHTBO over ~30 min. After most of the gas evolution has ceased, heat at 60℃ for 1.5 h until gas evolution has ceased completely.

(10) Add pentaerythrityl tetrabromide (20 g, 51.6 mmol, 206 mmol bromide) to the above mixture.

(11) Heat the mixture at 75℃ for 22 h under nitrogen. Cool the mixture to 25℃ and add dropwise to a flask containing 1 L of a well-stirred ice-water.

(12) Filter this mixture in a large Buchner funnel containing fast flow filter paper. Wash the white solid with deionized water (4×100 mL) and dry the solid at 40℃ under high vacuum overnight.

(13) Obtain NMR (^1H, ^{13}C), mass spectra, and SEC, as this product will be used in the next step.

Synthesis of poly(ether): [dendri-PE(OH)$_{12}$]

(14) Add PE(MBO)$_4$ (8 g, 11.4 mmol) to methanol (130 mL) in a two-necked round-bottomed flask (250 mL) equipped with a condenserand a dropping funnel.

(15) Add 1.2 mL of concentrated HCl to the latter reaction mixture. Gently reflux for 1 h.

(16) Add a Dean-Stark trap to the system (between the flask and condenser), distil methanol and methyl acetate until only about one-third of the solvent remains. Cool this mixture to 10℃.

(17) Filter the precipitate and wash it with methanol. Dry the final product under high vacuum overnight at 25℃.

(18) Obtain NMR (^{1}H, ^{13}C), mass spectra, and SEC, as this product will be used in the next step.

Synthesis of poly(ether): [dendri-PE(Tos)$_{12}$]

(19) Add dendri-PE-(OH)$_{12}$ (2 g, 3.29 mmol) to a flame-dried 500 mL three-necked flask equipped with a stirrer bar, a pressure-equalizing dropping funnel fitted with a rubber septum, a condenser, and a thermometer, and attached via the condenser to the dual manifold.

(20) Add 40 mL anhydrous pyridine via a cannula. Cool the mixture to 0℃ .

(21) In a flame-dried flask prepare a solution of p-toluenesulfonyl chloride (18.7 g, 9.8 mmol, 30 equiv. per dendri-PE-(OH)$_{12}$) in 100 mL anhydrous pyridine.e Cannula transfer this mixture to the dropping funnel. Add this solution from the dropping funnel to the PE-(OH)$_{12}$ solution maintaining the temperature at 0~5℃. Maintain the temperature at 0℃ and stir for 1 h.

(22) Seal the flask and leave the mixture at room temperature for 4 days.

(23) Pour this mixture into 500 mL ice-water and decant the solvent after the precipitate has agglomerated at the bottom of the beaker.

(24) Dry the crude product at 40℃ under high vacuum overnight. Dissolve this solid in 100 mL of chloroform and filter the precipitate. Dry the final product under high vacuum overnight at 25℃.

(25) Obtain NMR (^{1}H, ^{13}C), mass spectra, and SEC, as this product will be used in the next step.

From*polymer chemistry by Donald A. Tomalia*

New Words and Expressions

safety goggle 安全护目镜

rotary evaporator 旋转蒸发器

methyl acrylate [有化] 丙烯酸甲酯;丙烯酸酯

azeotropic [ˌeiziəˈtrɔpik] 共沸点的;恒沸点的

toluene [有化] 甲苯

nucleophilic displacement 亲核置换

alkoxide [ælˈkɔksaid] [有化] 醇盐;酚盐

tosylation 甲苯磺酰化

chronological sequencing 时间顺序

pressure-equalizing dropping funnel 恒压滴液漏斗

toluenesulfonyl chloride 对甲苯磺酰氯

dimethylformamid 二甲基甲酰胺

sodium hydride [无化] 氢化钠

pentaerythritol [有化] 季戊四醇

triethyl orthoacetate 原乙酸三乙酯

dioctylphthalate 邻苯二甲酸二辛酯

Notes

(1) **Donald A.Tomalia** 美国化学家,最早的关于"树枝状聚合物"的合成报告可以追溯到唐纳德·托马里亚(Donald A.Tomalia)于1987年获得的专利,当时被称为"棒型树枝状化合物"。

(2) **Dendrimer** 树枝状聚合物,又称树枝化聚合物,是每个重复单元上带有树枝化基元(dendron)的线状聚合物。树枝状聚合物最早由美国化学家 Tomalia DA 博士于20世纪80年代初发明并成功合成。树枝状聚合物在生物医学领域从简单的药物运送载体,到复杂的医疗成像等多个方面都得到了应用,包括纳米级生物传感器、纳米级催化剂等,树枝状聚合物具有精确的纳米构造,其合成方法有发散法和会聚法。由合成步骤决定了树枝状聚合物精确的代数(generations 或层数 layers)与体积。树枝状聚合物的直径范围从 G0 代到 G10 代分别为10~130nm。与普通高分子聚合物不同,树枝状聚合物具有低粘度、高溶解性、可混合性以及高反应性等特点。同时其体积和形态还可在合成过程中加以专一性的控制。比如,设计出具有巨大内部疏水空间(hydrophobic void spaces),而表面却是亲水性质的树枝状聚合物。

Unit 4 Surfactant Chemistry

4.1 History and Applications of Surfactants

4.1.1 Introduction

Surfactants (or 'surface active agents') are organic compounds with at least one lyophilic ('solvent-loving') group and one lyophobic ('solvent-fearing') group in the molecule. If the solvent in which the surfactant is to be used is water or an aqueous solution, then the respective terms 'hydrophilic' and 'hydrophobic' are used. In the simplest terms, a surfactant contains at least one non-polar group and one polar (or ionic) group and is represented in a somewhat stylised form shown in Fig 4.1.

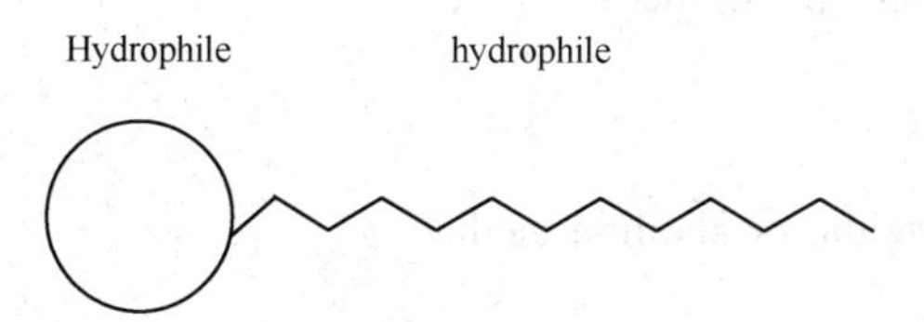

Fig.4.1 Simplified surfactant structure

Two phenomena result from these opposing forces within the same molecule: adsorption and aggregation. For example, in aqueous media, surfactant molecules will migrate to air/water and solid/water interfaces and orientate in such a fashion as to minimise, as much as possible, the contact between their hydrophobic groups and the water. This process is referred to as 'adsorption' and results in a change in the properties at the interface.

Likewise, an alternative way of limiting the contact between the hydrophobic groups and the water is for the surfactant molecules to aggregate in the bulk solution with the hydrophilic 'head groups' orientated towards the aqueous phase. These aggregates of surfactant molecules vary in shape depending on concentration and range in shape from spherical to cylindrical to lamellar (sheets/layers). The aggregation process is called and the aggregates are known as '*micelles*'. Micelles begin to form at a distinct and frequently very low concentration known as the '*critical micelle concentration*' or '*CMC*'.

In simple terms, in aqueous media, micelles result in hydrophobic domains within the solution whereby the surfactant may solubilise or emulsify particular solutes. Hence, surfactants will modify solution properties both within the bulk of the solution and at interfaces.

The hydrophilic portion of a surfactant may carry a negative or positive charge, both positive and negative charges or no charge at all. These are classified respectively as anionic, cationic, amphoteric (or 'zwitterionic') or non-ionic surfactant.

4.1.2 Properties and other criteria influencing surfactant choice

The principle 'surface active properties' exhibited by surfactants are

- Wetting
- Foaming/defoaming
- Emulsification/demulsification (both macro- and micro-emulsions)
- Dispersion/aggregation of solids
- Solubility and solubilisation (hydrotropic properties)
- Adsorption
- Micellisation
- Detergency (which is a complex combination of several of these properties)
- Synergistic interactions with other surfactants

Many surfactants possess a combination of these properties. In addition, depending on the chemical composition of a particular surfactant, some products may possess important ancillary properties including

- Corrosion inhibition
- Substantivity to fibres and surfaces
- Biocidal properties
- Lubricity
- Stability in highly acidic or alkaline media
- Viscosity modification

Thus, by defining the properties required to meet a specific application need, the choice of surfactant is narrowed down. It is further reduced by other criteria dictated by the end use, often underpinned by regulations or directives.

Table 4.1 illustrates a few application areas where other parameters have to be met.

Likewise other regional criteria may also come into play, e.g.

- Use of a surfactant may be banned in defined applications in one region of the world, such as the EU and not elsewhere.
- The surfactant must be listed on the regional inventory of approved chemicals, e.g. EINECS in Europe, TSCA in the United States, etc., to be used in that region.
- In cleaning applications and in uses resulting in discharges of effluent to the environment, the surfactant will be required to meet biodegradability criteria but test methods and biodegradability 'pass levels' vary worldwide.
- Customers may insist on compliance with specific regulations, usually governing permitted use levels (based on toxicological data), even though there is no legal requirement as such (e.g. compliance with FDA, EPA, BGA regulations).
- Specific by-products in some surfactants may give rise to concern on toxicological grounds and permitted by-product levels and/or use of levels of that surfactant may be restricted in particular formulations, e.g. nitrosamine levels in diethanolamides when used in personal care formulations.

Table 4.1　Additional criteria to be met in specific applications

Application	Criteria
Domestic, institutional and industrial cleaning products	Surfactants must be biodegradable
Toiletry and personal care products	Surfactants must be biodegradable Low skin and eye irritation Low oral toxicity
Crop protection formulations used in agriculture	Compliance with EPA regulations (not mandatory/customer requirement) Low phyto toxicity Low aquatic toxicity
Oil field chemicals (off shore) and oil spill chemicals	Must meet marine aquatic toxicity requirements in force in that location
Food grade emulsifiers	Must meet vigorous food additive standards for toxicity, etc.
Emulsion polymers for coatings, inks and adhesives	Emulsifier must comply with FDA or BGA regulations for some applications, e.g. direct/indirect food contact

4.1.3 Surfactant applications

The oldest surfactant is soap, which may be traced back to the ancient Egyptians and beyond. Synthetic surfactants had been produced in the first half of the 20th century but it was only after World War II, with the development of the modern petrochemical industry, that alternative feedstocks to oleochemicals became readily available. Hence chloroparaffins and/or alphaolefins and benzene were used to produce alkylbenzene (or 'alkylate'), processes were developed to produce a range of synthetic fatty alcohols and alkylene oxide chemistry resulted in ethylene oxide and propylene oxide building blocks becoming readily available.

Fatty alcohols derived from either oleochemical or petrochemical sources, they may be needed to produce several families of both non-ionic and anionic surfactants. Coconut oil and tallow have been traditional oleochemical raw materials for many years. However, the significant increase in the production of palm oil over recent decades has had a marked influence on the availability of such feedstocks. Methyl fatty esters, derived from oils and fats or fatty acids, are another key raw material for surfactant production.

On a global basis, the 11 million tonnes or so of surfactants produced each year (excluding soap) utilises approximately equal volumes of oleochemical and petrochemical feedstocks. However, it is interesting to note that, in the case of fatty alcohols, the balance has changed over the last 20 years.

In the mid-1990s, synthetic surfactant production finally overtook soap production, both of which were running at approximately 9 million tonnes per annum. Approximately 60% of all surfactants are used in detergents and cleaning products, ranging from household detergents and cleaners to personal care and toiletry products and a range of specialised hygiene products used in institutional and industrial applications. The other 40% finds application in a broad spectrum of agrochemical and industrial uses where 'detergency' is not required. The appendix attempts to illustrate many of these applications.

In 2003, the household and industrial cleaning market in Europe had a value of nearly 30 billion euro the distribution of which, in value terms (%), may be described as

- Household laundry products 41%
- Industrial and institutional cleaners 18%
- Hard surface household cleaners 11%
- Dishwash household products 11%
- Domestic maintenance products 10%
- Soaps 5%
- Domestic bleach products 4%

The industrial and institutional sector (value worth 5.7 billion euro) may be further broken down into

- Kitchen and catering 30%
- General surfaces 24%
- Industrial hygiene 17%
- Laundry 15%
- Others 14%

The source is AISE internal data/AISE collaboration with A.C.Nielsen.

From *Chemistry and Technology of Surfactants* by *Richard J.Farn*

New Words and Expressions

lyophilic[laɪə(ʊ)'fɪlɪk] [化学] 亲水的
lyophobic [laɪə'fəʊbɪk] [化学] 疏水的
bulk solution 本体溶液
amphoteric [æmfə'terɪk] [化学] 两性的(酸性的或碱性的)
cylindrical [sɪ'lɪndrɪkəl] 圆柱形的;圆柱体的
lamellar [lə'melə] 薄片状的,薄层状的;页片状
microemulsion [maɪkrəʊe'mʌlʃn] 微乳液
solubilisation [sɒljʊbɪlaɪ'zeɪʃən] 溶解,增溶(作用)
substantivity [sʌbstən'tiviti] 直接性;直染性;亲和力
zwitterionic [zwitərai'ɔnik] 两性离子的
chloroparaffin[klɔːrə'pærəfin] 氯化石蜡
biodegradable [ˌbaɪəʊdɪ'greɪdəbl] 生物所能分解的,能进行生物降解的
oleochemical [ˌbaɪəʊ'kemɪkl] 油脂化学
methyl fatty ester 甲基脂肪酸酯

Notes

(1) EINECS,化合物目录数据库(European Inventory of Existing commercial Chemical Substances,欧洲现有商业化学品目录,欧洲现有上市化学物质名录),即生产 ELINCS(European List of Notified Chemical Substances,欧盟委员会发布通报物质)。

(2) FDA,食品和药物管理局(Food and Drug Administration)的简称。FDA 有时也代表美国 FDA,即美国食品药物管理局,美国 FDA 是国际医疗审核权威机构,由美国国会即联邦政府授权,专门从事食品与药品管理的最高执法机关;是一个由医生、律师、微生物学家、药理学家、化学家和统计学家等专业人士组成的致力于保护、促进和提高国民健康的政府卫生管制的监控机构。其它许多国家都通过寻求和接收 FDA 的帮助来促进并监控其本国产品的安全。

4.2 Surfactants in Solution: Monolayers and Micelles

The amphiphilic nature of surfactants causes them to exhibit many properties that appear on first sight, to be contradictory. Because of their special molecular structures, they possess something of a "love-hate" relationship in most solvents, resulting in a tug of war among competing forces striving for a comfortable (energetically speaking) accommodation within a given environment. Surfactants, one might say, appear to feel to some extent that the grass is greener on the other side of the fence, and as a result, they spend much of their time sitting on that "fence" between phases. This chapter will begin the process of expanding on the specifics of how surfactant molecular structures affect their surface activity. Specific topics on the adsorption of surfactants at specific interfaces will be discussed in later chapters. At this point, it is important to understand some of the more important aspects of the solution behavior of surfactants and some of the circumstances that can affect that behavior.

4.2.1 Surfactant Solubility

The specific structures of surfactant molecules, having well-defined lyophilic and lyophobic components, is responsible for their tendency to concentrate at interfaces and thereby reduce the interfacial free energy of the system in which they are found. A molecule with the same elemental composition but a different structural distribution of its constituent atoms may show little or no surface activity. The primary mechanism for energy reduction in most cases will be adsorption at the available interfaces. However, when all interfaces are or begin to be saturated, the overall energy reduction may continue through other mechanisms as illustrated. The physical manifestation of one such mechanism is the crystallization or precipitation of the surfactant from solution—that is, bulk-phase separation such as that seen for a solution of any solute that has exceeded its solubility limit. In the case of surfactants, alternative options include the formation of molecular aggregates such as micelles and liquid crystal mesophases that remain in solution as thermodynamically stable, dispersed species with properties distinct from those of the monomeric solution. Before turning our attention to the subject of micelles, it is necessary to understand something of the relationship between the solubility of a surfactant or amphiphile in the solvent in question and its tendency to form micelles or other aggregate structures.

For most pure solutes, solubility is a more-or-less "yes or no" question. Under a given set of conditions of solvent and temperature, and sometimes pressure, the solute has a specific solubility limit which, when passed, results in the formation of crystals or at least a distinct separate phase that can hypothetically be separated from the solvent or supernatant liquid by physical means. While crystalline hydrates may be separated from water solutions, they will normally have specific compositions that make them unique and subject to characterization by chemical analysis, for example. Surfactants and other amphiphiles, on the other hand, can exhibit a number of intermediate or mesophases in going from a dilute solution of individual or "independent" molecules to crystalline hydrates or anhydrous structures.

A primary driving force for the industrial development of synthetic surfactants was the problem of the insolubility of the fatty acid soaps in the presence of multivalent cations such as calcium and magnesium or at low pH. While most common surfactants have a substantial solubility in water, that characteristic changes significantly with changes in the length of the hydrophobic tail, the nature of the head group, the electrical charge of the counterion, the system temperature, and the solution environment. For many ionic materials, for instance, it is found that the overall solubility of the material in water increases as the temperature increases. That effect is the result of the physical characteristics of the solid phase—that is, the crystal lattice energy and heat of hydration of the material being dissolved.

For ionic surfactants, the solubility of a material will often be observed to undergo a sharp, discontinuous increase at some characteristic temperature, commonly referred to as the Krafft temperature, T_K. Below that temperature, the solubility of the surfactant is determined by the crystal lattice energy and the heat of hydration of the system. The concentration of the monomeric species in solution will be limited to some equilibrium value determined by those properties. Above T_K, the solubility of the surfactant monomer increases to the point at which aggregate formation may begin, and the aggregated species (e.g., a micelle) becomes the thermodynamically favored or predominant form in solution. The micelle may be viewed, to a first approximation, as structurally resembling the solid crystal or a crystalline hydrate, so that the energy change in going from the crystal to the micelle will be less than the change in going to the monomeric species in solution. Thermodynamically, then, the formation of micelles favors an overall increase in solubility. The concentration of surfactant monomer may increase or decrease slightly at higher concentrations (at a fixed temperature), but micelles will be the predominant form of surfactant present above a critical surfactant concentration, the critical micelle concentration (cmc). The apparent solubility of the surfactant, then, will depend on not only the solubility of the monomeric material but also the solubility of the micelles or other aggregate structures. A schematic representation of the temperature-solubility relationship for ionic surfactants is shown in Fig. 4.2.

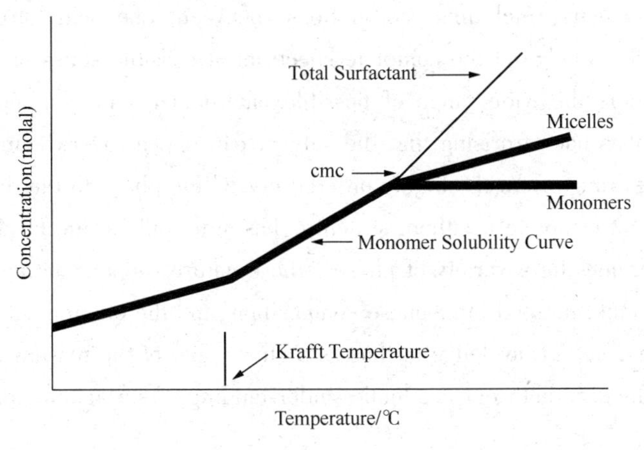

Fig.4.2 Temperature-solubility relationship for typical ionic surfactants

The Krafft temperature can be seen from the data that T_K can vary as a function of both the nature of the hydrophobic group and the character of the ionic interactions between the surfactant and its counterion. It should be noticed that no data are listed for nonionic surfactants. Nonionic surfactants, because of their different mechanism of solubilization, do not exhibit a Krafft temperature. They do, however, have a characteristic temperature-solubility relationship in water that causes them to become less soluble as the temperature increases. In some cases, phase separation occurs, producing a cloudy suspension of surfactant. The temperature (usually a range of temperature) at which the phases separate is referred to as the "cloud point" for that surfactant. This will be disscussed in more detail later. The intimate relationship between the Krafft temperature and the solid state of the surfactant is confirmed by the good correlation between T_K for a surfactant of a given chain length and the melting point of the corresponding hydrocarbon material. Such correlations can also be found for the appearance of other structural changes in surfactant solutions. As we shall see in later chapters, good practical use can be made of such temperature-related phenomena. Note that fluorinated surfactants have Krafft temperatures in roughly the same temperature range as hydrocarbon materials containing twice as many carbon atoms. That tendency is seen, not surprisingly, in comparing most surfactant properties of hydrocarbon versus fluorocarbon materials. As indicated above, an important characteristic of a surfactant in solution is its solubility relative to the critical concentration at which thermodynamic considerations result in the onset of molecular aggregation or micelle formation. Since micelle formation is of critical importance to many surfactant applications, the understanding of the phenomenon relative to surfactant structures constitutes an important element in the overall understanding of surfactant structure-property relationships.

4.2.2 The Phase Spectrum of Surfactants in Solution

Most academic discussions of surfactants in solution concern relatively low concentrations, so the system contains what may be called "simple" surfactant species such as monomers and their basic aggregates or micelles. Before entering into a discussion of micelles, however, it is important to know that although they have been the subject of exhaustive studies and theoretical considerations, they are only one of the several states in which surfactants can exist in solution. A complete understanding of surfactant solution systems, including correlations between chemical structures and surface properties, requires a knowledge of the complete spectrum of possible states of the surfactant.

When one considers the wide range of possible environments for surfactant molecules in the presence of solvents, it is not surprising that the subject can appear overwhelming to the casual observer. The possibilities range from the highly ordered crystalline phase to the dilute monomeric solution, which, although not completely without structure, has order only at the level of molecular dimensions. Between the extremes lie a variety of phases whose natures depend intimately on the chemical structure of the surfactant, the total bulk-phase composition, and the environment of the system (temperature, pH, cosolutes, etc.). Knowledge of those structures, and of the reasons for and consequences of their formation, influences both our academic understanding of surfactants and their technological application.

Pure, dry surfactants, like most materials, can be made to crystallize relatively easily. Because of their amphiphilic nature, however, the resulting structures always appear to be lamellar with alternating head-to-head and tail-to-tail arrangements. The energy of the surfactant crystal, as reflected by its melting point, for example, will be determined primarily by the chemical structure of the molecules. Terminally substituted, n-alkyl sulfates, for example, will have higher melting points than will the corresponding branched or internally substituted materials basically due to the more compact and ordered packing structures available to the straight-chain materials. Additionally, highly polar, small hydrophilic groups will provide enhanced crystal stability over bulky, more polarizable functionalities.

The packing of long hydrocarbon chains into a crystalline alignment is difficult because of the many possible variations in configuration for the units of the chain due to rotation about the four bonds to each carbon atom (rotational isomers). That difficulty is reflected in the relatively low melting points and poorly defined crystal structure of most hydrocarbons under normal conditions. When members of a homologous series or structural isomers are present as a mixture, a situation common to many important surfactant systems, the difficulty of crystal formation, is magnified. The crystallization of pure surfactants from a mixture, therefore, can be difficult, especially if the mixture is that of a series of homologs. For that reason, crystals of natural fatty acid soaps, commercial polyoxyethylene (POE) nonionic surfactants, and other surfactants containing homologous species or branched isomers are rare, and even relatively pure samples may exhibit a variety of crystal structures depending on the conditions of crystallization.

When surfactants are crystallized from water and other solvents that are strongly associated with the polar head group, it is common for the crystalline form to retain a small amount of solvent in the crystal phase. In the case of water, the material would be a hydrate. The presence of solvent molecules associated with the head group allows for the existence of several unique compositions and morphological structures that, although truly crystalline, are different from the structure of the anhydrous crystal.

Two general classes of liquid crystalline structures or mesophases are encountered whether one is considering surfactants or other types of material. These classes are the thermotropic liquid crystals, in which the structure and properties are determined by the temperature of the system, and lyotropic liquid crystals in which the structure is determined by specific interactions between the surfactant molecules and the solvent. With the exception of the natural fatty acid soaps, experimental data support the view that almost all surfactant liquid crystals are lyotropic in nature.

From *Surfactant Science and Technology* by *Drew Myers*

New Words and Expressions

amphiphilic[æmfəˈfɪlɪk][生化] 两亲的;两性分子的
supernatant liquid [化学] 澄清液,清液层,上层液体
mesophases[ˈmesəufeiz][化学] 中间相
crystal lattice energy 晶格能
crystalline hydrate 结晶水合物
cosolute[kəzəˈluːt][化学] 共存溶质
rotational isomer 旋转异构体
homologous [hɒˈmɒləgəs] 相应的;[生物] 同源的;类似的;一致的
polyoxyethylene[ˈpəʊljɒksjeθɪliːn] 聚氧乙烯
morphological structure 形态结构
amorphous[əˈmɔːfəs] 无定形的;无组织的;[物] 非晶形的
diffraction pattern 衍射图
optical birefringence 光学双折射
lyotropic [ljətˈrɒpɪk] 易溶的;溶致型的

Notes

(1) **Krafft temperature**,克拉夫特温度, 离子型表面活性剂在水中的溶解度随温度的变化与一般无机盐有些相似,即溶解度随温度的升高而增大。但有一个特点,即溶解度随温度的变化存在明显的转折点,这一突变温度称为克拉夫特点(Krafft Point)。在此温度以上,离子型表面活性剂的溶解度急剧变大。同系物的碳氢链越长,其 Kraff 点越高,因此,Kraff 点可以衡量离子型表面活性剂的亲水、亲油性 。

(2) **Cloud poin**,浊点,非离子型表面活性剂在水溶液中的溶解度随温度上升而降低,在升至一定温度值时出现浑浊,这个温度被称之为该表面活性剂的浊点,是反映聚氧乙烯型非离子表面活性剂亲水性的一个指标。

(3) **Liquid crystal**,液晶,一种介于液相和固态晶体的有机化合物,它既具有各向异性的晶体所特有的双折射性,又具有液体的流动性。一般可分热致液晶和溶致液晶两类。

4.3 Emulsion

4.3.1 Introduction

Emulsions are a class of disperse systems consisting of two immiscible liquids. The liquid droplets (the disperse phase) are dispersed in a liquid medium (the continuous phase). Several classes of emulsion may be distinguished, namely oil-in-water (O/W), water-in-oil (W/O) and oil-in-oil (O/O). The latter class may be exemplified by an emulsion consisting of a polar oil (e.g. propylene glycol) dispersed in a nonpolar oil (paraffinic oil), and vice versa. In order to disperse two immiscible liquids a third component is required, namely the emulsifier; the choice of emulsifier is crucial not only for the formation of the emulsion but also for its long-term stability. Emulsions may be classified according to the nature of the emulsifier or the structure of the system (see Table 4.2).

Table 4.2 Classification of emulsion types

Nature of emulsifier	Structure of the system
Simple molecules and ions	Nature of internal and external phase
Nonionic surfactants	O/W, W/O
Surfactant mixtures	Micellar emulsions (microemulsions)
Ionic surfactants	Macroemulsions
Nonionic polymers	Bilayer droplets
Polyelectrolytes	Double and multiple emulsions
Mixed polymers and surfactants	Mixed emulsions
Liquid crystalline phases	
Solid particles	

Several processes relating to the breakdown of emulsions may occur on storage, depending on:

① The particle size distribution and the density difference between the droplets and the medium;

② The magnitude of the attractive versus repulsive forces, which determines flocculation;

③ The solubility of the disperse droplets and the particle size distribution, which in turn determines Ostwald ripening;

④ The stability of the liquid film between the droplets, which determines coalescence; and Phase inversion.

The physical phenomena involved in each breakdown process is not simple, and requires an analysis to be made of the various surface forces involved. In addition, the above processes may take place simultaneously rather then consecutively, which inturn complicates the analysis. Model emulsions, with monodisperse droplets, cannot be easily produced and hence any theoretical treatment must take into account the effect of droplet size distribution. Theories that take into account the polydispersity of the system are complex, and in many cases only numerical solutions are possible. In addition, the measurement of surfactant and polymer adsorption in an emulsion is not simple, and such information must be extracted from measurements made at a planar interface.

A summary of each of the above breakdown processes is provided in the following sections, together with details of each process and methods for its prevention.

Creaming and Sedimentation This process results from external forces, usually gravitational or centrifugal. When such forces exceed the thermal motion of the droplets (Brownian motion), a concentration gradient builds up in the system such that the larger droplets move more rapidly either to the top (if their density is less than that of the medium) or to the bottom (if their density is greater than that of the medium) of the container. In the limiting cases, the droplets may form a close-packed array at the top or bottom of the system, with the remainder of the volume occupied by the continuous liquid phase.

Flocculation This process refers to aggregation of the droplets (without any change in primary droplet size) into larger units. It is the result of the van der Waals attractions which are universal with all disperse systems. Flocculation occurs when there is not sufficient repulsion to keep the droplets apart at distances where the van derWaals attraction is weak. Flocculation may be either strong or weak, depending on the magnitude of the attractive energy involved.

Ostwald Ripening (Disproportionation) This effect results from the finite solubility(et c.) of the liquid phases. Liquids which are referred to as being immiscible often have mutual solubilities which are not negligible. With emulsions which are usually polydisperse, the smaller droplets will have a greater solubility when compared to larger droplets (due to curvature effects). With time, the smaller droplets disappear and their molecules diffuse to the bulk and become deposited on the larger droplets. With time, the droplet size distribution shifts to larger values.

Coalescence This refers to the process of thinning and disruption of the liquid film between the droplets, with the result that fusion of two or more droplets occurs to form larger droplets. The limiting case for coalescence is the complete separation of the emulsion into two distinct liquid phases. The driving force for coalescence is the surface or film fluctuations; this results in a close approach of the droplets whereby the van der Waals forces are strong and prevent their separation.

Phase Inversion This refers to the process whereby there will be an exchange between the disperse phase and the medium. For example, an O/W emulsion may with time or change of conditions invert to a W/O emulsion. In many cases, phase inversion passes through a transition state whereby multiple emulsions are produced.

4.3.2 Industrial Applications of Emulsions

Several industrial systems consist of emulsions of which the following are worthy of mention:

① Food emulsions, such as mayonnaise, salad creams, deserts and beverages.

② Personal care and cosmetic products, such as hand-creams, lotions, hair-sprays and sunscreens.

③ Agrochemicals—self-emulsifiable oils which produce emulsions on dilution with water, emulsion concentrates (droplets dispersed in water; EWs) and crop oil sprays.

④ Pharmaceuticals, such as anesthetics of O/W emulsions, lipid emulsions, double and multiple emulsions.

⑤ Paints, such as emulsions of alkyd resins and latex emulsions.

⑥ Dry-cleaning formulations; these may contain water droplets emulsified in the dry cleaning oil, which is necessary to remove soils and clays.

⑦ Bitumen emulsions are prepared stable in their containers but, when applied the road chippings, they must coalesce to form a uniform film of bitumen.

⑧ Emulsions in the oil industry - many crude oils contain water droplets (e.g. North Sea oil); these must be removed by coalescence followed by separation.

⑨ Oil slick dispersants - oil spilled from tankers must be emulsified and then separated.

The emulsification of unwanted oil is a very important process in pollution control.

The above-described utilization of emulsions in industrial processes justifies the vast amount of basic research which is conducted aimed at understanding the origins of the instability of emulsions and developing methods to prevent their break down. Unfortunately, fundamental research into emulsions is not straightforward, as model systems (e.g. with monodisperse droplets) are difficult to produce. In fact, in many cases, the theoretical bases of emulsion stability are not exact and consequently semi-empirical approaches are used.

From *Emulsion Science and Technology* by *Tharwat F. Tadros*

New Words and Expressions

immiscible[ɪ'mɪsəbl] 不能混合的,不融和的
continuous phase 连续相(胶体系的外相)
disperse phase 分散质,分散相
vice versa [vaisi'vəːsə] 反之亦然
size distribution [冶] 粒度分布;粒径分布;大小分布;径谱
flocculation [flɔkju'leiʃən] 絮凝;絮结产物
anesthetic [ænɪs'θɛtɪk] 麻醉的;感觉缺失的
alkyd resin 醇酸树脂
Bitumen emulsion 沥青乳液,沥青乳浊液
sedimentation[sedɪmən'teɪʃən] [矿业][物化] 沉降,[化学] 沉淀
disproportionation[disprəpɔːʃə'neiʃən] 歧化作用;不相均

Notes

(1) **Microemulsion**,微乳液,是由两种或两种以上互不相溶液体经混合乳化后,分散液滴的直径介于 5~100nm,为透明分散体系,其形成与胶束的加溶作用有关,又称为“被溶胀的胶束溶液”或“胶束乳液”。通常由油、水、表面活性剂、助表面活性剂和电解质等组成的透明或半透明的液状稳定体系。其特点是分散相质点大小在 0.01~0.1μm,质点大小均匀,显微镜不可见;质点呈球状;微乳液呈半透明至透明,热力学稳定,如果体系透明,流动性良好,且用离心

机 100g 的离心加速度分离五分钟不分层即可认为是微乳液；与油、水在一定范围内可混溶。

（2）**Ostwald Ripening**，奥斯特瓦尔德熟化（或奥氏熟化）是一种可在固溶体或液溶胶中观察到的现象，其描述了一种非均匀结构随时间流逝所发生的变化：溶质中的较小型的结晶或溶胶颗粒溶解并再次沉积到较大型的结晶或溶胶颗粒上。奥氏熟化通常会出现在油包水乳剂中，而相对的在水包油乳剂中则会发生絮凝。

4.4 Surfactant Hydrophobic Feedstocks

By far the most common hydrophobic group used in surfactants is the hydrocarbon radical having a range of 8 ~ 22 carbon atoms. Commercially there are two main sources for such materials that are both inexpensive (relatively speaking) and available in sufficient quantity to be economically feasible: "natural" or biological sources such as agriculture and the petroleum industry (which is, of course, ultimately biological). There are, of course, alternative routes to the same materials, as well as other surfactant types that require more elaborate synthetic schemes. Those shown, however, constitute the bulk of the synthetic materials used today.

Many, if not most, surfactant starting materials are not chemically pure materials. In fact, for economic and technical reasons, most surfactant feedstocks are mixtures of isomers whose designations reflect some average value of the hydrocarbon chain length included rather than a "true" chemical composition. In some cases isomeric composition may be indicated in the surfactant name or description, while in others the user is left somewhat in the dark. The term "sodium dodecyl sulfate," for example, implies a composition containing only C_{12} carbon chains. The material referred to as "sodium lauryl sulfate," on the other hand, is nominally a C_{12}-surfactant, but will contain some longer-and shorter-chain homologues. Each source of raw materials may have its own local geographic or economic advantage, so that nominally identical surfactants may exhibit slight differences in surfactant activity due to the subtle influences of raw-materials variations. Such considerations may not be important for most applications, but should be kept in mind in critical situations.

4.4.1 The Natural Fatty Acids

One of the major sources of raw materials for the commercial production of surfactants is also the oldest source—agriculture. Fats and oils (oleochemicals), products of nature's ingenuity and human labor, are triglyceride esters of fatty acids, which can be readily hydrolyzed to the free fatty acids and glycerol. Naturally occurring plant or animal fatty acids usually contain an even number of carbon atoms arranged in a straight chain (no branching), so that groups symbolized by an R in abbreviated nomenclature will contain an odd number that is one less than that of the corresponding acid. The carbons are linked together in a straight chain with a wide range of chain lengths; those with 16 and 18 carbons are the most common. The chains may be saturated, in which case the R group has the formula C_nH_{2n+1}, or they may have one or more double bonds along the chain. Hydroxyl groups along the chain are uncommon, but not unknown, especially in soaps made using castor oil (recinoleic acid). Other substitutions are rare.

Commercially, the largest surfactant outlet for fatty acids is conversion to soap by neutralization with alkali. In a strict sense, this may be considered to be a synthetic process, and soap therefore a synthetic surfactant. However, common usage reserves the term "synthetic" for the more modern products of chemical technology, primarily petroleum-derived, that have been developed in the twentieth century and generally show important improvements over the older soap technology. The chemical processes required for the production of modern surfactants and detergents are also usually much more complicated than the simple neutralization involved in soap manufacture. In this context, the term "simple" is relative, as is well known to anyone who has ever tried to prepare homemade soap.

4.4.2 Saturated Hydrocarbons or Paraffins

The hydrophobic groups derived from petroleum are principally hydrocarbons, originating from the paraffinic or higher-boiling fractions of crude oil distillates. The chain lengths most suitable for surfactant hydrophobes, C_{10}–C_{20}, occur in the crude oil cuts boiling somewhat higher than gasoline, namely, kerosene and above. The main components of kerosene are saturated hydrocarbons ranging from $C_{10}H_{22}$ to $C_{15}H_{32}$, ordinarily containing 10%~25% of straight-chain homologs. There may be significant amounts of branched-chain isomers present, in addition to quantities of saturated cyclic derivatives, alkyl benzenes, and naphthalenes, and minor amounts of other polycyclic aromatics.

The paraffins have the disadvantage of being relatively chemically unreactive so that direct conversion to surfactants is difficult. As discussed above, substitution of one or more hydrogen atoms with halogen offers a pathway to some surfactant systems, but manufacturing complications can be an impediment. It is usually necessary to synthesize the surfactant by way of some more reactive intermediate structures, commonly olefins, alkyl benzenes, or alcohols. Such compounds contain reactive sites that are more easily linked to the required solubilizing groups.

4.4.3 Olefins

Olefins with the desired chain length are prepared by building up molecules from smaller olefins (oligomerization), by breaking down (cracking) larger molecules, or by direct chemical modification of paraffins of the desired chain length. An important historic example of surfactant-grade olefin production by the oligomerization process is the preparation of tetrapropylene, $C_{12}H_{24}$:

$CH_3CH_2{=}CH_2 \longrightarrow$ (mixed isomers of $C_{12}H_{24}$ olefins and higher and lower homologs)

This may be prepared by the oligomerization of propylene, a byproduct of refinery operations, under the influence of a phosphoric acid catalyst. The reaction conditions are drastic, and extensive random reorganization of the product molecules occurs with substantial formation of intermediate isomers in the C_{10}–C_{14} range. The final product is composed of a variety of highly branched isomers and homologs, with the double bond usually situated internally in the molecule.

A second type of built-up olefin is that obtained by the polymerization of ethylene using a Ziegler-Natta catalyst. Such materials are predominantly linear with even carbon numbers, although branched isomers are present in small amounts. The ethylene raw material historically has been more expensive than propylene.The catalyst is also more expensive and the reaction conditions more sensitive and critical than those for propylene oligomerization.The formula is:

$$CH_2=CH_2+Z-N \text{ catalyst} \longrightarrow \text{Predominantly } CH_2=CH-C_9H_{19} \text{ to } C_{19}H_{39}$$

Production of surfactant or detergent-class olefins from higher-molecular-weight precursors is accomplished by the cracking process, which uses high temperatures to split high-molecular-weight paraffins into smaller units. A catalyst may also be employed in the process. Basically, the reaction involves the splitting of a paraffin into two smaller molecules, a paraffin and an olefin. In practice, a wide range of products is obtained because the original molecules may split at any spot along the chain and the resulting products may themselves be cracked further. Each olefin molecule that undergoes such secondary cracking produces two more olefins so that the paraffin content becomes progressively smaller. The olefins produced are predominantly α-olefins, with the double bond located at the terminus of the molecule, as indicated above. If the original cracking stock is linear, the product olefins will be predominantly linear; if branched or cyclic structures were originally present, such structures will also appear in the product.

The third route to detergent olefins is from paraffins of the same chain length. In principle, it is necessary only to remove two hydrogen atoms from an adjacent pair of carbons along the chain to produce the desired olefin, but the difficulties of dehydrogenation are such that a two-step process of chlorination and dehydrochlorination has been developed. Ineither process the reaction easily proceeds past the desired stage to give polychlorinated paraffins and polyolefins, all undesirable byproducts.

4.4.4 Alkyl Benzenes

Alkyl benzenes first made their appearance as hydrophobic groups in the late 1940s as a result of new industrial processing capabilities related to the chemistry and chemical engineering of aromatic alkylation reactions. They were prepared by the more or less random chlorination of kerosene and the subsequent Friedel-Crafts alkylation of benzene. Their real dominance in the field began in the early 1950s with the appearance of tetrapropylene-based alkyl benzenes, a functionally better and cheaper product obtained in a one-step process involving addition of benzene across the double bond of the olefin. A variety of Friedel-Crafts catalysts may be employed including aluminum chloride or hydrogen fluoride.

The alkyl group of the alkyl benzene might be expected to have a carbon backbone identical to that of the olefin from which it was derived. This is true in most, but not all, cases because the alkylation process may cause rearrangement of the carbons in the chain. Furthermore, the reaction of the benzene ring with the double bond of the olefin involves a number of intermediate steps during which isomerization may occur, so that the benzene may finally link on at some position other that of the original double bond. Thus each of the many species that make up the olefin feedstock may give rise to several isomeric alkyl benzenes, and the resulting material will be an even more complex mixture than the original. Gas chromatographic analysis of typical products may show more than 100 at least partially resolved components.

Propylene-derived alkyl benzenes were pretty much phased out in the 1960s because of their exceptional (and undesirable) biological stability and were replaced by the linear alkyl benzenes prepared from linear olefins or intermediate chlorinated paraffins. These products, like the tetrapropylenes, have a chain length that may range from C_8 to C_{16}, variously distributed according to the specific preparation procedure and the properties desired in the final surfactant. The nominal carbon values for most commercial products are C_8 and C_9 alkyl groups.

4.4.5 Alcohols

Long-chain alcohols have been used as a source of surfactant hydrophobes since the earliest days of synthetic detergent manufacture. Linear alcohols have been used since the beginning; branched alcohols are a more recent addition to the chemical arsenal.

The classical route to a linear alcohol is by the reduction of the carboxyl group of a fatty acid. Actually, an ester of the carboxylic acid is usually employed, since the carboxyl group itself reacts rather sluggishly. Lauryl and tallow alcohols are two of the most commonly used substrates for surfactant synthesis. The first is derived from lauric acid, predominantly a C_{12} acid, but also usually containing some amounts of lower and higher homologs. The tallow alcohols average around C_{18}. Partially or completely hydrogenated fatty acids from oilseeds usually have carbon chains in the C_{16} ~ C_{18} range. Depending on the reduction process used, they may contain some unsaturated alcohols derived from the unsaturated acids in the original fatty acids:

$$CH_3(CH_2)_nCOOR+H_2 \longrightarrow CH_3(CH_2)_nCH_2OH$$

Since the early 1960s linear primary alcohols have been available from petroleum sources, namely, ethylene. The process for their preparation is similar to the Ziegler process for linear olefins, except that the last step is an oxidative one yielding the alcohol directly instead of an olefin. As oligomers of ethylene, the Ziegler-derived alcohols are produced in even-numbered chain lengths. The average chain length and distribution of homologs can be controlled somewhat by the reaction conditions, and completely by subsequent distillation. The —OH groups are at the end of the chain (terminal), so that they are identical to the alcohols derived from the natural fatty acids. However, the two products may differ slightly because of variations in the amounts and distribution of minor products and impurities.

Branched-chain alcohols were used extensively for surfactant manufacture prior to the changeover to the more readily biodegradable linear products. They were usually derived from polypropylenes by the "oxo" process, which involves catalytic addition of carbon monoxide and hydrogen to the double bond in a sequence of reactions. Thus the tetrapropylene derivative is nominally a C_{13} alcohol, as highly branched as the original raw material.

If a linear α-olefin is used in the oxo process, addition may occur at either end of the double bond to give a mixture of linear primary and methyl branched secondary alcohols. Substitution further down the chain occurs only in small amounts, and with the proper choice of reaction conditions the proportion of linear primary alcohol may reach 80% or more.

The development of linear paraffin feedstocks for the production of linear alkyl sulfates (LAS) also made many secondary alcohols feasible as surfactant hydrophobes. Here the —OH group may be introduced by reaction of the paraffin with oxygen, or by chlorination and subsequent hydrolysis. In either case all possible isomers are formed and the —OH group is found on any of the carbons along the chain.

4.4.6 Alkyl Phenols

The alkyl phenol hydrophobes are produced by addition of phenol to the double bond of an olefin. The alkyl group may be linked to the ring either ortho, meta, or para to the hydroxyl group, and the position can have a significant impact on the characteristics of the resulting surfactant. The earlier commercial products were derived from branched olefins such as octylphenols from diisobutylene and nonylphenols from tripropylene. More recently, linear alkylphenols have become available with the development of linear olefins for LAS.

4.4.7 Polyoxypropylenes

The polyoxypropylenes, oligomers of propylene oxide, can be cited as an example of nonhydrocarbon hydrophobes. A complete line of surfactants known commercially as the "pluronics" have been developed commercially. These are block copolymers of propylene oxide and ethylene oxide. By careful control of the relative amounts of each component incorporated into the polymer, it is possible to exercise a subtle control over the solubility and surfactant character of the product. The character of other hydrophobes such as alcohols and alkyl phenols may also be modified by addition of propylene oxide.

From *Surfactant Science and Technology* by *Drew Myers*

New Words and Expressions

geographic[dʒɪə'græfɪk] 地理的;地理学的
olefin['əʊlɪfɪn] 烯烃
sodium dodecyl sulfate 十二烷基硫酸钠
glycerol['glɪs(ə)rɒl] [有化] 甘油;丙三醇
chlorododecane 氯十二烷;氯代十二烷
castor oil 蓖麻油
crude oil distillate 原油馏分油
impediment[ɪm'pedɪm(ə)nt] 口吃;妨碍;阻止
lauric acid [有化] 月桂酸;十二烷酸
oligomerization [əuligəmərai'zeiʃən] 低聚;齐聚反应;寡聚化
diisobutylene[daiˌaisəu'bju:tili:n] [有化] 二异丁烯
nonylphenol['nɒnɪlfɪnɒl] 壬基苯酚,壬基酚

Notes

(1) **Friedel-Crafts alkylation**,傅瑞德-克拉夫茨烷基化反应(傅-克反应),是在强路易斯酸催化下,苯环上的氢原子被烷基所取代,由于傅-克烷基化反应是卤化烷、烯或醇在酸催化下形成的碳正离子引起的,如果形成的碳正离子不稳定,就会发生碳正离子的重排,从而得到重排产物。

(2) **Ziegler-Natta catalyst**,齐格勒-纳塔催化剂,一种有机金属催化剂,用于合成非支化、高立体规整性的聚烯烃,典型的齐格勒-纳塔催化剂是双组分:四氯化钛-三乙基铝[$TiCl_4$-$Al(C_2H_5)_3$],1953年德国人齐格勒在温度(60~90℃)和压力(0.2~1.5MPa)温和的条件下,得到了支链很少的高密度聚乙烯。意大利化学家居里奥·纳塔将这一催化剂用于聚丙烯生产,发现得到了高聚合度,高规整度的聚丙烯。1963年卡尔·齐格勒和居里奥·纳塔获诺贝尔化学奖。

(3) **"Oxo" process**,羰基合成,又称氢甲酰化反应,烯烃与一氧化碳和氢气在催化剂作用下,在烯烃双键上同时加上氢原子和甲酰基生成比原来烯烃多一个碳原子的两种异构醛的反应过程。

4.5 Anionic Surfactants

Anionic surfactants are the most commonly used class of surfactants in cleansing applications. These surfactants, in addition to their ability to emulsify oily soils into wash solutions, can lift soils, including particulates, from surfaces. This is because the negatively charged head group is repelled from most surfaces, which also tend to be slightly negatively charged—the reverse action to a cationic surfactant, where the positively charged head group is adsorbed onto a surface, giving an antistatic and conditioning effect.

The great majority of anionic surfactants will generate significant foaming in solutions above their critical micelle concentration (CMC), which is a desirable attribute in most cleansing applications, but can restrict the use of anionic surfactants in areas where foam is a problem. Anionics can be classified according to the polar group and the following will be considered: sulphonates, sulphates, phosphate esters, and carboxylates.

When specifying an anionic surfactant for an application, it is important to understand how the composition of the raw material (especially that of the hydrophobe) influences the performance of the surfactant and the properties of the formulated product. In looking at the properties of each surfactant type, the basic chemistry will be considered together with sources of hydrophobe and the manufacturing process used to functionalise them. How the composition of the surfactant affects its performance and physical properties will be examined together with how these properties lead to the applications of the surfactant.

4.5.1 Sulphonates

Sulphonation processes. Since many of the anionic surfactants to be discussed are made by the addition of SO_3 to an organic substrate, it is appropriate to consider, in overview, the main processes used and the contribution of the sulphonation process to the quality and performance of the surfactant. Sulphuric acid or oleum is probably the simplest and oldest sulphonating agent. Sulphuric acid is made by the reaction of gaseous sulphur trioxide with water, which is a very exothermic reaction. The ratio of SO_3 to water determines the acid strength, 96% or 98% being common commercial grades. Once the molar ratio of SO_3 to water is >1, the product is oleum, or fuming sulphuric acid. The strength of oleum is described in terms of the percentage of SO_3 added to 100% sulphuric acid. A product nominally consisting of 80% H_2SO_4 and 20% SO_3 would be called Oleum 20 but, in fact, the excess SO_3 reacts with the H_2SO_4 to form pyrosulphuric acid $HO(SO_2)O(SO_2)OH$.

(1) Alkylbenzene sulphonates

Alkylbenzene sulphonate is made from the reaction of an alkylbenzene with a sulphonating agent, to add a SO_3^- group to the aromatic ring, which forms the polar head group of the surfactant molecule. By far the most common sulphonating agent is gaseous sulphur trioxide which gives a clean reaction with minimal by-products. There is still some use of sulphuric acid and chlorosulphonic acid processes but high levels of impurities (sulphates and chlorides respectively) have made these products much less favored. Reaction of the LAB with a dilute solution of SO_3 in air (typically 10% SO_3) in a falling film sulphonator is the most common method of production.

(2) Petroleum sulphonates

The petroleum sulphonates vary in composition from being very similar to LAS to far removed from any detergent material (calcium dinonylnaphthalene sulphonate, molecular weight 708). The main effect of molecular weight is to influence the solubility of the product, lower molecular weights tending to better water solubility, higher weights to oil solubility. This general trend will apply within a structural type but should be treated with caution across differing systems, for example, alkylbenzene vs alkyl naphthalene.

The method of sulphonation will influence composition. Large scale commercial sulphonation of petroleum sulphonates is carried out using SO_3/liquid SO_2 technology to give a 'clean' product with good conversion and low levels of inorganic sulphates. Lower molecular weight (~350-400) acids have been manufactured using standard falling film reactors. As molecular weight increases, the products become more viscous and the reaction becomes difficult to control, leading to darker colors and increasing levels of sulphates. High molecular weight products are more difficult to manufacture and may use air/SO_3 reactors equipped to deal with higher temperatures and viscosity, or may use batch reaction with oleum. In the latter case, excess sulphuric acid leads to high levels of sodium sulphate on neutralization and, since it is insoluble in oil, it must be removed either by separation of waste sulphuric acid from the sulphonic acid or by a difficult filtration after neutralization.

4.5.2 Sulphates

Sulphates are the second most important class of anionic surfactants in terms of volumes and range of application and share many features with sulphonates in that they are manufactured in the same way. However, there is one very important difference between the two that is the chemical stability of the sulphate group compared to the sulphonate.

(1) Alkyl sulphates

Alkyl sulphates are a versatile and economic class of surfactants with applications in such diverse areas as polymerization and toothpaste.

Alkyl sulphates are prepared by reacting an alcohol with a sulphonating agent, normally sulphur trioxide (as illustrated in Fig.4.3).

$$ROH + SO_3 \longrightarrow ROSO_3H$$

Fig.4.3 Sulphation of an alcohol

The product is a sulphuric acid ester or 'sulpho acid' which is susceptible to acid hydrolysis, reverting to the alcohol and free sulphuric acid. Since the rate of hydrolysis is dependent on the concentration of the acid, the hydrolysis reaction is effectively self-catalyzing so that the sulpho acid must be neutralized as quickly as possible after manufacture to prevent reversion. This simple chemistry has a profound effect on the manufacture, use and economics of alkyl sulphates.

(2) Alkyl ether sulphates

This class of surfactants has possibly the widest range of use of any anionic surfactant. It is found in almost every product where foaming is desirable, in industrial, household and personal care applications. Alkyl ether sulphates are described in terms of their parent alcohol and the degree of ethoxylation. Thus, sodium laureth-2 is the sodium salt of a sulphated C_{12} alcohol, with an average of 2 mol of ethylene oxide added. Often, the alcohol is assumed to be the typical $C_{12\text{-}14}$ and the surfactant simply called a 2- or 3-mol ether sulphate.

4.5.3 Phosphate esters

They are a versatile surfactant type, with some properties analogous to those of ether sulphates. Unlike sulphate (which is a sulphuric acid mono alkyl ester), phosphate (Fig. 4.4) can form di- and triester, giving a wider range of structures and, with them, the ability to tailor the product to a greater number of application areas.

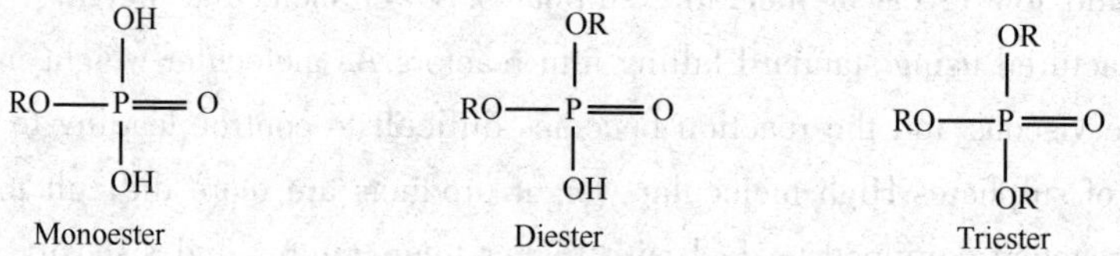

Monoester: RO—P(=O)(OH)—OH

Diester: RO—P(=O)(OR)—OH

Triester: RO—P(=O)(OR)—OR

Fig. 4.4 Structure of phosphate esters

As medium foaming/hydrotropic surfactants, PEs can be used in detergent cleansing applications with the short chain alkyl esters being effective hydrotropes for non-ionic surfactants. Application of PEs in detergents is relatively limited, due to their cost compared to a sulphate/sulphonate, or non-ionic but long chain diesters can be used as effective de-foamers in anionic systems.

Phosphate esters are widely used in metalworking and lubricants. A $C_{12\text{-}14}$ with 6 mol of ethylene oxide (diester) can be used as an emulsifier but also as an 'extreme pressure' additive, it can reduce wear where there is high pressure metal to metal contact. PEs can also show corrosion inhibiting properties, as with petroleum sulphonates and the emulsifying power of PEs with low foam is used in agrochemical formulations. PEs can act as dispersants or hydrotropes in plant protection formulations, allowing the development of easy-to-handle and dilute formulations of both poorly miscible and insoluble herbicides. The surfactant and anti-corrosion properties of PEs also find use in textile auxiliaries-low foaming gives further benefits and C_8 diesters are also reported to give antistatic effects on synthetic fibres.

4.5.4 Carboxylates

The final section of this chapter looks at anionic surfactants which derive their functionality from a carboxylate group. These include one of the earliest surfactants made by man (soap) to more complex 'interrupted soaps' where these structures give mild, hard-water-tolerant surfactants.

(1) Soap

The alkali-metal salts of fatty acids — soaps are the oldest synthetic surfactants and they have been prepared in various forms and in varying purity since pre-historic times. The technology has progressed from boiling animal fats with wood ash to an ultra-efficient high volume process, with a very extensive knowledge and literature base.

(2) Ether carboxylates

Ether carboxylates are a very versatile class of surfactants, used in diverse applications from mild personal care formulations to lubricants and cutting fluids. They are interrupted soaps, with the addition of a number of ethylene oxide groups between the alkyl chain and the carboxylate group. The additional solubility imparted by the EO groups gives much greater resistance to hardness and reduced irritancy compared to soap. Fig.4.5 shows how a carboxylate group is added to an ethoxylate by reaction with chloroacetic acid.

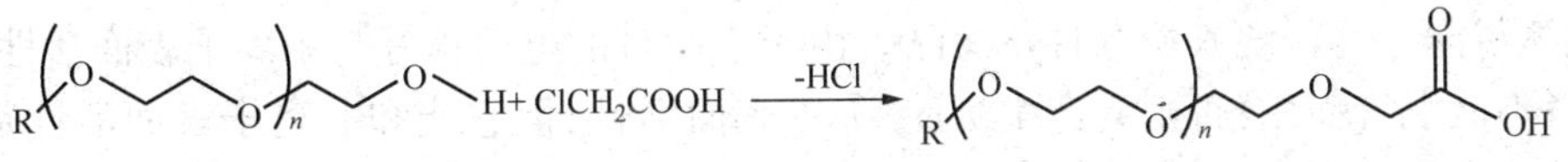

R, O, O, n, O, H + $ClCH_2COOH$ —(-HCl)→ R, O, O, n, O, O, OH

Fig.4.5 Preparation of ether carboxylate

With such a wide pallet of raw materials, it is possible to produce a very wide range of attributes in the surfactant and the HLB may be varied from 8 (low C number, low EO), to give surfactants soluble in organic media, to >20 (very good aqueous solubility). Ether sulphates (laureth-2 or laureth-3) would have HLB values of 20.

A key attribute of ether carboxylates is mildness which increases with EO number but this can also reduce detergency. Comparing sodium salts with a predominantly C_{12} alkyl chain, the 3-mol carboxylate would have a Zein score of ~150, while the 13-mol one would score ~80 (cf. laureth-2 sulphate at 270 and sodium lauryl sulphate at 490). The sodium salts show phase behavior similar to ether sulphates but the position and scale of the viscosity minimum can be varied with C chain, degree of ethoxylation and, unlike ether sulphates, by the degree of neutralization.

From *Chemistry and Technology of Surfactants* by *Richard J. Farn*

New Words and Expressions

anionic [ˌænaɪˈɒnɪk] 阴离子的,带负电荷的离子的
sulphonates[ˈsʌlfəneit] 磺酸盐; 使磺化
sulphur trioxide 三氧化硫
carboxylate [kɑːˈbɒksɪleɪt] [有化] 羧酸盐;羧化物
phosphate [ˈfɒsfeɪt] 磷酸盐
oleum [ˈəʊlɪəm] [无化] 发烟硫酸
naphthalene [ˈnæfθəliːn] 萘(球),卫生球
hydrotrope [ˈhaɪdrətrəʊp] [化学] 助水溶物,水溶物
betaine 甜菜碱
triethanolamine 三羟乙基胺
potassium oleate 油酸钾
hydrolysis [haɪˈdrɒlɪsɪs] 水解作用
cutting fluid[机] 切削液;乳化切削油
ethoxylation [eθɔksiˈleiʃən] 乙氧基化

Notes

(1) **Anionic surfactant**,阴离子表面活性剂,其分子溶于水发生电离后,亲水基是带阴离子电荷的表面活性剂。主要呈现为半透明黏稠液体,白色针状,白色粉状等形态,常用于生产洗发水、沐浴露、牙膏、洗衣粉等日化产品。由于它所带的电荷正好与阳离子表面活性剂相反,一般两者不能混合使用,否则会产生沉淀,失去表面活性;但它能和非离子表面活性剂或两性表面活性剂配合使用。

(2) **Petroleum sulphonate**,石油磺酸盐,是一种高品质的表面活性剂,主要用于油田三采期的驱油,石油磺酸盐是一种表面活性剂,它与表面活性剂助剂复配,其作用是降低油水界面张力,更好地提高洗油效率。而复配体系则利用聚合物和表面活性剂的化学协同效应,在提高水驱波及体积、降低油水流度比的基础上将滞留地层的残余油“强洗”出来,从而提高原油采收率。石油磺酸盐作为驱油用超级表面活性剂要达到以下技术指标:①有效降低油/水界面张力;②减少岩层吸附;③增溶、抗盐、耐温的一种新型驱油超级表面活性剂。

4.6 Cationic Surfactants

Cationic surfactants first became important when the commercial potential of their bacteriostatic properties was recognized in 1938. Since then, the materials have been introduced into hundreds of commercial products, although their importance does not approach that of the anionic materials in sheer quantity or dollar value. Currently, cationic surfactants play an important role as antiseptic agents in cosmetics, as general fungicides and germicides, as fabric softeners and hair conditioners, and in a number of bulk chemical applications. Many new applications for cationic surfactants have been developed since World War II, so that these compounds can no longer be considered to be specialty chemicals; rather, they truly fall into the category of bulk industrial surfactant products.

Relative to the other major classes of surfactants, namely, anionics and nonionics, the cationics represent a relatively minor part of worldwide surfactant production, probably less than 10% of total production. However, as new uses and special requirements for surfactants evolve, their economic importance can be expected to continue to increase.

Commercial cationic surfactants, like their anionic and nonionic counterparts, are usually produced as a mixture of homologs, a point that must always be kept in mind when discussing physical properties and applications of such materials. As previously noted, slight variations in the chemical structure or composition of the hydrophobic group of surfactants may alter their surface-active properties, leading to the possibility of important errors in the interpretation of results and in performance expectations. That possibility holds true for all surfactant classes and will be emphasized repeatedly throughout this work. When the sources of hydrophobic groups for cationic surfactants are natural fatty acids such as coconut oil or tallow, there may be significant variations in both chain length and the degree of unsaturation in the alkyl chain. When the alkyl group is derived from a petrochemical source, the components may be found to vary in molecular weight, branching, the presence of cyclic isomers, and the location of ring substitution in aromatic derivatives.

Pure cationic surfactants such as cetyltrimethylammonium bromide (CTAB) have been used extensively for research into the fundamental physical chemistry of their surface activity. Such investigations have led to a vast improvement in our basic understanding of the principles of surfactant action. Because of the significant differences in purity and composition between commercial- and research grade materials, however, care must be taken not to overlook the effects of such differences on the action of a given surfactant in a specific application.

Prior to the availability of straight-chain, petroleum-based surfactants, the sole sources of raw materials for cationic surfactants were vegetable oils and animal fats. All those materials could be considered to be derivatives of fatty amines of one, two, or three alkyl chains bonded directly or indirectly to a cationic nitrogen group. The most important classes of these cationics are the simple amine salts, quaternary ammonium compounds, and amine oxides:

$C_nH_{2n+1}NHR_2^+X^-$ (R = H or low-molecular-weight alkyl groups)

$C_nH_{2n+}NR_3^+X^-$ (R = low-molecular-weight alkyl groups)

$C_nH_{2n+}N^{\delta+}R_2 \longrightarrow O^{\delta-}$ (R = low-molecular-weight alkyl groups)

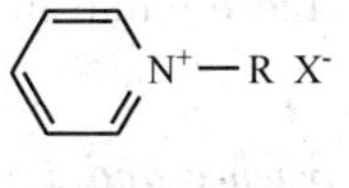

N-Alkylpyridinium salts ($N^+-R\ X^-$)

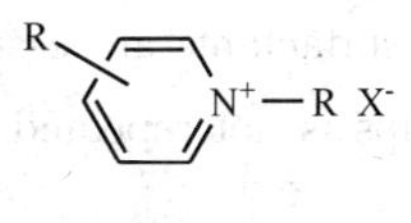

salts of alkyl-substituted pyridines (R on ring; $N^+-R\ X^-$)

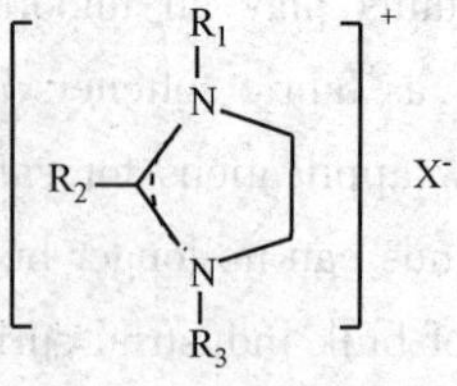

Imidazolinium derivatives 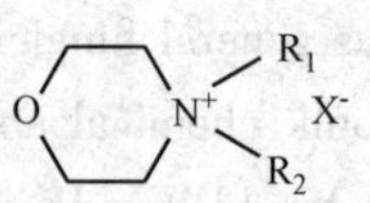　　　　morpholinium salts

There are two important categories of cationic surfactants that differ mainly in the nature of the nitrogen-containing group. The first consists of the alkyl nitrogen compounds such as simple ammonium salts containing at least one long-chain alkyl group and one or more amine hydrogen atoms, and quaternary ammoniun compounds in which all amine hydrogen have been replaced by organic substituents. The amine substituents may be either long-chain or short-chain alkyl, alkylaryl, or aryl groups. The counterion may be a halide, sulfate, acetate, or similar compound. The second category contains heterocyclic materials typified by the pyridinium, morpholinium, and imidazolinium derivatives. Other cationic functionalities are, of course, possible, but are much less common than these two major groups.

In the pyridinium and other heterocyclic amine surfactants, the surfactant properties are derived primarily from the alkyl group used to quaternize the amine. As a variation to that approach, however, it is possible to attach a surfactant-length alkyl (or fluoroalkyl) group directly to the heterocyclic ring and quaternize the nitrogen with a short-chain alkyl halide. Many commercial cationic surfactants with the general structures $R—(CH_2)_5NH^+R'X^-$ and $R—(CH_2)_5NH^+(R_2)'X^-$, where R is the surfactant-length hydrophobic group and R' is a short-chain alkyl or hydroxyalkyl chain, are available. As is the case with the anionic materials, the structures of such materials are limited mainly by the skill and imagination of the preparative organic chemist.

Some types of amphoteric surfactants (to be discussed below) in which the nitrogen is covalently bound to a group containing an anionic (e.g., $—CH_2CH_2SO_3^-$) or potentially anionic (e.g., —COOH) functionality are also classed as cationic in some publications; however, under the classification scheme employed in this work, such materials are covered in a separate category. The examples above represent the simplest types of cationic surfactants. Many modern examples contain much more complex linkages; however, the basic principles remain unchanged.

The economic importance of the cationic surfactants has increased significantly because of some of their unique properties. Most cationics are biologically active in that they kill or inhibit the growth of many microorganisms. They have also become extremely important to the textile industry as fabric softeners, waterproofing agents, and dye fixing agents. Because many important mineral ores and metals carry a net negative charge, the cationic surfactants are also useful in flotation processing, lubrication, and corrosion inhibition, and they are gaining importance as surface modifiers for the control of surface tribological properties, especially electrostatic charge control. Since the hydrophobic portions of the cationic surfactants are essentially the same as those found in the anionics, individual discussion of those groups is not repeated here.

From *Surfactant Science and Technology* by *Drew Myers*

New Words and Expressions

bacteriostatic property 抑菌性能
fungicide [ˈfʌndʒɪˌsaɪd] 杀真菌剂,抗真菌剂
germicide [ˈdʒɜːmɪsaɪd] 杀菌剂,杀菌物
hair conditioner 护发素,头发调理品
cetyltrimethylammonium bromide 溴化十六烷基三甲铵
alkylaryl [ɔːlkɪˈlærɪl] 烷基芳香基
pyridinium [piriˈdiniəm] 吡啶
imidazolinium [ɪmɪdəˈlɪnɪəm] [医] 咪唑啉
morpholinium [mɔːpˈhəʊlɪnɪəm] 吗啉正离子
hydroxyalkyl [maɪkrəʊˈɔːgənɪzəm] 微生物
fabric softener [化] 织物柔软剂
quaternize[kwəˈtɜːnaɪz] (使)季铵化
dye fixing agent 固定染料助剂,固色剂,染料固色剂
flotation processing 浮选工艺

Notes

(1) **Cationic surfactant,**阳离子表面活性剂,其分子溶于水发生电离后,与亲油基相连的亲水基是带阳电荷的。亲油基一般是长碳链烃基。亲水基绝大多数为含氮原子的阳离子,少数为含硫或磷原子的阳离子,根据氮原子在分子中的位置不同分为胺盐、季铵盐和杂环型三类。分子中的阴离子不具有表面活性,通常是单个原子或基团,如氯、溴、醋酸根离子等。阳离子表面活性剂带有正电荷,与阴离子表面活性剂所带的电荷相反,两者配合使用一般会形成沉淀,丧失表面活性。它能和非离子表面活性剂配合使用。

(2) **Flotation processing,**浮选法,是利用矿物表面物理化学性质的差异来选分矿石的一种方法,目前应用最广泛的是泡沫浮选法。矿石经破碎使各种矿物解离成单体颗粒,向矿浆加入各种浮选药剂并搅拌调和,使与矿物颗粒作用,以扩大不同矿物颗粒间的可浮性差别,然后送入浮选槽,搅拌充气。矿浆中的矿粒与气泡接触、碰撞,可浮性好的矿粒选择性地粘附于气泡并被携带上升成为气-液-固三相组成的矿化泡沫层,经机械刮取或从矿浆面溢出,再脱水、干燥成精矿产品。不能浮起的脉石等矿物颗粒,随矿浆从浮选槽底部作为尾矿产品排出,有时,将无用矿物颗粒浮出,有用矿物颗粒留在矿浆中,称为反浮选,如从铁矿石中浮出石英等。

4.7 Nonionic Surfactants

The term non-ionic surfactant usually refers to derivatives of ethylene oxide and/or propylene oxide with an alcohol containing an active hydrogen atom.However other types such asalkyl phenols, sugar esters, alkanolamides, amine oxides, fatty acids, fatty amines and polyols are all produced and used widely throughout the world in a multitude of industries. This chapter covers the production of these materials and how they can be modified to meet the desired end product use.

There are over 150 different producers and some 2 million tonnes of commercial nonionic surfactants manufactured worldwide of which at least 50% are alkoxylated alcohols. Ethoxylated nonylphenol production is falling and accounts for 20% of the market while alkoxylated fatty acids account for some 15%. Fatty acid amides and sugar esters account for another 10% and there are a large number of specialities making up the balance. In general, non-ionic surfactants are easy to make, relatively inexpensive and derived from a variety of feedstocks.

4.7.1 Fatty alcohol ethoxylates

With the slow demise of the nonylphenol ethoxylate market due to legislation, the fatty alcohol market has the chance to design alternatives by subtle changes to the hydrophobe chain lengths and alkoxylate levels. The effects must be achieved with biodegradability as a key parameter and fish toxicity as an up and coming extra requirement to keep in mind. There is no direct natural source of fatty alcohols: they all have to be synthesized and there are five major processes used. Historically the first alcohol ethoxylates were based on tallow or stearyl alcohol. The general processes for producing fatty alcohols are mentioned below.

1. Catalytic hydrogenation of fatty acids from natural fats and oils.

$$RCOOH \longrightarrow RCH_2OH$$

2. The OXO process from alpha olefins

$$RCH=CH_2 + CO + H_2 \rightarrow RCH_2CH_2CH_2OH + RCH(CH_3)CH_2OH$$

3. Ziegler process to give linear even numbered chains

4. From n-paraffins to give essentially linear even and odd numbered chains

5. From the Shell ‘SHOP’ process to give linear even and odd numbered chains From these various processes it is possible to produce a selection of alcohols from C_6 to C_{20}. From route 1, the ‘natural products’, one can have C_{10}, C_{12} and C_{14} from palm oil and C_{16}, C_{18} and C_{20} from tallow. It is also possible to have C_{18} from rapeseed. From roule 2, one gets the alcohols with the odd chain lengths C_9 through to C_{15} in various cuts determined by the alpha olefin used. From route 3 one can make both plasticizer C_6 to C_{10} and detergent alcohols C_{12} to C_{20}. Once again these are even chain numbers. From route 4 and 5 one gets mixtures of odd and even chain numbers in roughly equal proportions.

It can be seen that it is possible to vary the alcohol chain greatly and the alkoxylate chain can be varied in the same way as with nonylphenol to produce both water and oil soluble products. The major difference between nonylphenol and alcohol ethoxylates is the distribution of ethoxylate chains. The rate constants for the addition of ethylene oxide to primary alcohols are comparable and are essentially the same as for the 1-mol or the 2-mol adduct. These addition products, of course, are still primary alcohols. Thus, if one were making a 2 mol adduct of the alcohol, there would be a fair proportion of free alcohol still present of the order of 10%~20%. Chain growth starts well before all the starting alcohol has reacted and alcohol ethoxylates have therefore a much broader ethoxylate chain distribution than the comparable nonylphenol ethoxylate. It has been shown that ethylene oxide consumption becomes constant after 8 or 9 mol of ethylene oxide per mole of alcohol has been added.

4.7.2 Polyoxyethylene esters of fatty acids

Probably the third largest group of ethoxylated products is the esters of fatty acids. The fatty acids used are almost entirely derived from natural products by fat splitting in which the triglyceride, (fat or oil) is reacted with water to form $CH_2OH—CHOH—CH_2OH$ (glycerol) plus 3 mol of fatty (e.g. stearic) acid $C_{17}H_{35}COOH$. These are homogeneous reactions taking place in the fat or oil phase because water is more soluble in fat than fat is in water. Continuous fat splitting plants usually have counter current oil and water phases and operate at high temperatures and pressures, which reduce reaction times. The acids quite often contain some unsaturation, e.g. oleic acid, and this, in particular, should be stored and transported under nitrogen to prevent oxidation if the later ethoxylation product is to be of good odor, color and quality. Peroxide values of starting acids should in particular be measured. A very cheap source of acid used for some products is tall oil fatty acid obtained from the pulp and paper processes.

Ethoxylation is carried out in the same plants and manner as all the alcohol ethoxylates described earlier. Initially, the ethylene oxide reacts with the acid to produce ethylene glycol monoester, $RCOO(CH_2CH_2O)H$, and then reacts rapidly with further ethylene oxide to produce the polyethoxylated product $RCOO(CH_2CH_2O)_nH$. However, the reaction conditions are ideal for ester interchange and the final product contains free polyethylene glycol, the monoester and the diester [$RCOO(CH_2CH_2O)_nOCR$] in the ratio 1 : 2 : 1. An alternative method of preparation of these products is to react polyethylene glycol of desired molecular weight and esterify it with acid in an ester kettle. Reaction temperatures and catalysts vary but are in the region 100-200℃. An equimolar ratio of fatty acid to polyethylene glycol results in a mixture similar to the product via ethoxylation, i.e. dominant in monoester. If high excesses of polyethylene glycol are used, monoester dominates but even purified monoester products revert to the mixture on storage, with an adverse effect on wetting properties.

The esters formed in this process are hydrolyzed in both acid and base conditions and are much less stable than alcohol ethoxylates. This limits the applications in detergents but they have many industrial uses. In the textile industry they have good emulsifying, lubricating, dispersive and antistatic properties. They are also used widely in personal care, institutional and industrial cleaning, crop protection, paints and coatings and adhesives.

4.7.3 Glycerol esters and polyglycerol esters

Mono- and diglycerides are the most commercially important members of this series being used extensively as emulsifiers in the food and cosmetic industries. They can be prepared from individual fatty acids, but more commonly, directly from oils or fats by direct glycerolysis. Thus, oil or fat is heated directly with glycerol at 180~230℃ in the presence of an alkaline catalyst. Ideally, 1mol of coconut oil plus 2mol of glycerol will yield 3mol of monoglyceride. In practice, the reaction product is 45% monoester, 44% diester and 11% triester and any unreacted glycerol is removed by washing with water.

Polyglycerol esters are produced in a two-step process from glycerol and fatty acids. The first step is a controlled polymerization of glycerol into a polymeric form by heating the glycerol in the presence of an alkaline catalyst, such as 1% caustic soda, at a temperature of 260~270℃:

$$HOCH_2CH(OH)CH_2OH + HOCH_2CH(OH)CH_2OH \longrightarrow HOCH_2CH(OH)CH_2OCH_2CH(OH)CH_2OH$$

As the reaction proceeds, the material becomes more viscous such that most commercial products are only 2~10 units long.

Esters of oligomers can be made, with or without more catalyst addition, by reacting with any fatty acid. The addition of more hydroxyl groups with each additional glycerol means that a large range of polyglycerol esters can be made with various fatty acids from C_{10} to C_{18} and the hydrophile/lipophile balance (HLB) range of these products can vary from 3 to 16 making a very good series of emulsifiers.

4.7.4 Anhydrohexitol esters

The only hexitol-derived surfactants toachieve commercial importance are those where a portion of the polyol has been anhydrised. They are manufactured by the direct reaction of hexitols with fatty acids during which internal ether formation as well as esterification occurs.

The plant used for these reactions is a typical ester kettle which can have internal or external heating, recirculation of the reactants, low shear or high shear stirring of the immiscible reactants, condensation for return of free acids, water removal facilities, etc. The quality of ester obtained depends not only on reactor geometry and mixing abilities but, particularly, on how the heat input is achieved. Local 'hot spots' must be avoided and it is essential that high quality acids with low peroxide values, where applicable, are used. Being insoluble, the reactants need good mixing to achieve faster reaction times and an alkaline catalyst is used which, in effect, is the sodium salt of the acid. Acidic catalysts such as phosphoric acid are also used.

Esterification takes place between 180℃ and 240℃. Although esterification and dehydration occur simultaneously, esterification takes place faster and the reaction is cooked down at high temperatures to complete the anhydride ring formation and meet the hydroxyl specification. The total reaction time is about 6 h and, after completion of the reaction, the product is neutralized and filtered. The majority of the commercially available products are monoesters although one or two triesters are sold. The esterification occurs mainly on the primary hydroxyl group but small quantities of monoester do occur on the three secondary hydroxyl groups together with some *di-* and triesters. The anhydrisation can also be carried out further to yield isosorbides and their derivatives.

Span surfactants are lipophilic and are generally soluble or dispersible in oil, forming water in oil emulsions. They are used for their excellent emulsification properties in personal care, industrial cleaning, fibre finish, crop protection, water treatment, paints and coatings, lubricant and other industrial applications.

From *Chemistry and Technology of Surfactants* by *Richard J. Farn*

New Words and Expressions

alkanolamide [ɔːlkænəʊˈləmaɪd] 烷醇酰胺
fatty alcohol ethoxylate 脂肪醇乙氧基化物
polyol [ˈpɒlɪɒl] 多羟基化合物，多元醇
biodegradability[ˈbaɪəʊdɪˌgreɪdəˈbɪlətɪ] [生物] 生物降解能力
stearyl alcohol 十八烷醇；[有化] 硬脂醇（等于 octadecanol）
oleic acid 十八烯酸，[有化] 油酸
polyoxyethylene ester 聚氧乙烯酯
esterification [εˌstεrəfəˈkeʃən] 酯化反应
tall oil fatty acid 妥尔油脂肪酸
glycerolysis [化]甘油解
caustic soda [无化] 苛性钠；氢氧化钠
anhydrohexitol [ænhaɪdrəʊˈheksɪtɒl] [化] 失水山梨醇
polyglycerol esters 聚甘油酯
fibre finish 纤维上油，纤维整理

Notes

(1) **Nonionic surfactant**，非离子表面活性剂，其溶于水时不发生解离，其分子中的亲油基团与离子型表面活性剂的亲油基团大致相同，其亲水基团主要是由具有一定数量的含氧基团（如羟基和聚氧乙烯链）构成。由于非离子表面活性剂在溶液中不是以离子状态存在，所以它的稳定性高，不易受强电解质存在的影响，也不易受酸、碱的影响，与其他类型表面活性剂能混合使用，相容性好，在各种溶剂中均有良好的溶解性，在固体表面上不发生强烈吸附。按亲水基团分类，有聚氧乙烯型和多元醇型两类。具有良好的洗涤、分散、乳化、润湿、增溶、匀染、防腐蚀、和保护胶体等优良性能。

(2) **Span surfactant**，斯盘型非离子表面活性剂，是由脂肪酸与山梨醇在碱催化剂的作用下加热到 190℃脱水酯化，得到脂肪酸山梨醇酯。再加热到 250℃时，山梨醇分子内部脱水，生成脂肪酸失水山梨醇酯。是重要的乳化剂，也是纤维油剂的重要配料。在羟基上与环氧乙烷加成后，得到吐温（Tween）型非离子表面活性剂。

4.8 Fluorinated Surfactants

4.8.1 Introduction

Commercially, the production and use of surfactants is dominated by modified hydrocarbon-based chemicals. In a number of instances, however, a hydrocarbon-type surfactant will not provide the desired product attributes or performance and, in such cases, two options are presented. One involves reformulation of the product to accommodate a hydrocarbon-type surfactant and the other is the use of a fluorosurfactant. Fluorosurfactants behave typically as would a hydrocarbon type except that properties such as surface tension reduction are larger in magnitude. Furthermore, the presence of fluorine in the hydrophobic portion of the molecule causes them to differ from their hydrocarbon counterparts in more subtle ways that have commercial importance. An example of a difference would be the reduced dielectric constant or index of refraction of a fluorosurfactant compared to its hydrocarbon analog. While this may be of no consequence when formulating cleaners, it most certainly exists in a number of electronics applications.

The large majority of surfactants can be classified as hydrocarbon types, which means that the hydrophobe is a hydrocarbon. There is another class of surfactants, differentiated by the name fluorosurfactant, that uses a fluorocarbon instead of a hydrocarbon as the hydrophobe. Typically, the fluorocarbon is based on $—(CF_2)_nF$ where the number-averaged value of $n \approx 8$ and effectiveness and efficiency of fluorosurfactants are sensitive functions of n. Commercially, a value for $n \approx 8$ is chosen to give maximum effectiveness and efficiency. When compared to hydrocarbon surfactants, similar fluorocarbonsurfactants have a higher efficiency and effectiveness. The interested reader is referred to an excellent review on the structure and properties of fluorosurfactants.

4.8.2 Uses

Clearly, due to the price differential between a hydrocarbon and analogous fluorocarbon surfactant (~10–100×), the latter is used often as a "last resort" when nothing else will perform adequately. Not only does the fluorosurfactant provide a lower surface tension than a hydrocarbon analog on a molecule-to-molecule basis, but many other important differences are used advantageously. Often, the user is searching for a material that not only will dominate the surface but also impart unique properties to the material. Several such uses are worthy of recognition. Hydraulic fluid used in aircraft contains a fluorosurfactant and, although there is disagreement about the actual mechanism, it is claimed that the presence of the fluorosurfactant is necessary for the proper functioning of valves in the air craft hydraulic system. Very thin layers (~5–10 Å) of fluorosurfactants are used as antireflection layers in the photolithographic process in microelectronics fabrication. Typical photoresists have a relatively high index of refraction and, as the light used to process the photoresist reflects off the substrate, standing wave patterns are exposed in the photoresist due to multiple, coherent internal reflections. These standing wave patterns affect the critical tolerance of the desired pattern. The lowered index of refraction of a thin layer of fluorosurfactant on the photoresist negates internal reflection of the light off the substrate, therefore, allowing for greater control of critical tolerances. Gelatin is used in large quantities in the photographic film industry as a stabilizer for the colloid responsible for the latent image. The completed film is then wound rapidly on metal spools causing a great deal of triboelectric charging. Discharging exposes the film and is undesirable and the addition of a fluorosurfactant to the gelatin layer mitigates cathodic charging on the native material. Advanced fire fighting foams (AFFF) are another example that exploits the inherent properties of fluorosurfactants versus hydrocarbon surfactants. AFFF materials are aqueous-based products used to combat fire in critical applications such as aircraft. The material must foam, contain the fire from spreading, not become fuel itself, help to extinguish the fire by preventing oxidant from entering the combustion zone and not damage sensitive components. Another application that exploits the differences between hydrocarbon and fluorocarbon surfactants is electroplating. Here, the problem lies in the very aggressive conditions of low pH and air-sparging of the bath. Fluorosurfactants, typically, have a much larger pH usage range than hydrocarbon analogs and can tolerate exposure to pH values in the 1~2 range that would be representative of an electroplating bath. In addition, the fluorosurfactant-rich foam present on an air-sparged electroplating bath is a more effective barrier to evaporation and aerosolization of a corrosive mist in a manufacturing environment.

4.8.3 Applied theory

There is nothing magical about fluorocarbons or, specifically, the $—(CF_2)_nF$ group. The $—(CF_2)_nF$ is similar to $—(CH_2)_nH$ in many ways. These include dipole moments and polarizabilities that are related to intermolecular forces and, hence, surface tension. Where they do differ is in size, specifically diameter, and a relative comparison for a typical hydrocarbon and similar fluorocarbon surfactant is shown. The terminal $—CF_3$ group ≈ 20% larger than the $—CH_3$ group and the same rationale applies for the $—CF_2—$ group versus the $—CH_2—$ group. This means that intermolecular forces/unit volume for fluorocarbons are less than that for hydrocarbons of similar structure.

A term to describe the aforementioned quotient is cohesive energy density (CED; heat of vaporization/unit volume). To a first approximation, the lower the CED, the lower will be the surface tension and this is the source of the increased efficiency in surface tension reduction of fluorosurfactants versus hydrocarbon surfactants. Therefore, fluorosurfactants are often the choice for applications demanding ultimately low surface tension. Furthermore, fluorosurfactants are far less compatible with water than are hydrocarbon surfactants. This is the origin of the increased effectiveness compared to hydrocarbon surfactants.

How surface tension translates to commercial applications will now be examined. Surfactants are often added to reduce surface tension of a liquid enabling it to wet a surface and the equation governing this phenomenon is attributed to Young:

$$\gamma_{lv}\cos\theta = \gamma_{sv} - \gamma_{sl} \tag{1}$$

where γ is interfacial tension of liquid/vapor (lv), solid/vapor (sv) and solid/liquid (sl) interfaces and θ is the contact angle of a liquid droplet on a surface. This equation states that if a liquid has a higher interfacial tension than a solid, the liquid will not wet the solid. A familiar case would be a water droplet resting on a surface of Teflon where the water 'beads' on the surface and, for the curious, this is the basis for repellency. In most coating systems, for example, one desires to have the surface tension of the coating to be lower than that of the substrate surface. This ensures that the coating wets the surface and the proper choice of surfactant can aid the coating in wetting the substrate. If the substrate has a very low surface tension (e.g., polyolefin), then an even lower surface tension surfactant, such as a fluorosurfactant, must be used. An example of this effect is shown in Fig. 4.6.

As an example, consider a clean floor tile with a surface tension of 32 $mN \cdot m^{-1}$ and the same tile that has been soiled (27$mN \cdot m^{-1}$). This is an example of a very realistic possibility. One can observe clearly that the contact angle increases dramatically once the surface tension of the liquid (for example, floor polish with surfactant) increases above that of the substrate. Shown also in Fig. 4.6 are bars that span typical surface tension expected at use conditions for solutions containing fluorocarbon and hydrocarbon surfactants. The anticipated contact angles would be seen from the intersection of the vertical lines with the two curves. It is obvious that the fluorosurfactant would provide a better guarantee of the liquid (floor polish in this case) wetting (contact angle $\rightarrow 0$) the substrate (floor tile whether clean or soiled) than would a hydrocarbon surfactant.

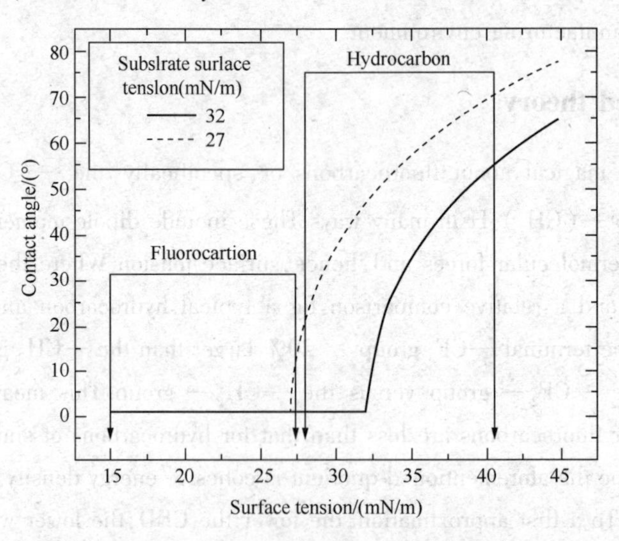

Fig. 4.6 Contact angles as a function of liquid surface tension for substrates with surface tension of 32 (—) and 27 (- - -) mN/m

Interfacial rheology is another issue. Much like the bulk, interfaces (specifically, the liquid/air interface) have their own rheological issues. For example, foaming is a consequence of interfacial rheology and low surface tension is a necessary, but not sufficient cause for foaming. In order to support a foam, an interface must have some elasticity. Consider a balloon and imagine immersing the balloon in liquid nitrogen and then adding a great deal of pressure to expand further. The balloon breaks. At liquid nitrogen temperatures, the polymer constituting the balloon is no longer elastic, but rather a rigid solid so it breaks rather than expands. Flow and leveling is an even more complicated issue and, according to first principles, leveling stress, λ, is given by(2):

$$\lambda = -\gamma / r \quad (2)$$

where γ is surface tension and$r$ is radius of interfacial curvature. Equation 2 states that leveling stress is directly proportional to surface tension and implies that lowering the surface tension of a coating will make it level even less with added surfactant than without. This seems contrary to what is observed in reality and obviously, the situation is not that simple. Many factors, as yet understood poorly in practical coating systems, are at play and a complete theoretical description has not been forwarded. In summary, flow and leveling in coatings is a dynamic event that defies description by a single static variable such as surface tension. Surface tension alone is not a very good indicator of the performance of materials in most situations and applications and must be measured as some facsimile of the application to get meaningful information.

For many uses, hydrocarbon surfactants often behave satisfactorily. There simply is no need to use expensive fluorosurfactants. However, one of the basic requirements of a coating is that it wets the substrate and secondly is that it flows and levels to give required optical properties such as gloss or distinctness of image. This is where fluorosurfactants come into play. Some coating systems inherently wet, flowing and level well without fluorosurfactants whilst others do not. Wetting is troublesome particularly when trying to coat low tension surfaces such as polyolefins (polyethylene or polypropylene, for example) or surfaces contaminated with low tension materials such as silicone oils or greases. The relevant equation governing wetting is given in Equation (1) and is well known. The surface tension of the coating has to be less than that of the substrate in order to wet and fluorosurfactants afford lower surface tension than comparable hydrocarbon surfactants. Thus, if wetting of a low tension surface is desired, particularly from a relatively high tension aqueous coating, then the use of a fluorosurfactant is warranted. In contrast, flow and leveling is much less understood without a detailed understanding of all the material parameters. To know this a priori would take far too long to be practical and the formulator is faced with a simple question: does the coating flow and level satisfactorily or not? If it does, then there is no need for a fluorosurfactant. If it does not, then a fluorosurfactant may well provide the required flow and leveling attributes desired.

From *Chemistry and Technology of Surfactants* by *Richard R. Thomas*

New Words and Expressions

fluorosurfactant 含氟表面活化剂
dielectric constant [电] 介电常数;电容率
hydraulic fluid [haɪˈdrɔːlɪk] 液压的;水力的;水力学的
antireflection 防反射;抗反射
microelectronics fabrication 微电子制造
photolithographic process 光刻工艺
photoresist [电子] 光刻胶;[印刷] 光致抗蚀剂;[光] 光阻材料
gelatin [ˈdʒelətɪn] 明胶;动物胶;胶制品
triboelectric [trɪbɔɪˈlɛktrɪk]摩擦电的
electroplating [ɪˈlektrəʊ,pleɪtɪŋ] 电镀;电镀术
aerosolization [ˈɛərəu,sɔlaiˈzeiʃən] 烟雾化
dipole moment 偶极矩
intermolecular force [物化] 分子间作用力
contact angle [矿业][机] 接触角;交会角
surface tension 表面张力
interfacial rheology 界面流变学;界面流变

polyolefin [高分子] 聚烯烃

priori [praɪˈɔːraɪ] 先验的;优先的

pharmaceutical [fɑːməˈsuːtɪk(ə)l] 制药(学)的;药物

urethane coating 氨基甲酸乙酯涂层

perfluorooctanoic acid 全氟辛酸

Notes

Fluorinated Surfactant 是以氟碳链为非极性基团的表面活性剂,即以氟原子部分或全部取代碳氢链上的氢原子。热稳定性高。全氟磺酸盐能在350~400℃不发生分解,全氟羧酸在400℃环境下能稳定存在,全氟羧酸盐也能应用在250℃的高温体系中;化学稳定性好,氟碳表面活性可在强酸、强碱、强氧化介质等特殊应用体系中稳定有效地发挥其表面活性剂作用,不会与体系发生反应或分解。如全氟磺酸盐在含氧化铬(10G/L)的98%硫酸溶液中于90℃温度下存放28天其性能不发生任何变化;相容性好。高的化学稳定性就意味着高的化学惰性,氟碳表面活性剂能与其它各类活性剂很好地相容,并可应用于几乎所有配方体系。

4.9 Gemini Surfactants

Milton J.Rosen[a] and David J.Tracy[b]

[a]Surfactant Research Institute, Brooklyn College of the City University of New York, Brooklyn, New York 11210, and [b]Surfactants and Performance Ingredients, Rhodia, Inc., Cranbury, New Jersey 08512-7500

ABSTRACT: The literature, including patents, describing the emerging area of gemini surfactants is reviewed. The differences in structure/property relationships between gemini and comparable conventional surfactants are described and discussed in terms of their predicted performance properties. Supportive performance data are numerated.

KEY WORDS: Critical micelle concentration, dispersion, emulsification, foaming, gemini, irritation, surfactants, synergism, wetting.

Gemini surfactants contain two hydrophilic groups and two (sometimes three) hydrophobic groups. Although patents on surfactants of this structure have existed since 1935, interest has been intense only in the past several years as a result of a report which pointed out that these surfactants could be more surface-active by orders of magnitude than comparable conventional surfactants containing a similar single hydrophilic group and a single hydrophobic group. As a result, numerous papers have appeared in the chemical literature describing the fundamental properties of gemini surfactants as well as, especially in the last few years, a flurry of patents covering all the electrical charge types of surfactants: anionics, cationics, zwitterionics, and nonionics.

1. Gemini VS.Conventional Surfactants: Differences in Fundamental properties

Compared to their conventional analogs, gemini surfactants possess: (i) much lower critical micelle concentration (CMC) values (a measure of their tendency to form micelles), (ii) much lower C_{20} values (a measure of their tendency to adsorb at an interface), (iii) closer packing of the hydrophobic groups, and (iv) stronger interaction with oppositely charged surfactants at the aqueous solution/air interface. In addition, they are more soluble in water and some geminis exhibit unique rheological properties.

1.1 CMC and C_{20} values

CMC values in aqueous media and C_{20} values [the surfactant concentration in the aqueous phase that produces a 20 dyne/cm reduction in the surface tension of the solvent] are shown in Table 4.3, together with some values for comparable conventional surfactants. (Note: Methylene groups between two hydrophilic groups, such as an oxygen atom up to four carbon atoms away from a second hydrophilic group, generally are equivalent to one-half of a methylene group in a straight alkyl chain.) It is apparent from the data in Table 4.3 that geminis can have CMC values one to two orders of magnitude smaller and C_{20} values two to three orders of magnitude smaller than those of comparable conventional surfactants. This is claimed also for zwitterionic geminis of structure $[C_{11}H_{23}C(O)NHCH_2CH_2N^+H(CH_2CH_2COO^-)CH_2]_2$.

The length and structure of the linkage between the two hydrophilic groups also determines the CMC and C_{20} values. For the cationic series $C_{10}H_{21}N^+(CH_3)_2(CH_2)_mN^+(CH_3)_2C_{10}H_{21}\cdot 2Br^-$ and $C_{12}H_{25}N^+(CH_3)_2(CH_2)_mN^+(CH_3)_2C_{12}H_{25}\cdot 2Br^-$ it has been found that the CMC value increases with m to a maximum at m equals ca.4 and then decreases. The cause of the decrease, it is suggested, is that once the hydrophobic polyethylene group reaches sufficient length, it penetrates into the interior of the micelle and thus is removed from contact with the aqueous phase. The smallest CMC and C_{20} values appear to arise with linkages that are short, slightly hydrophilic (just capable of hydrogen bonding with water), and, if hydrophobic, flexible. Thus, for geminis with similar alkyl chains and the following linkages, the CMC values increase in the order: $—CH_2CH_2OCH_2CH_2— < —CH_2CHOHCH_2CH_2— < —CH_2CHOHCHOHCH_2— < —CH_2CH_2CH_2CH_2— < p—CH_2C_6H_4CH_2—$ (Table 4.3), and both the CMC and C_{20} values increase in the order: $O < —O(CH_2CH_2O)_3—$. The rigidity of the hydrophobic $—CH_2C_6H_4CH_2—$ linkage prevents its removal from contact with the aqueous phase by looping into the interior of the micelle [as is possible with the flexible $—(CH_2)_m—$ linkage], and this inhibits micelle formation (increases the CMC value). Addition of $—OCH_2CH_2—$ units to the linkage also leads to an increase in the CMC and C_{20} values as a result of an increase in the size and hydrophilicity of the molecule.

Table 4.3 Critical Micelle Concentration (CMC) and C_{20} Values of Some Gemini Surfactants in H_2O at 25℃

Compound	CMC(mM)	C_{20}(mM)
$[C_{10}H_{21}OCH_2CH(SO_4^-Na^+)CH_2OCH_2]_2$	0.013	0.001
$C_{12}H_{25}SO_4^-Na^+$	8.2	3.1
$[C_{10}H_{21}OCH_2CH(OCH_2CH_2CH_2SO_3^-Na^+)CH_2]_2O$	0.033	0.008
$C_{12}H_{25}SO_3^-Na^+$	9.8	4.4

Continned

Compound	CMC(mM)	C_{20}(mM)
$[C_{10}H_{21}OCH_2CH(OCH_2COO^-Na^+)CH_2]_2O$	0.084	0.004
$[C_{10}H_{21}OCH_2CH(OCH_2COO^-Na^+)CH_2OCH_2CH_2]_2OO-CH-COO-Na^+$	0.26	0.04
$(C_{11}H_{23})_2CO-CH-COO-Na^+$	0.090	
$[C_{14}H_{29}CH(SO_3^-Na^+)COOCH_2]_2$	0.072	
$C_{14}H_{29}(SO_3^-Na^+)COOCH_2CH_2OH$	0.30	
$C_{11}H_{23}COO^-Na^+$	20	5.0
$[C_{12}H_{25}N^+(CH_3)_2CH_2CHOH]_2 \cdot 2Br^-$	0.7	0.13
$[C_{12}H_{25}N^+(CH_3)_2CH_2CH_2]_2O \cdot 2Cl^-$	0.5	0.25
$C_{12}H_{25}N^+(CH_3)_3Br^-$	16	8
$[C_{12}H_{25}N^+(CH_3)_2CH_2CH_2]_2 \cdot 2Br^-$	1.2	
$[C_{12}H_{25}N^+(CH_3)_2CH_2]_2C_6H_4 \cdot 2Br^-$(50℃)	1.0	
$[C_{12}H_{25}N+(CH_3)_2CH_2]2CHOH \cdot 2C^-$(20°C)	0.65	

1.2 Packing of the hydrophobic groups at the interface

When the linkage between the two hydrophilic groups is small or hydrophilic, the area/hydrophobic chain at the aqueous solution/air interface has been found to be smaller in geminis than in comparable conventional surfactants. This means that the hydrophobic groups are packed more closely in these geminis than in the conventional surfactants, making for a more coherent interfacial film. Thus, in the series $[C_{10}H_{21}OCH_2CH(OCH_2CH_2COO^-Na^+)CH_2]_2Y$, the area/molecule at the aqueous solution interface in 0.1 M NaCl at 20℃ (pH 11) for the following Y linkages is: O <—OCH_2CH_2O—<—$O(CH_2CH_2O)_2$—<—$O(CH_2CH_2O)_3$—, but even for the linkage —$O(CH_2CH_2O)_3$, the area/hydrophobic chain is smaller than that for $C_{11}H_{23}COO^-Na^+$ under the same conditions. Closer packing of the hydrophobic chains has also been observed for disulfonate geminis having three hydrophobic groups in the molecule and for diquaternary ammonium geminis. Some representative data for areas are shown in Table 4.4.

Table 4.4 Area/Hydrophobic Chain of Gemini and Other Surfactants at the Air/Aqueous Solution Interface[a]

Surfactant	Temperature/℃	Area/chain/Å^2
$[C_{10}H_{21}OCH_2CH(OCH_2CH_2CH_2SO_3Na)CH_2]_2O$	25	26
$C_{10}H_{21}OCH_2CH_2SO_3Na$	25	43
$[C_{10}H_{21}OCH_2CH(OCH_2COONa)CH_2]_2O$	20	41
$C_{11}H_{23}COONa$	20	69

[a]Medium: 0.1 M NaCl.

1.3 Water solubility, Krafft points

The water solubility of some dicarboxylate gemini surfactants in hard water has been reported to be greater than that of comparable sodium carboxylates. The Krafft points of several series of disulfonate, disulfate, and diphosphate gemini surfactants have been reported to be below 0℃, or considerably below those of comparable conventional surfactants.

1.4 Interaction with oppositely charged surfactants

Ionic geminis carry twice the charge of conventional ionic surfactants and therefore can interact much more strongly than the latter with oppositely charged surfactants. The strength of the interaction between different types of surfactant molecules is measured by the so-called "β parameter," which is a measure of the nonideality of the system and is a major factor in determining whether the mixture will exhibit synergy. The more negative the value of the β parameter, the stronger is the attractive interaction between the components of a mixture and the greater the probability of a synergistic interaction. Some β parameter values are listed in Table 4.5 for interaction at the aqueous solution/air interface (β^{σ}) and in mixed micelles (β^{m}) in the aqueous phase. Also listed are the mole fractions of the gemini in the surfactant mixture, on a surfactant-only basis, at the aqueous solution/air interface (X^{σ}) and in the mixed micelle (X^{m}). Synergistic interaction leading to a 10-fold decrease in the CMC is claimed when the zwitterionic gemini, $[C_{11}H_{23}C(O)NHCH_2CH_2N^+H(CH_2CH_2COO^-)CH_2]_2$ is mixed with a sodium lauryl ether sulfate.

Table 4.5 Interaction Parameters (β) for Some Gemini-Containing Binary Mixtures of Surfactants at 25℃[a]

System (solvent)	β^{σ}	β^{m}	X^{σ}	X^{m}
$[C_8H_{17}N^+(CH_3)_2CH_2CHOH]_2 \cdot 2Br^- C_{10}H_{21}SO_3^-Na^+$ (0.1M NaBr)	−26	−12	0.49	0.51
$[C_8H_{17}N^+(CH_3)_2CH_2CHOH]_2 \cdot 2Br^- C_{12}H_{25}SO_3^-Na^+$ (0.1M NaBr)	−31	−14	0.45	0.45
$[C_{10}H_{21}N^+(CH_3)_2CH_2CHOH]_2 \cdot 2Br^- C_{10}H_{21}SO_3^-Na^+$ (0.1 M NaBr)	−34	−14	0.48	0.50
$[C_{10}H_{21}N^+(CH_3)_2CH_2CHOH]_2 \cdot 2Br^- C_{12}H_{21}SO_3^-Na^+$ (0.1 M NaBr)	−34	−17	0.50	0.55
$C_8H_{17}SO_4^-Na^+ - C_8H_{17}N^+(CH_3)_3 \cdot Br^-$	−14	−10	—	—
$(C_{10}H_{21})2C_6H_2(SO_3^-Na^+)OC_6H_4SO_3^-Na^+C_{14}H_{29}N(CH_3)_2O$ (0.1M NaCl)	−9.9	−2.4	—	—
$C_{10}H_{21}C_6H_3(SO_3^-Na^+)OC_6H_4SO_3^-Na^+C_{14}N_{29}N(CH_3)_2O$ (0.1 N NaCl	−8.3	−7.5	—	—
$C_{10}H_{21}C_6H_3(SO_3^-Na^+)OC_6H_5C_{14}H_{29}N(CH_3)_2O$ (0.1 M NaCl)	−4.7	−3.2	—	—
$(C_{10}H_{21})_2C_6H_2(SO_3^-Na^+)OC_6H_4SO_3^-Na^+C_{12}H_{25}(OC_2H_4)_7OH$ (0.1M NaCl)	−6.9	−0.8	—	—
$C_{10}H_{21}C_6H_3(SO_3^-Na^+)OC_6H_4SO_3^-Na^+C_{12}H_{25}(OC_2H_4)_7OH$ (0.1 M NaCl)	−1.8	−0.9	—	—

a. values not available.

In addition to the much larger β^{σ} values for gemini-containing mixtures compared to the conventional mixtures listed, there are two other noteworthy features of the interaction between diquaternary gemini and conventional anionic surfactant mixtures: (i) the much weaker interaction between the gemini and the second surfactant in the mixed micelle (β^{m}) than in the mixed monolayer at the aqueous solution/air interface (β^{σ}) and (ii) the mole fraction values of approximately 0.5, both in the mixed monolayer at the aqueous solution/air interface (X^{σ}) and in the mixed micelle (X^{m}) in the aqueous phase. The weaker interaction of geminis in the mixed micelle compared to that in the mixed monolayer is believed to be due to the greater difficulty of incorporating the two hydrophobic groups of the geminis into a convex micelle compared to that of accommodating them at a planar interface. The mole fraction values of approximately 0.5 indicate that the two surfactants are interacting in a 1 : 1 molar ratio, rather than in the 2 : 1 molar ratio that would be expected of a divalent gemini and a monovalent conventional surfactant. This 1 : 1 molar interaction leaves a net charge on the interaction product and allows it to be water-soluble, in contrast to the usual anionic-cationic interaction product of conventional surfactants which has no net charge and generally precipitates from aqueous solution.

The synergism in surface tension reduction resulting from this very strong interaction is illustrated in Fig. 4.7. The cationic gemini-conventional anionic mixture is both more efficient and more effective at reducing surface tension than either component by itself.

Gemini surfactants also exhibit some unusual behavior that may be unique to them. When the number of methylene groups in the alkyl chain of the hydrophobic group of a conventional surfactant is increased, there is a monotonic increase in the surface activity of the compound. This is shown by the commonly observed linear relationship between the log of either the CMC or the C_{20} value and the number of carbon atoms in the straight alkyl chain. These linear relationships are also shown by gemini surfactants up to some critical carbon number. When this critical carbon number is exceeded (about 14 per chain, the exact value depending upon the structure of the gemini and the molecular environment, e.g., temperature, electrolyte content), then geminis start to exhibit less than their expected surface activity. Thus, the log CMC and log C_{20} values start to deviate from their linear relationship with the number of carbon atoms in the alkyl chains in the direction of decreased surface activity. The deviation increases with an increasing number of carbon atoms until a value is reached where surface activity decreases with increase in the length of the alkyl chains, i.e., the CMC and C_{20} values increase with this change. This has been attributed to the formation of small, nonsurface-active aggregates below the observed CMC. Calculation of equilibrium constants for this self-association indicates that the aggregates are probably small oligomers (dimers, trimers, and tetramers). This phenom-

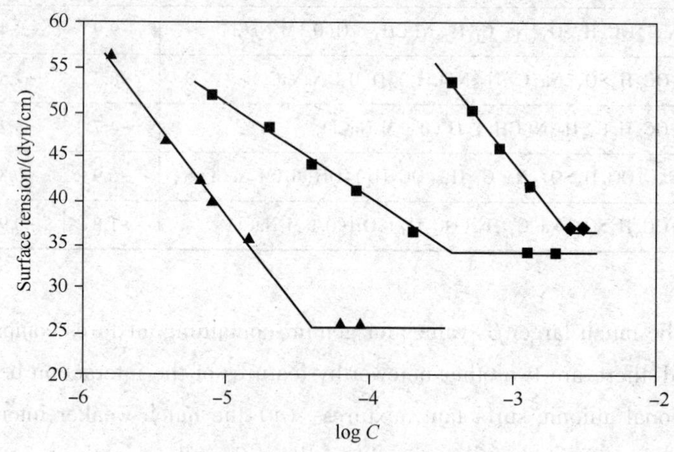

Figure 4.7 Synergism in the system $[C_{10}H_{21}N^{+}(CH_3)_2CH_2CHOH]_2 \cdot 2Br^{-}$, $C_{12}H_{25}SO_3^{-}Na^{+}$, and theirmixture at 0.67 mole fraction of the diquaternary gemini in 0.1 M NaBr at 25°C. Surface tension vs. log molar surfactant concentration (C) plots of ◆, $[C_{10}H_{21}N^{+}(CH_3)_2CH_2CHOH]_2 \cdot 2Br^{-}$; ■, $C_{12}H_{25}SO_3^{-}Na^{+}$; ▲, their mixture.

enon has also been observed in geminis with three hydrophobic groups in the molecule. This unusual behavior may be due to the exceptionally large free energy decrease resulting from hydrophobic bonding between molecules containing two long hydrophobic groups each.

Still another unusual phenomenon shown by ionic geminiswith alkyl chains longer than a critical length is their interaction with oppositely charged surfactants. As described above, ionic geminis can interact very strongly with oppositely charged surfactants to increase the surface activity of the system synergistically. Above a critical combined alkyl chain length for the two interacting surfactants, however, their interaction produces a marked decrease in the surface activity of the system. This is shown in Fig 4.8. As in the self-association described above, this decrease in surface activity is attributed to the formation of small, nonsurface-active aggregates, but in this case involving two different surfactants. Calculation of the equilibrium constants for the interaction between the two surfactants indicates that they form a 1 : 1 molar interaction product, as in the interactions of the shorter-chain cationic geminis with similar conventional anionics shown in Table 4.5. However, in those cases the interaction products are more surface active than the component surfactants, whereas with these longer-chain geminis the interaction products are nonsurface active. As before, the 1 : 1 molar interaction permits them to remain water soluble.

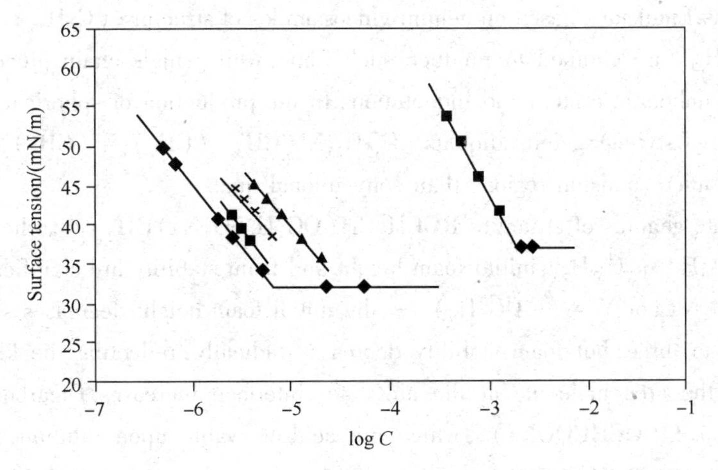

Fig 4.8 Surface tension vs. log molar surfactant concentration (C) plots of $[C_{12}H_{25}N^+(CH_3)_2CH_2CHOH]_2 2Br^-$, $C_{12}H_{25}SO_3^-Na^+$, and their mixtures in 0.1 M NaBr at 25℃ at various mole fractions (α) of the diquaternary gemini. ◆, $[C_{12}H_{25}N^+(CH_3)_2CH_2CHOH]_2 \cdot 2Br^-$; ■, $C_{12}H_{25}SO_3^-Na^+$; ▲, mixture 1, $\alpha = 0.33$; ×, mixture 2, $\alpha = 0.50$; ×, mixture 3, $\alpha = 0.67$

Still another unusual property of geminis is their formation of long worm-like micelles at low concentrations in aqueous solution. This accounts for the unusually high viscosities at low surfactant concentrations in aqueous solution shown by some geminis. Aqueous solutions of cationic geminis of structure $C_{12}H_{25}N^+(CH_3)_2(CH_2)_2N^+(CH_3)_2C_{12}H_{25} \cdot 2Br^-$ show shear thickening at concentrations of 1.4% and become viscoelastic above 2%.

2. Performance Properties

As a result of the fundamental physicochemical properties already described, gemini surfactants should exhibit some interesting, useful performance properties. Their greater water solubility and higher surface activity, compared to analogous conventional surfactants, make them potential candidates for use in new laundry washing machines which limit the use of water. Their very low CMC values imply low skin irritation, which is associated with monomeric surfactant concentrations. The maximum possible concentration of monomeric surfactants in the system decreases as the CMC decreases. Low CMC values also promise greater efficiency in the solubilization of water-insoluble material, since solubilization occurs only above the CMC. Closer packing of hydrophobic chains at the interface, compared to conventional surfactants, means a more laterally cohesive interfacial film, implying better foaming, dispersing, and emulsifying properties, whereas the 1 : 1 molar interaction of ionic geminis with oppositely charged surfactants makes them more compatible than comparable conventional surfactants. Many of these potential performance properties have been observed in actual practice.

2.1 Dispersing, Emulsifying, and Foaming

Drug particles with enhanced bioavailability are claimed to be obtained by wet milling a stable dispersion to an average particle size of less than 400 nm in the presence of the gemini surfactant of structure $[R'NHCH_2CH_2N(R)CH_2]_2$, where R′ is lactobionoyl.

Gelatin emulsions for photographic purposes must produce perfectly smooth coatings free of craters or indentations. Emulsions based on gemini glucosamides of structure $(C_6H_{13})_2C[CH_2NHC(O)(CHOH)_4CH_2OH]_2$ are claimed to produce such films, while single-chain glucosamides produce films with large numbers of craters and indentations. In the production of spherical latex particles in the polymerization of styrene, gemini diquats $[C_{12}H_{25}N^+(CH_3)_2(CH_2)_xN^+(CH_3)_2C_{12}H_{25}]$ give rise to more extensive microemulsion regions than conventional quats.

For disulfonate geminis of structure $ROCH_2CH(OC_3H_6SO_3Na)CH_2]_2Y$, when Y is small (—O—) and R is $C_{10}H_{21}$ or $C_{12}H_{23}$, initial foam height and foam stability are significantly greater than those of $C_{12}H_{25}SO_4Na$. For Y =—$(OC_2H_4)_n$—, the initial foam height decreases slightly as n is increased from one to three, but foam stability decreases radically, reflecting the known decrease in foam stability as the area/molecule at the air/water interface increases. Dicarboxylate geminis of structure $(C_{11}H_{23})_2C(OCHCOONa)_2$, which are acid-cleavable upon dilution, were observed to foam more strongly than $C_{11}H_{23}COONa$. Compounds of structure $[C_{10}H_{21}OCH_2CH(OCH_2COO^-Na^+)CH_2]_2Y$, where Y is —O— or —$OCH_2CH_2O$—, also display significantly better foaming properties than $C_{11}H_{23}COO^-Na^+$. The disodium phosphated geminis $[C_{10}H_{21}OCH_2CH[OP(O)(ONa)(OH)]CH_2]_2Y$, where Y=$(OCH_2CH_2)$—, showed excellent foaming properties, but the analogous tetrasodium phosphates$[C_{10}H_{21}OCH_2CH\{OP(O)(ONa)_2\}CH_2]_2Y$ showed almost no foaming. All of these compounds were tested at 0.1% concentration.

Dicarboxylates of structure $[C_{10}H_{21}N(COCH_2CH_2COONa)CH_2]_2CHOH$ are claimed to show much improved foaming behavior at 0.1% concentration than sodium laurate. Conventional alkyl trimethylammonium cationic surfactants show very little foam at 0.1% concentration. In contrast, C_{12} and C_{14} cationic geminis of structure $[RN^+(CH_3)_2]_2Y \cdot 2Cl^-$ show very high foam when the spacer is short (three carbon atoms), in some cases higher even than that of sodium dodecyl sulfate. When the spacer length is increased to four carbon atoms, however, the initial foam decreases somewhat, but foam stability decreases dramatically, similar to what was observed above with increasing spacer length of disulfonate geminis.This may account for the observed poor foam stability in the gemini series, $RCH(SO_3Na)CO_2(C_2H_4O)_nOCCH(SO_3Na)R$, with a considerable spacer length between the hydrophilic head groups.In this series, foam stability was poorer than for the comparable conventional surfactants.

2.2 Wetting

Geminis in which the alkyl chains are short and branched and thelinkage between the hydrophilic groups is short show excellent dynamic wetting properties. The acetylenic glycols of structure $RR'C(OH)C \equiv CC(OH)R'R$, where R′ is —$CH_3$ and R is a C_2-C_4 alkyl group, are commercially available geminis that have been utilized for decades.Compounds of structure $\{RN[(C_2H_4O)_xH]C(O)\}_2Y$, where Y is —$CH_2CH_2$— or —CH=CH— and R is 2-ethyl—hexyl, have been reported to be excellent hydrophobic soil wetting and rewetting agents. For the dicarboxylate geminis $\{C_{10}H_{21}OCH_2CH(OCH_2COO^-Na^+)CH_2\}_2Y$, where Y is —O— or —$OCH_2CH_2O$—, Draves skein wetting times at 0.1% concentration were 6 and 16 s, respectively, compared to 226 s for $C_{11}H_{23}CO_2Na$.

2.2.1 Solubilization of Water-insoluble Surfactants and Enhancement of Wetting

An alkylated diphenyl ether sulfonate of the gemini type (disodium didecyl diphenyl ether disulfonate) has been shown to solubilize water-insoluble nonionic surfactants of the alcohol ethoxylate, alkylphenol ethoxylate, and N-alkylpyrrolidone type more efficiently and more effectively than its monodecyl disulfonate, monodecyl monosulfonate, or didecylmonosulfonate analogs. As a result of this solubilization, mixtures of these water-insoluble surfactants containing 20% of this gemini at 0.1% concentration in water showed wetting times in the Draves skein wetting test of less than 15 s, while the water-insoluble surfactants and the gemini by itself showed very poor wetting under the same conditions.

2.2.2 Solubilization of Hydrocarbons and Other Water-insoluble Material

Cationic geminis of structure $C_{12}H_{25}N^+(CH_3)_2(CH_2)_sN^+(CH_3)_2C_{12}H_{25} \cdot 2Br^-$ are reported to solubilize more toluene and more n-hexane than the comparable conventional surfactants.This is most pronounced when s = 2.Possibly owing to their enhanced solubilization of normally water-insoluble material, the use of gemini surfactants of the type $(RC_6H_3SO_3M)_2O$ is claimed to result in improved removal of stains from inks, paints, and other coloring materials from skin and clothing.

2.2.3 Skin and Eye Irritation

As far back as 1967, a patent claimed that gemini betaines of structure $[C_9H_{19}NH(CH_2)_2N^+(CH_3)_2(CH_2COO^-)]_2(CH_2)_m$ were less irritating (Draize test) than their single-chain analogs. A recent patent claims that a zwitterionic gemini of structure $[C_{11}H_{23}CONHC_2H_4N^+H(CH_2COO^-)CH_2]_2$, is a minimal irritant (Eytex protocol) and reduces the irritancy of a sulfated alcohol ethoxylate.

Dicationic geminis of structure $[C_{12}H_{25}N^+(CH_3)_2CH_2CONH]_2Y$, where Y is —$(CH_2)_4$— or —$(CH_2)_2SS(CH_2)_2$— werefound to be nonirritating (Draize test) at concentrations <0.5%.They also conferred antimicrobial resistance to wool fabric.

Gemini sulfoesters of structure $[C_{12}H_{25}CH(SO_3Na)COOCH_2]_2$, tested both on human arms and by protein denaturation, are claimed to be superior to the single-chain sulfomonoester.The gemini sulfonates gave lower protein denaturation by a factor of 10, and a 5% solution applied to the arms of human subjects for 5 d resulted in no cases of redness or skin rashes. Sulfated geminis of structure $[C_{12}H_{25}N(COCH_2CH_2CH_2OSO_3Na)CH_2]_2CHOH$, tested on the abdominal region of guinea pigs, are claimed to show no irritancy (triethanolamine lauryl sulfate resulted in erythema), while methylene bis(alkylphenol) sulfates of structure $\{[C_8H_{17}C_6H_3O(C_2H_4O)_7SO_3Na]\}_2CH_2$, tested (Eytex protocol) as mild surfactants.

2.3 Antimicrobial Properties

Diquaternary ammonium geminis of structure $[C_{12}H_{25}N^+(CH_3)_2CH_2CONH]_2Y \cdot 2Cl^-$, where $Y = -(CH_2)_4-$ or $-(CH_2)_2SS(CH_2)_2-$, were shown to have greater antimicrobial activity against both gram-positive and gram-negative organisms and *Candida albicans* than hexadecyltrimethylammonium bromide. There was no significant difference in the antimicrobial activity of the two compounds with different linkages.

A nonionic gemini based upon dioctanoyl lysine, of structure $C_7H_{15}CONH(CH_2)_4CH(NHCOC_7H_{15})CON[(C_2H_4O)_2CH_3]_2$, was found to be nonirritating and nonhemolytic but highly surface-active and hence suitable for personal-care or pharmaceutical formulations.

2.4 Hair Conditioning

The diquaternary ammonium chloride, $[C_{18}H_{37}N^+(CH_3)_2CH_2]_2CHOH \cdot 2Cl^-$, is claimed to exhibit hair conditioning properties superior to those of stearalkonium chloride. The Rubine dye test demonstrated its higher substantivity to hair, presumably owing to the double charge on the gemini surfactant.

REFERENCES

(omitted)

Author

Professor Rosen is Director of the Surfactant Research Institute at Brooklyn College of the City University of New York. He has published six books and well over 100 research papers in the area of surfactants. He acts as consultant and frequently invited lecturer in the field and has served on the advisory or editorial boards of several leading journals.

Dr. David Tracy is a Scientist Fellow at Rhodia Inc. where heis responsible for research and development of new products, including gemini surfactants. He graduated with honors, receiving a B. A. in chemistry from Thomas More College and an M.S. and Ph.D. in organic chemistry from the University of Illinois. Throughout his thirty-year career he also held various R&D positions at GAF before joining Rhodia Inc. in 1990.

Unit 5 Oilfield Chemistry

5.1 Drilling Muds

Drilling fluids are mixtures of natural and synthetic chemical compounds used to cool and lubricate the drill bit, clean the bottom of hole, carry cuttings to the surface, control formation pressures, and improve the function of the drill string and tools in the hole. They are divided into two general types: water-based fluids and oil-based fluids. The type of fluid base used depends on drilling and formation needs, as well as the requirements for disposition of the fluid after it is no longer needed. Drilling muds are a special class of drilling fluids used to drill most deep wells. Mud refers to the thick consistency of the formulation. The functions of a drilling mud are as follows: (1) to remove rock bit cuttings from the bottom of the hole and carry them to the surface, (2) to overcome the fluid pressure of the formation, (3) to avoid damage of the producing formation, (4) to cool and lubricate the drill string and the bit, (5) to prevent drill pipe corrosion fatigue, (6) to allow the acquisition of information about the formation being drilled.

5.1.1 Classification of Muds

The classification of drilling muds is based on their fluid phase alkalinity, dispersion, and the type of chemicals used. The classification of muds are as follows: freshwater muds, inhibited muds, emulsions and oil-based muds. Freshwater muds and inhibited muds are belong to dispersed noninhibited systems. The drilling fluids are used in the sections of upper hole. They are formulated from freshwater and may contain bentonite. The flow properties are controlled by a flocculant or thinner, and the fluid loss is controlled with bentonite and carboxymethylcellulose. Low-solids muds is belong to nondispersed systems. In the nondispersed systems, no special agents are added to deflocculate the solids in the fluid. The main advantages of these systems are the higher viscosities and the higher yield point-to-plastics viscosity ratio. These alterated flow properties provide a better cleaning of the bore hole, allow a lower annular circulating rate, and minimize wash out of the bore hole.

Quebracho-treated freshwater mudsis one of dispersed noninhibited systems. Quebracho is a natural product extracted from the heartwood of the Schinopsis trees that grow in Argentina and Paraguay. Quebracho is a well characterized polyphenolic and is readily extracted from the wood by hot water. Quebracho is widely used as a tanning agent. It is also used as a mineral dressing, as a dispersant in drilling muds, and in wood glues. Quebracho is commercially available as a crude hot water extract, either in lump, ground, or spray-dried form, or as a bisulfite treated spraydried product that is completely soluble in cold water. Quebracho is also available in a bleached form, which can be used in applications where the dark color of unbleached quebracho is undesirable. Quebracho-treated freshwater muds were used in shallow depths.It is also referred to as red mud because of the deep red color.Quebracho acts as a thinner.Polyphosphates are also added when quebracho is used.Quebracho is active at low concentrations and consists of tannates.

Lignosulfonate freshwater muds is also belong to dispersed noinhibited systems. Lignosulfonate freshwater muds contain ferrochrome lignosulfonate for viscosity and gel-strength control.These muds are resistant to most types of drilling contamination because of the thinning efficiency of the lignosulfonate in the presence of large amounts of salt and extreme hardness.

Low-solids freshwater muds is one of nondispersed systems.Clear freshwater is the best drilling fluid in terms of penetration rate.Therefore it is desirable to achieve a maximal drilling rate using a minimal amount of solid additives.Originally low-solids mud formulations were used in hard formations,but they now also tend to find use in other formations.Several types of flocculants are used to promote the settling of drilled solids by flocculation

5.1.2 Mud Compositions

Components for water-based muds are as follows: glycol based alkali silicates, acrylamide homopolymer, carboxymethylcellulose, carboxymethylcellulose zinc oxide, acrylamide copolymer, and polypropylene glycol.Various modification methods for lignosulfonates have been described, for example, by condensation with formaldehyde or modification with iron salts.It has been found that chromium- modified lignosulfonates, as well as mixed metal lignosulfonates of chromium and iron, are highly effective as dispersants and therefore are useful in controlling the viscosity of drilling fluids and in reducing the yield point and gel strength of the drilling fluids.Because chromium is potentially toxic, its release into the natural environment and the use thereof is continuously being reviewed by various government agencies around the world.Therefore less toxic substitutes are desirable.

Diesel oilof oil-based muds is harmful to the environment, particularly the marine environment in offshore applications.The use of palm oil derivative could be considered as an alternative oil-based fluid that is harmless to the environment.Poly-α-olefins (PAOs) are biodegradable and nontoxic to marine organisms; they also meet viscosity and pour point specifications for formulation into oilbased muds.The hydrogenated dimer of 1-decene can be used instead of conventional organic-based fluids, as can n−1-octene.Oil-based muds are being replaced now by synthetic muds.

Synthetic-based muds are mineral oil muds in which the oil phase has been replaced with a synthetic fluid, such as ether, ester, PAO, or linear alkylbenzene, and are available from major mud companies. The mud selection process is based on the mud's technical performance, environmental impact, and financial impact.Synthetic muds are expensive.Two factors influence the direct cost: unit or per- barrel cost and mud losses. Synthetic muds are the technical equivalent of oil-based muds when drilling intermediate hole sections. They are technically superior to all water-based systems when drilling reactive shales in directional wells.However, with efficient solids-control equipment, optimized drilling, and good housekeeping practices, the cost of the synthetic mud can be brought to a level comparable with oil-based mud.

Other base materials proposed areethers of monofunctional alcohols, branched didecyl ethers, α-sulfofatty acids, oleophilic alcohols, oleophilic amides, hydrophobic side chain poly amides from N, N-didodecylamine and sodium polyacrylate or polyacrylic acid, Poly ether poly amine, Phosphate ester of a hydroxy polymer. Quaternary oleophilic esters of alkylol amines and carboxylic acids improve the wettability of clay. Nitrates and nitrites can replace calcium chloride in inverted emulsion drilling muds.

5.1.3 Characterization of Drilling Muds

Important parameters to characterize the properties of a drilling mud are specific weight, viscosity, gel strength, and filtration performance. Viscosity is measured by means of a Marsh funnel. The funnel is dimensioned so that the outflow time of 1qt (926 mL) freshwater at 21°C (70°F) is 26s. Viscosity also is measured with a rotational viscometer. The mud is placed between two concentric cylinders. One cylinder rotates with constant velocity. The other cylinder is connected with a spring. The torque on this cylinder results in a deviation of the position from rest, which may serve as a measure of viscosity. Gel strength is obtained with the rotational viscometer when the maximal deflection of the pointer is monitored when the motor is turned on with low speed, the liquid being at rest for a prolonged time before, for example, for 10 minutes. This maximal deflection is referred to as a 10-minute gel.

From *oil field chemicals by Johannes Karl Fink*

New Words and Expressions

lubricate [ˈluːbrɪkeɪt] 润滑,使润滑
alkalinity [ˌælkəˈliniti] 碱度,碱性
carboxymethylcellulose [kɑːbɒksiːmeθɪlˈseljʊləʊ] 羧甲基纤维素
ferrochrome lignosulfonate 铁铬木质素磺酸盐
quebracho [keɪˈbrɑːtʃəʊ] 白雀树,白坚木
unbleached [ʌnˈbliːtʃt] 未漂白的,原色的
lignosulfonate [lɪgnəʊsʌlˈfɒneɪtiː] 木质素磺酸盐
polypropylene glycol 聚丙二醇
Oil-based muds 油基泥浆
schinopsis trees 红坚木

Notes

(1) **Drilling muds** 钻井液泥浆,被公认为油田钻井的血液,主要是把岩屑从井底携带至地面,除此之外,还有以下作用:清洗井底,悬浮携带岩屑,保持井眼清洁;平衡地层压力,稳定井壁、防止井塌、井喷、井漏;传递水功率、以帮助钻头破碎岩石;为井下动力钻具传递动力和冷却钻头、钻具。

（2）**Marsh funnel** 马氏漏斗是一种用于日常测量钻井液粘度的仪器。采用 Q/02NXL 002-2007 企业标准结合美国 API 标准制造。以定量钻井液从漏斗中流出的时间来确定钻井液的粘度。结构简单、使用方便。广泛应用于石油,地质勘探等部门。

5.2 Recovery

5.2.1 Introduction

Recovery, as applied in the petroleum industry, is the production of oil from a reservoir. There are several methods by which this can be achieved, which range from recovery because of reservoir energy (i.e., the oil flows from the well hole without assistance) to enhanced recovery methods in which considerable energy must be added to the reservoir to produce the oil. However, the effect of the method on the oil and on the reservoir must be considered before application.

Thus, once the well is completed, the flow of oil into the well commences. For limestone reservoir rock, acid is pumped down the well and out the perforations. The acid dissolves channels in the limestone that lead oil into the well. For sandstone reservoir rock, a specially blended fluid containing proppants (sand, walnut shells, aluminum pellets) is pumped down the well and out the perforations. The pressure from this fluid makes small fractures in the sandstone that allow oil to flow into the well, while the proppants hold these fractures open. Once the oil starts flowing, the oil rig is removed from the site and production equipment is set up to extract the oil from the well.

A well is always carefully controlled in its flush stage of production to prevent the potentially dangerous and wasteful gusher. This is actually a dangerous condition, and is (hopefully) prevented by the blowout preventer and the pressure of the drilling mud. In most wells, acidizing or fracturing the well starts the oil flow.

As already noted, crude oil accumulates over geological time in porous underground rock formations called reservoirs that are at varying depths in the earth's crust, and in many cases elaborate, expensive equipment is required to get it from there. The oil is usually found trapped in a layer of porous sandstone, which lies just beneath a dome-shaped or folded layer of some nonporous rock such as limestone. In other formations the oil is trapped at a fault or break in the layers of the crust.

Generally, crude oil reservoirs exist with an overlying gas cap, in communication with aquifers, or both. The oil resides together with water and free gas in very small holes (pore spaces) and fractures. The size, shape, and degree of interconnection of the pores vary considerably from place to place in an individual reservoir. Below the oil layer, the sandstone is usually saturated with salt water. The oil is released from this formation by drilling a well and puncturing the limestone layer on either side of the limestone dome or fold. If the peak of the formation is tapped, only the gas is obtained. If the penetration is made too far from the center, only salt water is obtained. Oil wells may be either on land or under water. In North America, many wells are offshore in the shallow parts of the oceans. The crude oil or unrefined oil is typically collected from individual wells by small pipelines.

The oil in such formation is usually under such great pressure that it flows naturally, and sometimes with great force, from the well. However, in some cases, this pressure later diminishes so that the oil must be pumped from the well. Natural gas or water is sometimes pumped into the well to replace the oil that is withdrawn. This is called repressurizing the oil well.

Conventional crude oil is a brownish green to black liquid of specific gravity in a range from about 0.810 to 0.985, with a boiling range from about 20℃ to above 350℃, above which active decomposition ensues when distillation is attempted. The oils contain from 0% to 35% or more of gasoline, as well as varying proportions of kerosene hydrocarbons and higher boiling constituents up to the viscous and nonvolatile compounds present in lubricant oil and in asphalt. The composition of the crude oil obtained from the well is variable and depends not only on the original composition of the oil in situ, but also on the manner of production and the stage reached in the life of the well or reservoir.

5.2.2 Primary Recovery (Natural Methods)

If the underground pressure in the oil reservoir is sufficient, then the oil will be forced to the surface under this pressure. Gaseous fuels or natural gas are usually present, which also supply needed underground pressure. In this situation, it is sufficient to place a complex arrangement of valves (the Christmas tree) at the well head to connect the well to a pipeline network for storage and processing. This is called primary oil recovery.

Thus, primary oil production (primary oil recovery) is the first method of producing oil from a well and depends upon natural reservoir energy to drive the oil through the complex pore network to producing wells. If the pressure on the fluid in the reservoir (reservoir energy) is great enough, the oil flows into the well and up to the surface. Such driving energy may be derived from liquid expansion and evolution of dissolved gases from the oil as reservoir pressure is lowered during production, expansion of free gas, or a gas cap, influx of natural water, gravity, or combination of these effects.

Crude oil moves out of the reservoir into the well by one or more of three processes. These processes are: dissolved gas drive, gas cap drive, and water drive. Early recognition of the type of drive involved is essential to the efficient development of an oil field.

In dissolved gas drive, the propulsive force is the gas in solution in the oil, which tends to come out of solution because of the pressure released at the point of penetration of a well. Dissolved gas drive is the least efficient type of natural drive as it is difficult to control the gas-oil ratio; the bottom-hole pressure drops rapidly, and the total eventual recovery of petroleum from the reservoir may be less than 20%.

If gas overlies the oil beneath the top of the trap, it is compressed and can be utilized (gas cap drive) to drive the oil into wells situated at the bottom of the oil-bearing zone. By producing oil only from below the gas cap, it is possible to maintain a high gas-oil ratio in the reservoir until almost the very end of the life of the pool. If, however, the oil deposit is not systematically developed, so that bypassing of the gas occurs, an undue proportion of oil is left behind. The usual recovery of petroleum from a reservoir in a gas cap field is 40% to 50%.

Usually the gas in a gas cap (associated natural gas) contains methane and other hydrocarbons that may be separated out by compressing the gas. A well-known example is natural gasoline that was formerly referred to as casinghead gasoline or natural gas gasoline. However, at high pressures, such as those existing in the deeper fields, the density of the gas increases and the density of the oil decreases until they form a single phase in the reservoir. These are the so-called retrograde condensate pools, because a decrease (instead of an increase) in pressure brings about condensation of the liquid hydrocarbons. When this reservoir fluid is brought to the surface and the condensate is removed, a large volume of residual gas remains. The modern practice is to cycle this gas by compressing it and inject it back into the reservoir, thus maintaining adequate pressure within the gas cap, and condensation in the reservoir is prevented. Such condensation prevents recovery of the oil, for the low percentage of liquid saturation in the reservoir precludes effective flow.

The most efficient propulsive force in driving oil into a well is natural water drive, in which the pressure of the water forces the lighter recoverable oil out of the reservoir into the producing wells. In anticlinal accumulations, the structurally lowest wells around the flanks of the dome are the first to come into water. Then the oil-water contact plane moves upward until only the wells at the top of the anticline are still producing oil; eventually, these also must be abandoned as the water displaces the oil.

In a water drive field, it is essential that the removal rate be adjusted so that the water moves up evenly as space is made available for it by the removal of the hydrocarbons. An appreciable decline in bottom-hole pressure is necessary to provide the pressure gradient required to cause water influx. The pressure differential needed depends on the reservoir permeability; the greater the permeability, the lesser the difference in pressure necessary. The recovery of petroleum from the reservoir in properly operated water drive pools may run as high as 80%. The force behind the water drive may be hydrostatic pressure, the expansion of the reservoir water, or a combination of both. Water drive is also used in certain submarine fields.

For primary recovery operations, no pumping equipment is required. If the reservoir energy is not sufficient to force the oil to the surface, then the well must be pumped. In either case, nothing is added to the reservoir to increase or maintain the reservoir energy or to sweep the oil toward the well. The rate of production from a flowing well tends to decline as the natural reservoir energy is expended. When a flowing well is no longer producing at an efficient rate, a pump is installed.

The recovery efficiency for primary production is generally low when liquid expansion and solution gas evolution are the driving mechanisms. Much higher recoveries are associated with reservoirs with water and gas cap drives and with reservoirs in which gravity effectively promotes drainage of the oil from the rock pores. The overall recovery efficiency is related to how the reservoir is delineated by production wells. Thus, for maximum recovery by primary recovery, it is often preferable to sink several wells into a reservoir, thereby bringing about recovery by a combination of the methods outlined here.

5.2.3 Secondary Recovery

Over the lifetime of the well the pressure will fall, and at some point there will be insufficient underground pressure to force the oil to the surface. If economical, and itoften is, the remaining oil in the well is extracted using secondary oil recovery methods. It is at this point that secondary recovery methods must be applied.

Secondary oil recovery methods use various techniques to aid in recovering oil from depleted or low-pressure reservoirs.Sometimes pumps on the surface or submerged (electrical submersible pumps (ESPs)),are used to bring the oil to the surface.Other secondary recovery techniques increase the reservoir's pressure by water injection and gas injection,which injects air or some other gas into the reservoir.Together,primary recovery and secondary recovery allow 25% to 35% of the reservoir's oil to be recovered.

Primary (or conventional) recovery can leave as much as 70% of the petroleum in the reservoir.Such effects as microscopic trapping and bypassing are the more obvious reasons for the low recovery.There are two main objectives in secondary crude oil production.One is to supplement the depleted reservoir energy pressure,and the second objective is to sweep the crude oil from the injection well toward and into the production well.In fact,secondary oil recovery involves the introduction of energy into a reservoir to produce more oil.For example,the addition of materials to reduce the interfacial tension of the oil results in higher recovery of oil.

There are also secondary oil recovery operations that involve the injection of water or gas into the reservoir.When water is used the process is called a waterflood; with gas, a gasflood.Separate wells are usually used for injection and production.The injected fluids maintain reservoir pressure or repressure the reservoir after primary depletion and displace a portion of the remaining crude oil to production wells.In fact,the first method recommended for improving the recovery of oil was probably the reinjection of natural gas,and there are indications that gas injection was utilized for this purpose before 1900.These early practices were implemented to increase the immediate productivity and are therefore classified as pressure maintenance projects.Recent gas injection techniques have been devised to increase the ultimate recovery,thus qualifying as secondary recovery projects.

In secondary recovery,the injected fluid must dislodge the oil and propel it toward the production wells.Reservoir energy must also be increased to displace the oil.Using techniques such as gas and water injection does not change the state of oil.Similarly,there is no change in the state of the oil during miscible fluid displacement technologies.The analogy that might be used is that of a swimmer (in water) in which there is no change to the natural state of the human body.

Thus,the success of secondary recovery processes depends on the mechanism by which the injected fluid displaces the oil (displacement efficiency) and on the volume of the reservoir that theinjected fluid enters (conformance or sweep efficiency).In most proposed secondary projects, water does both these things more effectively than gas.It must be decided whether the use of gas offers any economic advantages because of availability and relative ease of injection.In reservoirs with high permeability and high vertical span,the injection of gas may result in high recovery factors as a result of gravity segregation,as described in a later section.However,if the reservoir lacks either adequate vertical permeability or the possibility for gravity segregation, a frontal drive similar to that used for water injection can be used (dispersed gas injection).Thus,dispersed gas injection is anticipated to be more effective in reservoirs that are relatively thin and have little dip.Injection into the top of the formation (or into the gas cap) is more successful in reservoirs with higher vertical permeability (200 md or more) and enough vertical relief to allow the gas cap to displace the oil downward.

Vaporization is another recovery mechanism used to inject gas into oil reservoirs. A portion of the oil affected by the dry injection gas is vaporized into the oil and transported to the production wells in the vapor phase. In some instances, this mechanism has been responsible for a substantial amount of the secondary oil produced.

During the withdrawal of fluids from a well, it is usual practice to maintain pressure in the reservoir at or near the original levels by pumping either gas or water into thereservoir as the hydrocarbons are withdrawn. This practice has the advantage of retarding the decline in the production of individual wells and considerably increasing the ultimate yield. It also may bring about the conservation of gas that otherwise would be wasted, and the disposal of brines that otherwise might pollute surface and near-surface potable waters.

In older fields, it was not the usual practice to maintain reservoir pressure, and it is now necessary to obtain petroleum from these fields by means of secondary recovery projects. Thus, several methods have been developed to obtain oil from reservoirs where previous economic policies dictated that ordinary production systems were no longer viable.

Considerable experimentation has been carried on in the use of different types of input gas. Examples are wet casinghead gas, enriched gas, liquefied petroleum gas (LPG), such as butane and propane, high-pressure gas, and even nitrogen. High-pressure gas not only pushes oil through the reservoir, but may also produce a hydrocarbon exchange so that the concentration of liquid petroleum gases in the oil is increased.

Water injection is still predominantly a secondary recovery process (waterflood). Probably, the principal reason for this is that reservoir formation water is ordinarily not available in volume during the early years of an oil field and pressure maintenance water from outside the field may be too expensive. When a young field produces considerable water, it may be injected back into the reservoir primarily for the purpose of nuisance abatement, but reservoir pressure maintenance is a valuable by-product.

From *the Chemistry and Technology of Petroleum* by *Heinz Heinemann*

New Words and Expressions

reservoir energy 储层能量

limestone reservoir rock 石灰岩储集岩

proppant 支撑剂

walnut shell 核桃壳

geological time 地质时间;地质年代

aquifer [ˈækwɪfə](美)蓄水层;含水土层

asphalt[ˈæsfælt] 沥青;柏油

in situ 在原地,就地;在原来位置

directional drilling [油气] 定向钻进;定向钻井;定向钻孔

ocean rig 海洋钻井平台
primary oil recover 一次采油
casinghead gasoline 井口汽油
natural water drive 天然水驱
hydrostatic pressure [流] 液体静压力
secondary recovery 二次采油
sucker-rod pump 有杆泵;杆式泵

Notes

(1) **Primary oil recover** 仅依靠天然能量开采原油的方法。天然能量包括:天然水驱、弹性能量驱、溶解气驱、气驱及重力驱等。人类对油藏的作用只限于钻出油井,为油流提供通道。

(2) **Secondary recovery** 依赖地层天然压力采油称为一次采油。随着地层压力的下降,需要用注水补充地层压力的办法来采油,称为二次采油。据计算,一次采油的采收率在20%~30%,二次采油可使采收率达到40%。

5.3 Enhanced Oil Recovery

Approximately 60% to 70% of the oil in place cannot be produced by conventional methods. Enhanced oil-recovery methods gain importance in particular with respect to the limited worldwide resources of crude oil. The estimated worldwide production from enhanced oil-recovery projects and heavy-oil projects at the beginning of 1996 was approximately 2.2 million barrels per day (bpd). This is approximately 3.6% of the world's oil production. At the beginning of 1994, the production had been 1.9 million bpd.

Enhanced oil-recovery processes include chemical and gas floods, steam, combustion, and electric heating. Gas floods, including immiscible and miscible processes, are usually defined by injected fluids (carbon dioxide, flue gas, nitrogen, or hydrocarbon). Steam projects involve cyclic steam (huff and puff) or steam drive. Combustion technologies can be subdivided into those that autoignite and those that require a heat source at injectors.

Chemical floods are identified by the chemical type that is injected. The most common processes are polymers, surfactants, and alkalis, but chemicals are often combined. For example, polymer slugs usually follow surfactant or alkaline slugs to improve sweep efficiency. Injection of materials that plug permeable channels may be required for injection profile control and to prevent or mitigate premature water or gas breakthrough. Crosslinked or gelled polymers are pumped into injectors or producers for water shutoff or fluid diversion. Cement squeezes often can effectively fix near-wellbore water channeling problems. The design of chemical injection-enhanced oil recovery projects can be more complicated than that of waterflood projects. Down-hole conditions are more severe than are those for primary or secondary recovery production. Well injectivity is complicated by chemicals in injected waters, so in addition to precautions used in waterfloods, chemical interactions, reduced injectivity, deleterious mixtures at producers, potential for accelerated corrosion, and possible well stimulations to counter reduced injectivity must be considered. Monographs on enhanced oil-recovery technologies have been presented by Green and Willhite, Sorbie, and Littmann.

5.3.1 Polymer waterflooding

The polymer in a polymer waterflooding processacts primarily as a thickener. It decreases the permeability of the reservoir and thus improves the vertical and lateral sweep efficiency. Associative copolymers of acrylamide with N-alkylacrylamides, terpolymers of acrylamide, N-decylacrylamide, and sodium-2-acrylamido-2-methylpropane sulfonate (NaAMPS), sodium acrylate (NaA), or sodium-3-acrylamido-3-methylbutanoate (NaAMB) have been shown to possess the required rheologic behavior to be suitable for enhanced oil-recovery processes. Other copolymers of acrylamide with the zwitterionic 3-(2-acrylamido-2-methylpropyl- dimethyl ammonio)-1-propane sulfonate (AMPDAPS) monomer also have been examined.

The low-tension polymer flood technique consists of combining low levels of polymer-compatible surfactants and a polymer with a waterflood. This affects mobility control and reduces front-end and total costs.

The viscosity and non-Newtonian flooding characteristics of the polymer solutions decrease significantly in the presence of inorganic salts, alkali silicates, and multivalent cations. The effect can be traced back to the repression of the dissociation of poly electrolytes, to the formation of a badly dissociating polyelectrolyte metal complex, and to the separation of such a complex from the polymer solution. A modified acrylamide polymer that is hydrophobically associating has remarkably improved the properties of salt resistance and temperature resistance, compared with high-molecular-weight poly acrylamide.

Pseudozan is an exopolysaccharide produced by a *Pseudomonas* species. It has high viscosities at low concentrations in formation brines, forms stable solutions over a wide pH range, and is relatively stable at temperatures up to 65°C. The polymer is not shear degradable and has pseudoplastic behavior. The polymer has been proposed for enhanced oil-recovery processes for mobility control.

5.3.2 Combination flooding

Combination flooding comprises the combination of at least two basic techniques from gas flooding, caustic flooding, surfactant flooding, polymer flooding, and foam flooding. There may be synergisms between the various chemical reagents used. There are specific terms that clarify the individual combination of the basic methods, such as surfactant-enhanced alkaline flooding, alkalineassisted thermal oil recovery, and others.

Coinjection of a low-concentration surfactant and a biopolymer, followed by a polymer buffer for mobility control, leads to reduced chemical consumption and high oil recovery. There may be synergistic effects between the surfactant and the polymer in a dynamic flood situation. The chromatographic separation of surfactant and polymer is important to obtain good oil recovery and low surfactant retention. In buffered surfactant-enhanced alkaline flooding, it was found that the minimum in interfacial tension and the region of spontaneous emulsification correspond to a particular pH range, so by buffering the aqueous pH against changes in alkali concentration, a low interfacial tension can be maintained when the amount of alkali decreases because of acids, rock consumption, and dispersion.

The effectiveness of alkaline additives tends to increase with increasing pH. However, for most reservoirs, the reaction of the alkaline additives with minerals is a serious problem for strong alkalis, and a flood needs to be operated at the lowest effective pH, approximately 10. The ideal process by which alkaline agents reduce losses of surfactants and polymers in oil recovery by chemical injection has been detailed in the literature.

5.3.3 Foam flooding

A process for enhancing the recovery of oil in a subterranean formation comprises injecting a foam having oil-imbibing and transporting properties. A foam having such properties is selected either by determination of the lamella number or by micro-visualization techniques. The method for selecting a surfactant capable of forming a foam functional to both imbibe and transport an oil phase in a subterranean formation comprises: (1) Determining the surface tension of the foaming solution, (2) Measuring the radius of a foam lamella plateau border where it initially contacts the oil or of an emulsified drop, (3) Determining the interfacial tension between the foaming solution and the oil, (4) Correlating these measurements with a mathematical model to obtain a value indicative of the oil imbibing properties of the foam.

The foam, having a viscosity greater than the displacing medium, will preferentially accumulate in the well-swept and/or higher permeability zones of the formation. The displacing medium is thus forced to move into the unswept or underswept areas of the formation. It is from these latter areas that the additional oil is recovered. However, when a foam is used to fill a low oil content area of the reservoir, the oil contained therein is, for all practical purposes, lost. This is because the foam functions to divert the displacement fluid from such areas.

A foam drive method comprises the following steps: (1) Injecting into the reservoir an aqueous polymer solution as preceding slug, (2) Periodically injecting simultaneously or alternately a noncondensable gas and a foaming composition solution containing alkalis, surfactants and polymers to form combined foam or periodically injecting the gas and the foam previously formed from the solution, (3) Injecting a polymer solution as a protecting slug and then continuing with waterflooding.

5.3.4 Steam flooding

When the temperature of a carbonate reservoir that is saturated with high viscosity oil and water increases to 200℃ or more, chemical reactions occur in the formation, resulting in the formation of considerable amounts of CO_2. The generation of CO_2 during thermal stimulation of a carbonate reservoir results from the dealkylation of aromatic hydrocarbons in the presence of water vapor, catalytic conversion of hydrocarbons by water vapor, and oxidation of organic materials. Clay material and metals of variable valence (e.g., nickel, cobalt, iron) in the carbonate rock can serve as the catalyst. An optimal amount of CO_2 exists for which maximal oil recovery is achieved. The performance of a steam-flooding process can be improved by the addition of CO_2 or methane.

The reactivity of steam can be reduced via pH control. The injection or addition of a buffer such as ammonium chloride inhibits the dissolution of certain mineral groups, controls the migration of fines, inhibits the swelling of clays, controls chemical reactions in which new clay minerals are formed, and helps prevent the precipitation of asphaltenes and the formation of emulsions as a result of steam injection The reaction of sulfate with sulfide is strongly pH dependent, and the oxidation potential of sulfate in the neutral pH region is very low. At atmospheric pressure and temperatures up to the boiling points of the inorganic and organic media (e. g., water, alkanes, alkyl-substituted arenes), no reaction takes place within 100 hours. However, the reaction may proceed very slowly over geochemical time periods. During the steamflooding processes, huge amounts of H_2S are produced together with CO_2 and small amounts of elemental hydrogen.

In the producing zones, the temperatures lie in the range of 250℃ to 270℃, which is significantly below the conditions described in the literature. The H_2S production rises from 50 ppm to up to 300,000 ppm, causing enormous corrosion and health safety risks.

5.3.5 Microbial-enhanced oil-recovery techniques

Microbial-enhanced oil recovery (MEOR) was first proposed in 1926 by A. Beckman. Between 1943 and 1953, C.E. Zobell laid the foundations of MEOR techniques. The results were largely dismissed in the United States because there was little interest in finding methods to enhance the recovery of oil at this time. However, in some European countries, the interest for MEOR increased and several field trials were conducted. The first MEOR water flood field project in the United States was initiated in 1986. The site selected was in the Mink Unit of Delaware-Childers Field in Nowata County, Oklahoma.

Microbiologists laid the foundations of MEOR. After the petroleum crisis in 1973, the interest in MEOR generally increased. Monographs about the underlying ideas and the practice of MEOR are available.

The most widely practiced technique for applying MEOR involves cyclic stimulation treatments of producing wells. Improvements in oil production can result from removal of paraffinic or asphaltic deposits from the near-wellbore region or from mobilization of residual oil in the limited volume of the reservoir that is treated. An alternate method involves applying microbes in an ongoing waterflood to improve oil recovery. In the laboratory, microorganisms have been shown to produce chemicals such as surfactants, acids, solvents (alcohols and ketones), and gases (primarily CO_2) that can be effective in mobilizing crude oil under reservoir conditions. Microbial growth and polymer production in porous media have been shown to improve the sweep efficiency by permeability modification. In general, cost-effective MEOR methods are best applied in shallow, sandstone reservoirs in mature producing fields.

The function of aerobic MEOR is based on the ability of oil-degrading bacteria to reduce the interfacial tension between oil and water. This process implies pumping water containing oxygen and mineral nutrients into the oil reservoir to stimulate growth of aerobic oil-degrading bacteria. Based on core flood experiments, the amount of bacterial biomass responsible for dislodging the oil can be calculated. The process is limited by the amount of oxygen available to the bacteria to degrade the oil. The bacterial biomass is more efficient than synthetic surfactants in dislodging the oil.

Experiments have shown that bacterial cells may penetrate a solid porous medium with at least 140-mD permeability and that a bacterial population may be established in such a medium if suitable substrates are supplied. Enhanced oil-recovery organism suitability is governed by parameters such as capacity to produce a surfactant/cosurfactant, cell morphology and relationship of bacterial size to pore size, and pore size distribution of the porous rock. The activity of the organism is directly affected by conditions in the reservoir, such as oxygen availability, temperature, pressure, and substrate availability. A physical model to predict the large-scale application for MEOR has been developed. This model simulates both the radial flow of fluids toward the wellbore and bacteria transport through porous media.

From *oil field chemicals* by *Johannes Karl Fink*

New Words and Expressions

combustion technologies 燃烧技术
N-alkylacrylamides *N*-烷基丙烯酰胺
zwitterionic [ˌtsvɪtəraɪˈɒnɪk] 两性离子的
polyelectrolyte [pɒlɪɪˈlektrəʊlaɪt] 聚合(高分子)电解质,聚电解质
exopolysaccharide [eksəˈpɒlɪsækærɪd] 胞外多糖类
alkalineassisted thermal oil recovery 碱性辅助热力采油
alkaline [ˈælkəlaɪn] 碱性的,碱的;含碱的
subterranean [ˌsʌbtəˈreɪniən] 地表下面的,地下的
underswept [ˈʌndəˈswept] 扫描不足的,扫描线少的
thixotropic [θɪksəˈtrɒpɪk] 触变的,具有触变作用的
dealkylation [diːælkɪˈleɪʃən] 脱烷作用
geochemical [dʒiːɒˈkemɪkl] 地球化学的
microbial-enhanced oil recovery 微生物驱油

Notes

(1) Foam flooding 泡沫驱是指将碱、表面活性剂、聚合物和气有机地组合到一起,使其具有超低界面张力、良好的水相发泡性能、良好的上浮能力。该项技术是由大庆油田有限责任公司王德民等提出的,该项技术取得了中国发明专利权、同时获得了美国、加拿大、俄罗斯和英国获得了发明专利,在挪威和印尼两个国家已进入实质性审查阶段。

（2）**Microbial-enhanced oil recovery techniques** 微生物驱油技术是技术含量较高的一种提高采收率技术，包括微生物在油层中的生长、繁殖和代谢等生物化学过程，还包括微生物菌体、微生物营养液、微生物代谢产物在油层中的运移，以及与岩石、油、气、水的相互作用引起的岩石、油、气、水物性的改变。微生物采油施工简单、成本低，是一种廉价有效的采油技术，有望成为未来油田开发后期稳油控水、提高采收率的主要技术之一。

5.4 Acid Stimulation

5.4.1 Introduction

About 50% of all oil reservoirs worldwide are based on carbonate minerals (limestone/chalk/dolomite) and about 50% are sandstone (quartz, feldspar, etc.), although they can also contain a small percentage carbonate minerals. Acid stimulation is used to increase permeability both in production and injector wells, carbonate or sandstone, by dissolving various acid-soluble solids naturally present in the rock matrix or as formation damage. There are many types of formation damage only some of which can be treated with acids; for example, organic deposits such as wax and asphaltenes cannot be treated. This section on acidizing should also be read together with the section on scale dissolvers on scale control since low-pH acidizing will also remove carbonate and sulfide scale deposits. To chemically remove sulfate scale, high pH chelates such as salts of polyaminocarboxylic acids are used.

Stimulation by acidizing is an old production enhancement technique dating as far back as the nineteenth century. Several books describe the fundamentals of acid stimulation, including a good introduction to modern techniques. Shorter reviews are also available.

Acid stimulation needs to be carried out with a full knowledge of the history of the well to determine the best course of action since there have been many cases of acid stimulation causing temporary or permanent formation damage, includingturning oil-producing wells into 100% water producers. This probably stems from the complex, heterogeneous nature of formation minerals and the unpredictability of their response to conventional oilfield acid formulations. Here, we will give just a brief summary of the techniques used and concentrate on the chemicals involved in the various acidizing treatment strategies. The various books and articles in the list of references at the end of this chapter should be consulted for a more comprehensive understanding of acid stimulation. There are two basic methods of using acids to stimulate production: ① Fracture acidizing; ② Matrix acidizing

5.4.2 Facture Acidizing of Carbonate Formations

Fracturing can be done hydraulically with proppants or with acids. In both cases, the goal is to create long, open channels from the wellbore penetrating deep into the formation. In fracture, acidizing some or all of the acid treatment is pumped in above the fracturing pressure. Fracture acidizing is usually carried out on carbonate reservoirs, which have lower permeability than sandstone reservoirs. Fracture acidizing of carbonate formations (chalk, limestone, and dolomite) can be used to either remove formation damage or stimulate undamaged formations. Once fractures have been formed by the overpressure, the acid is needed to etch out the fractures, leaving high and low points along the channel. This produces a conductive channel within the fracture where oil or gas can migrate. The acids used are the same as in carbonate matrix acidizing discussed below.

A problem with fracture acidizing is that as the acid is injected, it tends to react with the most reactive rock and/or the rock with which it first comes into contact. Thus, much of the acid is used up near the wellbore and is not available for etching of the fracture faces farther from the wellbore. Furthermore, the acidic fluid follows the paths of least resistance, which are, for example, either natural fractures in the rock or areas of more permeable or more acid-soluble rock. This process creates typically long-branched passageways in the fracture faces leading away from the fracture, usually near the wellbore. These highly conductive microchannels are called "wormholes" and are very deleterious because later-injected fracturing fluid tends to leak off into the wormholes rather than lengthening the desired fracture. To block the wormholes, techniques called "leak-off control techniques" have been developed. This blockage should be temporary because the wormholes are preferably open to flow after the fracturing treatment; fluid production through the wormholes adds to total production. Commonly, the same methods may be used for leak-off control in acid fracturing and for "diversion" in matrix acidizing. Thus, an acid-etched fracture can be created using either viscous fingering (pad acid) or viscous acid fracturing. With viscous fingering, a fracture is first formed using a viscous gelled water pad. Acid with lower viscosity is then injected, which fingers through the viscous pad in the fracture, etching out an uneven pattern as it goes. Viscous acid fracturing uses viscous acid systems such as gelled, emulsified, or foamed systems, or chemically retarded acids.

5.4.3 Matrix Acidizing

In matrix acidizing, the acid treatment is pumped into the production well at or below the formation fracturing pressure. Matrix acidizing is useful for stimulating both sandstone and carbonate reservoirs. A useful state-of-the-art review was published in 2003.

In carbonate matrix acidizing, the objective is to allow the acid to dissolve channels called wormholes in the near-wellbore region, reaching as far as possible into the formation. If the formation is undamaged, the production rate can be doubled at best; however, with damaged formations, higher production rates can be obtained. It should be noted that carbonate matrix acid stimulations are also useful for treating carbonate-cemented sandstones and damage from acid-soluble species such as calcium carbonate ($CaCO_3$), lost circulation material, or carbonate or sulfide scales.

5.4.4 Acids used in Acidizing

(1) Acids for Carbonate Formations

The most common acid used in carbonate fracture or matrix acidizing is hydrochloric acid (HCl). Organic acids such as acetic acid (CH_3COOH) and formic acid (HCOOH) are sometimes used particularly for high-temperature applications. Concentrations of HCl used in the field vary: 15 wt.% is common, but a concentration as high as 28 wt.% may be used (commercial HCl is usually sold as a 37 wt.% aqueous solution).Lower concentrations can be used as pickling acids to clean up the well in a preflush (to remove scale and rust) or an afterflush. HCl acid blends formulated with calcium chloride or bromide salts have been used to stimulate high-pressure, high-temperature wells.

Calcium carbonate rock (limestone or chalk) dissolves in the acid to release carbon dioxide and form a calcium chloride solution. The reaction with HCl is given below.

$$CaCO_3 + 2HCl \longrightarrow CaCl_2 + CO_2 + H_2O$$

Dolomite rock, which contains magnesium and calcium, will release both cations on treatment with acids. Strong acids such as HCl form predominantly unbranched wormholes, whereas weaker organic acids and so-called retarded acids form more branched wormholes. In some high-temperature applications, HCl does not produce acceptable stimulation results because of a lack of penetration or surface reactions. Organic acids, such as formic acid and acetic acid, were introduced to offer a slower-reacting and thus deeper-stimulating acid. These "retarded" acids have shortcomings due to solubility limitations of acetate or formate salts and also corrosivity at high temperatures. Maleic or lactic acid have also been proposed as their salts (e.g., calcium maleate) are more water soluble. Corrosion problems with organic acids is less than with HCl. Alkanesulfonic acids (e.g., methanesulfonic) have also been proposed as carbonate-acidizing agents with low corrosivity. Esters and other chemicals that hydrolyze in water to produce acid downhole are discussed in other Sections.

High-pH chelating agents for matrix acidizing, such as salts of EDTA or hydroxyaminocarboxylic acids have also been proposed. By adjusting the flow rate and pH of the fluid, it becomes possible to tailor the slower-reacting chelate solutions to the well conditions and achieve maximum wormhole formation with a minimum amount of solvent. In addition, use of high-pH solvents significantly reduces corrosion problems. Carbonate-acidizing treatments with chelates at temperatures of 177℃ have been reported. Chelates such as EDTA are considerably more expensive than HCl and organic acids. Long-chained carboxylic acids have also been investigated, offering low-corrosion rates, good dissolving power at high temperature, high biodegradability, and easy and safe handling.

(2) Acids for Sandstone Formations

In sandstone matrix acidizing, the primary purpose is to remove acid-soluble damage in the well and near-wellbore area, thus providing a better pathway for the flow of oil or gas. Treating an undamaged sandstone well with matrix acids does not usually lead to stimulation unless the reservoir is naturally fractured. Some carbonate-based damage can be removed with the same acids used in carbonate matrix acidizing. However, in a sandstone reservoir, which is composed mainly of quartz and aluminosilicates (such as feldspars), migration of particles (fines) into the pores of the near-wellbore area can cause reduced production. These fines will not dissolve in strong acids such as HCl but will dissolve in hydrofluoric (HF) acid.

Although highly corrosive, HF is classed as a weak acid owing to its low ionization in water. HF is also very toxic. HF, or more usually HF-releasing chemicals such as ammonium bifluoride (NH_4HF_2), is used for sandstone matrix acidizing combined with HCl or organic acids. HF will also dissolve clays left behind after drilling operations, such as bentonite. An aqueous HF/HCl blend is often called a "mud acid." A preflush and overflush of an ammonium salt is often used to remove incompatible ions such as Na^+, K^+, and Ca^{2+} that could lead to precipitation of insoluble fluorosilicate salts (e.g., Na_2SiF_6). The concentrations of the acids used in sandstone and carbonate matrix-acidizing treatments vary somewhat according to the service companies who carry out such operations. Guidelines to the concentrations have been documented. Generally, an HF concentration with a maximum of 3% has been used owing to the fear of deconsolidation of the near-wellbore of sandstone reservoirs. HCl/HF ratios usually vary from 4 : 1 to 9 : 1.

In sandstone acidizing, one has to be particularly careful of reprecipitation of reaction products, which could cause new formation damage. They occur mostly if the well is shut-in for a long period. The chemistry is complicated, but basically, HF reacts first with aluminosilicates to generate fluorosilicates, which react further with clays to form insoluble sodium or potassium fluorosilicates. Overflushes of dilute HCl or NH_4Cl can be used to push these potentially precipitous solutions away from the critical near-well area and deeper into the formation. Another method to control this precipitation problem is by using delayed acid formulations that generate HF slowly. Examples are clay acid (tetrafluoroboric acid, HBF_4) and self-generating acids, which can be esters that hydrolyze to acids at elevated temperatures. Buffered acid systems that allow for deeper penetration can also be used. For example, a buffered HF acid solution of pH 1.9~4 containing organic acids mixed with salts of organic acids and a phosphonate to alleviate the formation of siliceous precipitates can be used. High-pH-buffered systems have been used successfully in single-stage sandstone-acidizing treatments, eliminating the need for preflushesand overflushes. Silica precipitation inhibitors such as polycarboxylates, phosphonates, or an organosilane can be used. improved performance of HF or HF-releasing stimulation packages can be accomplished by adding alkali metal complexing agent, such as crown ethers.

The aluminum in the clays reacts later on with HF after the silicates have reacted. Aluminum fluoride salts are soluble in the spent acid unless diluted or the pH is raised by postflushes. Chelating agents such as polyaminocarboxylic acids or buffered organic acids can be added to the acid itself to prevent this precipitation from happening. Such blends can be used in single-stage treatments, compared with multistage mud acid treatments, and are especially useful for high-temperature wells. In addition, insoluble calcium fluoride can precipitate in the spent acid if too much calcium carbonate is present in the sandstone reservoir. In such cases, HCl treatment alone will suffice. Insoluble iron(Ⅲ) salts can also cause problems if the pH of the spent acid is raised above approximately. It has been proposed that the HCl-based preflush used in sandstone acidizing may be sufficient to remove much of the formation damage in certain cases, for example, calcium carbonate scales. This avoids potential formation damage caused by products of HF acidizing. A phosphonate scale inhibitor can also be added to avoid reprecipitation of carbonate scales.

5.4.5 Potential Formation Damage from Acidizing

There are several other ways that acidizing, both for sandstone and carbonate reservoirs, can lead to formation damage if not carried out correctly. These include

① Loss of near-wellbore compressive strength due to using too much HF either in volume or concentration

② Formation of emulsions or asphaltic sludge due to incompatibility between the acid and formation fluids

③ Water-blocking and wettability-alteration damage (this can be repaired with mutual solvent treatments (mixed with water or hydrocarbon solvent) containing surfactants)

④ Fines migration after acidizing (this is fairly common in sandstone acidizing; bringing the well on slowly after treatment can minimize this damage)

Experience has shown that for sandstone formations, oil wells respond to matrix acidizing in a different manner as compared with gas wells. For oil wells, the improvement in permeability resulting from the stimulation treatment peaks at a certain acid volume and then drops as the volume of acid injected increases. For gas wells, however, the resulting improvement in permeability is roughly proportional to the volume of acid injected, and is normally better than that obtained with oil wells. It is, therefore, expected that stimulation of oil wells in sandstone formations could be improved by displacing the oil in the zone to be treated with gas. Gas injection before acidizing is sought to minimize the formation of emulsions or sludge resulting from reactions between the spent acid products and the oil that otherwise would be contacted.

5.4.6 Acidizing Additives

The acid main flush contains several additives to bring control to the treatment. These almost always include: Corrosion inhibitor; Iron control agent; Water-wetting surfactant. Many other additives can also be used. The three classes listed above will be discussed first followed by other classes of additives.

From *Production Chemicals for the Oil and Gas Industry* by *Malcolm A. Kelland*

New Words and Expressions

acid stimulation 酸刺激，酸化增产

formation damage 地层损害，油层损害；生产层损坏

scale control 防垢；结垢控制

fracture acidizing 压裂酸化

matrix acidizing 基质酸化

carbonate reservoir ［地质］碳酸盐岩储层

deleterious ［ˈdɛləˈtɪrɪəs］有毒的，有害的

retarded acid 缓速酸

state-of-the-art 最先进的；已经发展的；达到最高水准的

formic acid ［有化］甲酸；［有化］蚁酸

pickling acid 酸洗用酸

afterflush 后处理

dolomite rock 白云岩

lactic acid ［有化］乳酸

maleic acid ［有化］马来酸；［有化］顺丁烯二酸
methanesulfonic acid ［有化］甲磺酸，［有化］甲基磺酸
spent acid ［化工］废酸

Notes

（1）**Fracture acidizing** 压裂酸化，是在足以压开地层形成裂缝或张开地层原有裂缝的压力下对地层挤酸的酸处理工艺。可分为前置液酸压和普通酸压（或一般酸压）。压裂酸化主要用于堵塞范围较深或者低渗透区的油气井。注酸压力高于油（气）层破裂压力的压裂酸化，人们习惯称之为酸压。酸化液压是国内外油田灰岩油藏广泛采用的一项增产增注措施。现已开始成为重要的完井手段。

（2）**Matrix acidizing** 基质酸化，又称常规酸化，即施工时井底压力低于岩层破裂压力，把酸液注入地层。作用是解除近井堵塞、扩大和延伸缝缝洞洞恢复和提高地层渗透率。特点是用量一般，浓度较低（10%左右）。

5.5 Sand Control

5.5.1 Introduction

Sand (or "fines") production is common in many oil and gas wells throughout the world. The flow of abrasive sand through wells and production lines causes unwanted erosion of equipment, and its production may also exacerbate oil-water separation in the process facilities. There are a number of ways to reduce sand production mechanically, including the use of screens, gravel packing, frac-packing, and modification to the perforation technique usually carried out at the well-completion stage. For poorly consolidated reservoirs, which are still producing excessive sand, chemical sand control can be an option. This can be especially rewarding for subsea wells if expensive intervention costs can be avoided.

5.5.2 Chemical Sand Control

1. Consolidation With resins

Chemical sand control has been carried out for many years with resins or epoxy, which harden unconsolidated sand. Typical systems are based on bisphenol A-epichlorohydrin resin, polyepoxde resin, polyester resin, phenol-aldehyde resin, urea-aldehyde resin, furan resin, urethane resin, acrylic resins, and glycidyl ethers. Sand consolidation fluid and field application in shallow gas reservoirs has been reported.

If the resin comprises a bisphenol A-epichlorohydrin polymer, a preferred curing agent is 4,4-methylenedianiline. If the resin comprises a polyurethane, the curing agent is preferably a diisocyanate. The furan resin system is one of the most common: the key chemical is furfuryl alcohol and does not require a curing agent, as it is self-polymerizing in the presence of acid catalysts (Fig. 5.1). These systems are designed to maintain sufficient permeability of the formation to allow production. Self-diverting resin-based sand consolidation fluids have been claimed that allow a greater interval to be treated than conventional resin treatments. Most resin-based chemicals are not considered to be very environmentally friendly.

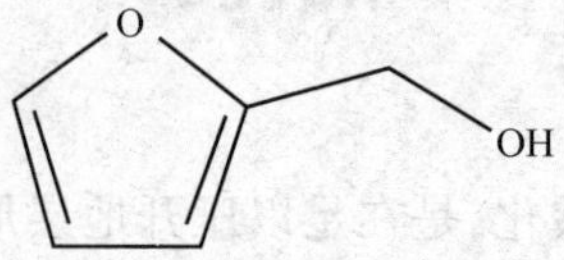

Fig 5.1 Furfuryl alcohol.

Various aqueous and nonaqueous tackifying chemicals, including silyl-modified polyamides that impart a sticky character to sand particles, hindering their movement, have been claimed. A broader range of tackifying agents, including both water-soluble and oil-soluble products, such as resins, gels, silicates, emulsions, polymers, and monomers, have been claimed.

Both foamed and nonfoamed aqueous treatments have been applied in the field successfully. Curable epoxy resins dispersed in brine with a water-soluble activator have been used in the North Sea to treat poorly consolidated formations. A permeability above 500 Md is recommended to avoid loss of productivity. The curable consolidation component is emulsified to form a water-external emulsion so that the active material can be delivered in a brine-based solution. Foam diverters have also been used for better placement of the treatment fluids.

Several polymer formulations have shown good results in sand pack laboratory tests. The materials used are amino aldehyde resins with water-soluble resins with dispersants and inorganic crosslinkers. Commercial sand-fixing products such as styrene/acrylic copolymers were also used.

Cross-linked polymer gels, similar to those used in water shut-off treatments, have also been proposed for sand consolidation. Polymer gel systems, such as those based on polyacrylamides, are claimed to impart a lower probability of failure tothe formation compared with resin treatments.

Inorganic chemical systems for sand consolidation have also been developed. For example, a system based on an insoluble silica source and a source of calcium hydroxide (e.g., aqueous solutions of calcium chloride and sodium hydroxide) has been claimed. The components of the aqueous system react to produce a calcium silicate hydrate gel having cementitious properties within the pores of the formation. An enzyme-based process for consolidation of sand with calcium carbonate has also been proposed. It requires calcium chloride, urea, and urease enzyme. The enzyme catalyzes the decomposition of urea to ammonia and carbon dioxide, raising the pH. In the presence of soluble calcium ions, insoluble calcium carbonate is formed that deposits on the sand and core, binding them together.

A non-enzymatic method for consolidating an underground formation with a consolidating mineral has been reported. The consolidating material is a carbonate and is produced from an alkaline treatment fluid which can contain environmentally-friendly and inexpensive components. The system utilises novel decarboxylation reactions to deposit calcium carbonate from an aqueous solution at a controlled rate over a range of conditions to achieve and improve cementation between sand grains. The majority of these reactions involve amino acid salts, for example glycine.

A modification of this quasinatural consolidation technology using calcium nitrate (nor calcium chloride) has been successfully applied in the North Sea to improve well performance by containing sand production.

2. Consolidation with Organosilanes

Since about 2005, a new sand consolidation method based on organosilane chemistry has been developed and used in the field. In comparison with other treatments, this method only increases the residual strength of the formation by a small amount. The treatment is oil soluble and will therefore not alter the relative permeability in the oil-bearing zones, thereby reducing the risk of increased skin due to changes in saturation. This system is especially beneficial for fields with low reservoir pressure. The method is employed by simple bullheading and can have self-diverting properties. In laboratory studies, the organosilane treatment was shown to give better overall performance with regard to sand consolidation and moderate permeability reduction compared with other treatments such as water-soluble gelling polymers and the $CaCl_2$/urea/enzyme system discussed earlier.

Preferred oil-soluble organosilanes that can be used are 3-aminopropyltriethoxysilane and bis (triethoxy silylpropyl) amine or mixtures thereof (Fig 5.2). They are usually mixed in diesel and bullheaded into the well. The authors suggest that the presence of the amine function appears to result in better adsorption of the organosilane to the sand grains. It is also believed that the presence of an amine group may contribute to the formation of a gel-like structure having viscoelastic properties. The authors suggest that the organosilane compounds react with water and hydrolyze.

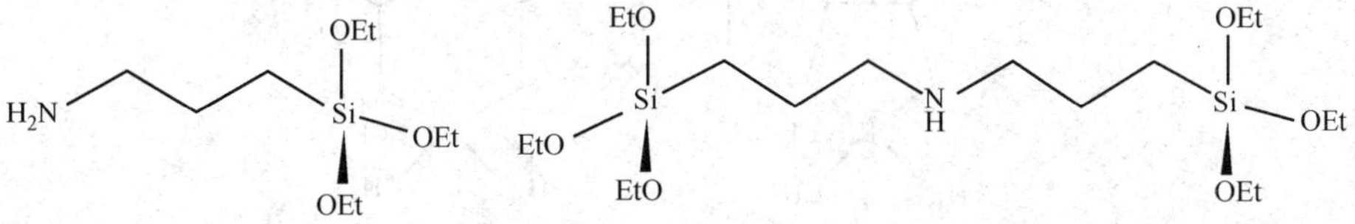

Fig. 5.2 3-Aminopropyltriethoxysilane and bis(triethoxy silylpropyl) amine.

The resulting chemicals then react with siliceous surfaces in the formation (e.g., the surface of silica sand), coat any sand particles, and bind them in place by the formation of silicate bridges restricting their movement. The advantage of bifunctional organosilanes, such as bis(triethoxy silylpropyl) amine, is their ability to bind two particles together. The organosilanes are claimed to be environmentally acceptable with low bioaccumulation potential and high biodegradation. Several types of wells have been treated with the organosilane system at a chemical concentration in the range of 5-7 vol.% of the active components. The first results in terms of sand production reduction were mixed, with the subsea well responding best to the treatment. A moderate reduction in permeability was observed in some wells, which reduced the production index (PI) of the well by 10%~15%. However, the production of the wells was limited by sand production levels, so such a reduction in PI was acceptable. Correct placement, especially in horizontal wells, was shown to be critical when it comes to performance with regard to increasing the maximum sand-free rate (MSFR). Organosilanes used at higher concentrations than those for sand consolidation have also been claimed as chemicals for water shut-off treatments. Consolidation of particulates using a hardenable resin and an organosilane coupling agent have been claimed.

Silylated polymers for sand consolidation have been claimed. The center of the polymer is carbon based, such as a poly(meth)acrylate, poly(oxyalkylene), or polyurethane; A is an alkoxy linkage and the R groups are preferably small alkyl groups (Fig. 5.3).

$$(RO)_{3-n}R_nSi—A—B—A—Si(OR)_{3-n}R_n$$

Fig. 5.3 Silylated polymers for sand consolidation.

3. Other Chemical Consolidation Methods

Another claimed sand consolidation method, which imparts small incremental forces or a relatively weak residual strength to the formation, is by using a positively charged water-soluble polymer. Examples are polyamino acids, such as polyaspartate and copolymers comprising aspartic acid and proline and/or histidine, and poly(diallyl ammonium salts) such as polydimethyldiallylammonium chloride and mixtures thereof (Fig. 5.4). It is thought that by virtue of its length and multiple positive charges, the polymer may interact electrostatically with a number of different particles of the formation, thereby holding or binding them together. In so doing, the polymer chain is likely to span the interstitial space between sand particles of the formation. The result is simply the formation of a "mesh-like" or "net-like" structure that does not impair fluid flow. Hence, the permeability of a subterranean formation treated according to the method described by the present invention is largely unchanged after treatment.

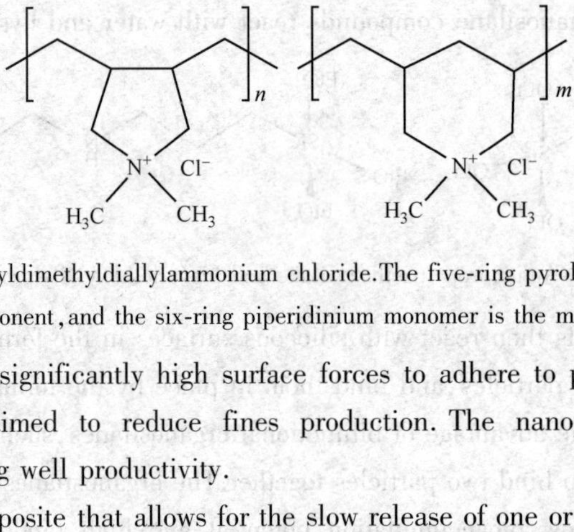

Fig. 5.4 Polydiallyldimethyldiallylammonium chloride. The five-ring pyrolidinium monomer is the major component, and the six-ring piperidinium monomer is the minor component.

Nanoparticles having significantly high surface forces to adhere to proppant used in frac-pack operations have been claimed to reduce fines production. The nanoparticles adsorb migrating formation fines, maintaining well productivity.

A well treatment composite that allows for the slow release of one or more well treatment agents has a nano-sized calcined porous substrate (adsorbent) of high surface area onto which the well treatment agent is applied. The composites are suitable for use in sand control operations as well as in hydraulic fracturing.

From *Production chemicals for the oil and Gas industry* by *Malcolm A. Kelland*

New Words and Expressions

abrasive sand [机] 研磨砂

gravel packing 砾石充填

polyester resin［树脂］聚酯树脂
urea-aldehyde resin 脲醛树脂
furan resin［树脂］呋喃树脂
urethane resin 聚氨酯树脂
glycidyl ether 缩水甘油醚
curing agent［助剂］固化剂;［助剂］硬化剂
diisocyanate［dai,aisəˈsaiəneit］［有化］二异氰酸盐(酯)
furfuryl alcohol［有化］糠醇;呋喃甲醇
water shut-off 堵水
glycine n.甘氨酸;［有化］氨基乙酸
oil-bearing zone 含油层
production index 生产指标
polyaspartate 聚天冬氨酸
acrylic resin［树脂］丙烯酸树脂

Notes

(1) **Chemical Sand Control** 化学防砂,化学防砂是利用化学药剂的化学反应把地层中的砂砾或充填到地层的砂石胶结起来,稳定地层结构或形成具有一定强度和渗透率的人工井壁,从而达到防止地层出砂的目的。防砂按工艺可分为两种方法:一是先压裂后,再向井眼内充填用固砂剂涂敷的石英砂、陶粒砂等颗粒物质。二是注入固砂剂,使地层中疏松的颗粒物质胶结;两者都是靠化学固砂剂在地层下固化形成一道渗透性良好的筛网以允许烃类、水等流通,而防止疏松砂粒侵入井筒。

(2) **Organosilane** 有机硅烷,它可以跟许多无机和有机材料反应,是个非常多功能的产品。有机硅烷作为偶联剂、交联剂和表面改性剂的独特性令其广泛地应用于许许多多的领域,从胶粘剂,到涂料、到铸造粘合剂等等。

5.6 Crude Oil Emulsions

Crude oils are typically water in crude oil (W/O) emulsions, which are often very stable. Among the indigenous natural surfactants contained in the crude oils, asphaltenes and resins are known to play an important role in the formation and stability of W/O emulsions. Asphaltenes are the most polar and heaviest compounds in the crude oil. They are composed of several poly nuclear aromatic sheets surrounded by hydrocarbon tails, and form particles whose molar masses are included between 500 and 20,000 g/mol. Resins are molecules defined as being soluble in light alkanes (pentane, hexane, or heptane), but insoluble in liquid propane. Resins are effective as dispersants of asphaltenes in crude oil. It was postulated that asphaltenes stabilize W/O emulsions in two steps. First, disk-like asphaltene molecules aggregate into particles or micelles, which are interfacially active. Then, these entities upon adsorbing at the W/O interface aggregate through physical interactions and form an interfacial network.

5.6.1 Types of emulsions

Emulsions have long been of great practical interest due to their widespread occurrence in everyday life. They may be found in important areas such as food, cosmetics, pulp and paper, pharmaceutical and agricultural industry. Petroleum emulsions may not be as familiar but have a similar long-standing, widespread, and important occurrence in industry, where they are typically undesirable and can result in high pumping costs, pipeline corrosions, reduced throughput and special handling equipment. Emulsions may be encountered at all stages in the petroleum recovery and processing industry (drilling fluid, production, process plant, and transportation emulsions. Emulsions are defined as the colloidal systems in which fine droplets of one liquid are dispersed in another liquid where the two liquids otherwise being mutually immiscible. Oil and water produce emulsion by stirring; however, the emulsion starts to break down immediately after stirring is stopped.

Depending upon the nature of the dispersed phase, the emulsions are classified as, O/W emulsion or oil droplets in water and W/O emulsion or water droplets in oil. Recently, developments of W/O/W type emulsion or water dispersed within oil droplets of O/W type emulsion and O/W/O type. (1) Oil-in-water emulsions (O/W): The emulsion in which oil is present as the dispersed phase and water as the dispersion medium (continuous phase) is called an oil-in-water emulsion. (2) Water-in-oil emulsion (W/O): The emulsion in which water forms the dispersed phase, and the oil acts as the dispersion medium is called a water-in-oil emulsion.

5.6.2 Properties of emulsions

The emulsions satisfy the following criteria: (1) Emulsions show all the characteristic properties of colloidal solution such as Brownian movement, Tyndall effect, electrophoresis etc. (2) These are coagulated by the addition of electrolytes containing polyvalent metal ions indicating the negative charge on the globules. (3) The size of the dispersed particles in emulsions in larger than those in the sols. It ranges from 1000 Å to 10 000Å. However, the size is smaller than the particles in suspensions. (4) Emulsions can be converted into two separate liquids by heating, centrifuging, freezing etc. This process is also known as demulsification.

Many advances have been made in the field of emulsions in recent years. Emulsion stability depends on presence of adsorbed structures on the interface between the two liquid phases. Emulsion behavior is largely controlled by the properties of the adsorbed layers that stabilize the oil-water surfaces. The knowledge of surface tension alone is not sufficient to understand emulsion properties, and surface rheology plays an important role in a variety of dynamic processes. When a surface active substance is added to water or oil, it pontaneously adsorbs at the surface, and decreases the surface tension γ. In the case of small surfactant molecules, a monolayer is formed, with the polar parts of the surface-active molecules in contact with water, and the hydrophobic parts in contact with oil. The complexity of petroleum emulsions comes from the oil composition in terms of surface active molecules contained in the crude, such as low molecular weight fatty acids, naphthenic acids and asphaltenes. These molecules can interact and reorganize at oil/water interfaces. These effects are very important in the case of heavy oils because this type of crude contains a large amount of asphaltenes and surface-active compounds. In petroleum industry, water-in-oil (W/O) or oil-in-water (O/W) emulsions can lead to enormous financial losses if not treated correctly. Knowing the particular system and the possible stability mechanisms is thus a necessity for proper processing and flow assurance.

5.6.3 Emulsion formation

There are three main criteria that are necessary for formation of crude oil emulsion: (1) Two immiscible liquids must be brought in contact; (2) Surface active component must present as the emulsifying agent; (3) Sufficient mixing or agitating effect must be provided in order to disperse one liquid into another as droplets.

During emulsion formation, the deformation of droplet is opposed by the pressure gradient between the external (convex) and the internal (concave) side of an interface. The pressure gradient or velocity gradient required for emulsion formation is mostly supplied by agitation. The large excess of energy required to produce emulsion of small droplets can only be supplied by very intense agitation, which needs much energy.

A suitable surface active component or surfactant can be added to the system in order to reduce the agitation energy needed to produce a certain droplet size. The formation of surfactant film around the droplet facilitates the process of emulsification and a reduction in agitation energy by factor of 10 or more can be achieved. A method requiring much less mechanical energy uses phase inversion. For instance, if ultimately a W/O emulsion is desired, then an O/W emulsion is first prepared by the addition of mechanical energy. Then the oil content is progressively increased. At some volume fraction above 60%~70%, the emulsion will suddenly invert and produce a W/O emulsion of smaller water droplet sizes than were the oil droplets in the original O/W emulsions.

5.6.4 Emulsion breakdown

In general there are three coupled sub-processes that will influence the rate of reakdown processes in emulsions. These are aggregation (Flocculation), coalescence and phase separation. They will be discussed in some details.

Flocculation:

It is the process in which emulsion drops aggregate, without rupture of the stabilizing layer at the interface. Flocculation of emulsions may occur under conditions when the van der Waals attractive energy exceeds the repulsive energyand can be weak or strong, depending on the strength of inter-drop forces. The driving forces for flocculation can be: (1) Body forces, such as gravity and centrifugation causing creaming or sedimentation, depending on whether the mass density of the drops is smaller or greater than that of the continuous phase. (2) Brownian forces or. (3) Thermo capillary migration (temperature gradients) may dominate the gravitational body force for very small droplets, less than 1μm.

Coalescence

It is an irreversible process in which two or more emulsion drops fuse together to form a single larger drop where the interface is ruptured. As already mentioned, for large drops approaching each other (no background electric field), the interfaces interact and begin to deform. A plane parallel thin film is formed, which rate of thinning may be the main factor determining the overall stability of the emulsion. The film thinning mechanism is strongly dependent on bulk properties (etc. viscosity) in addition to surface forces. The interaction of the two drops across the film leads to the appearance of an additional disjoining pressure inside the film,

Phase separation

The processes of flocculation and coalescence are followed by phase separation, i. e. emulsion breakdown.

5.6.5 Stabilization of the emulsion

There are many factors that usually favor emulsion stability such as low interfacial tension, high viscosity of the bulk phase and relatively small volumes of dispersed phase. A narrow droplet distribution of droplets with small sizes is also advantageous, since polydisperse dispersions will result in a growth of large droplets on the expense of smaller ones. The potent stabilization of the emulsion is achieved by stabilization of the interface .

1. Steric stabilization of the interface

The presence of solids at interfaces may give rise to repulsive surface forces which thermodynamically stabilize the emulsion. As concluded (28 ~ 30), many of the properties of solids in stabilizing emulsion interfaces can beattributed to the very large free energy of adsorption for particles of intermediate wettability (partially wetted by both oil and water phases). This irreversible adsorption leads to extreme stability for certain emulsions and is in contrast to the behavior of surfactant molecules which are usually in rapid dynamic equilibrium between the oil: water interface and the bulk phases. According to the asphaltene stabilization mechanism, coalescence requires the solid particles to be removed from the drop-drop contact region. Free energy considerations suggest that lateral displacement of the particles is most likely, since forcing droplets into either phase from the interface require extreme energies. The asphaltenes stabilization effect for water droplets has already been pictured where droplet contact is prevented by a physical barrier around the particles.

2. Electric stabilization of the interface

Electrical double layer repulsion or charge stabilization by polymers and surfactants with protrudingmolecular chains may prevent the droplets to come into contact with each other. Also, polymers, surfactants or adsorbed particles can create a mechanically strong and elastic interfacial film that act as a barrier against aggregation and coalescence. A film of closed packed particles has considerable mechanical strength, and the most stable emulsions occur when the contact angle is close to 90°, so that the particles will collect at the interface. Particles, which are oil-wet, tend to stabilize W/O emulsions while those that are water-wet tend to stabilize O/W emulsions. In order to stabilize the emulsions the particles should be least one order of magnitude smaller in size than the emulsion droplets and in sufficiently high concentration. Nevertheless, stable W/O emulsions have been generally found to exhibit high interfacial viscosity and/or elasticity modulus. It has been attributed to physical cross-links between the naturally occurring surfactants in crude oil (i.e. asphaltenes particles) adsorbed at the water-oil interface.

3. Composition of crude oil

The complexity comes mostly from the oil composition, in particular from the surface-active molecules contained in the crude. These molecules cover a large range of chemical structures, molecular weights, and HLB (Hydrophilic-Lipophilic Balance) values; they can interact between themselves and/or reorganize at the water/oil interface. Oil-water emulsions are fine dispersions of oil in water (O/W) or of water in oil (W/O), with drop sizes usually in the micron range. In general, emulsions are stabilized by surfactants. In some cases multiple emulsions such as water in oil in water (W/O/W) or oil in water in oil (O/W/O) can be found. Emulsions can be stabilized by other species, provided that they adsorb at the oil-water interface and prevent drop growth and phase separation into the original oil and water phases. After adsorption, the surfaces become visco-elastic and the surface layers provide stability to the emulsion. Crude oils contain asphaltenes (high molecular weight polar components) that act as natural emulsifiers. Other crude oil components are also surface active: resins, fatty acids such as naphthenic acids, porphyrins, wax crystals, etc, but most of the time they cannot alone produce stable emulsions. However, they can associate to asphaltenes and affect emulsion stability. Resins solubilize asphaltenes in oil, and remove them from the interface, therefore lowering emulsion stability. Waxes co-adsorb at the interface and enhance the stability. Naphthenic and other naturally occurring fatty acids also do not seem able to stabilize emulsions alone. However, they are probably partly responsible for the important dependence of emulsion stability upon water pH.

4. Asphaltenes

Asphaltenes stabilize the crude oil emulsion by different modes of action. When asphaltenes disperse on the interface, the film formed at a water/crude oil interface behaves as a skin whose rigidity can be shown by the formation of crinkles at interface when contracting the droplet to a smaller drop size. They can also aggregate with resin molecules on the interfaces and prevent droplet coalescence by steric interaction. Some authors suggest that asphaltenes stabilize the emulsion by formation of hydrogen bonding between asphaltenes and water molecules .

Singh et al. postulated that asphaltenes stabilize W/O emulsions in two steps. First, disk-like asphaltenes molecules aggregate into particles or micelles, which are interfacially active. Then, these entities upon adsorbing at the W/O interface aggregate through physical interactions and form an interfacial network.

5. Resins

Resins are thought to be molecular precursors of the asphaltenes. The polar heads of the resins surround the asphaltenes, while the aliphatic tails extend into the oil. Resins may act to stabilize the dispersion of asphaltene particles and can be converted to asphaltenes by oxidation. Unlike asphaltenes, however, resins are assumed soluble in the petroleum fluid. Pure resins are heavy liquids or sticky (amorphous) solids and are as volatile as the hydrocarbons of the same size. Petroleum fluids with high-resin content are relatively stable. Resins, although quite surface-active, have not been found to stabilize significantly water-in-oil emulsions by themselves in model systems. However, the presence of resins in solution can destabilize emulsions via asphaltenes solvation and/or replacement at the oil/water interface.

6.Saturates

Saturates are nonpolar and consist of normal alkanes (n-paraffins), branched alkanes (isoparaffins) and cyclo-alkanes (also known as naphthenes). Saturates are the largest single source of hydrocarbon or petroleum waxes, which are generally classified as paraffin wax, microcrystalline wax, and/or petrolatum. Of these, the paraffin wax is the major constituent of most solid deposits from crude oils.

7.Aromatics

Aromatics are hydrocarbons, which are chemically and physically very different from the paraffins and naphthenes. They contain one or more ring structures similar to benzene. The atoms are connected by aromatic double bonds.

From *Crude Oil Emulsions- Composition Stability and Characterization* by *Manar El-Sayed Abdel-Raouf*

New Words and Expressions

asphaltene [æs'fæltiːn] [油气] 沥青质
resin ['rezɪn] 胶质
cosmetic [kɒz'metɪk] 美容的;化妆用的 化妆品;装饰品
colloidal system 胶体体系
dispersed phase [物化] 分散相
dispersion medium 分散介质
Tyndall effect 丁达尔效应
spontaneous [spɒn'teɪnɪəs] 自发的;自然的;无意识的
flocculation 絮凝;絮结产物
coalescence [kəuə'lesns] 合并;联合;接合
emulsion breakdown 破乳
sedimentation [矿业][物化] 沉降,[化学] 沉淀
interfacial film 界面膜
Hydrophilic-Lipophilic Balance values 亲水亲油平衡值

Notes

(1) **Emulsion** 乳状液,一种液体以液珠形式分散在与它不相混溶的另一种液体中而形成的分散体系。乳状液一般不透明,呈乳白色。液滴直径大多在 100nm~10μm 之间,可用一般光学显微镜观察。乳状液可分水包油和油包水两种类型。橡胶乳汁、原油乳状液等均属此种分散体系。

(2) **Asphaltene** 沥青质,是一种由多种复杂高分子碳氢化合物及其非金属衍生物组成的复杂混合物。一般把石油中不溶于分子($C_5 \sim C_7$)正构烷烃,但能溶于热苯的物质称为沥青质。研究中常用的是正戊烷沥青质和正庚烷沥青质。沥青质外形为固体无定形物,黑色,相对密度略大于1,加热时不熔化,温度升高到300~350℃以上时,分解为气态、液体产物及缩合生焦。沥青质集中于减压渣油中。

5.7 Demulsifiers

In the production of crude oil, the greatest part of the crude oil occurs as a water-in-oil emulsion. The composition of the continuous phase depends on the water/oil ratio, the natural emulsifier systems contained in the oil, and the origin of the emulsion. The natural emulsifiers contained in crude oils have a complex chemical structure, so that, to overcome their effect, petroleumemulsion demulsifiers must be selectively developed. As new oil fields are developed, and as the production conditions change at older fields, there is a constant need for demulsifiers that lead to a rapid separation into water and oil, as well as minimal-residual water and salt mixtures

The emulsion must be separated by the addition of chemical demulsifiers before the crude oil can be accepted for. The quality criteria for a delivered crude oil are the residual salt content and the water content. For the oil to have a pipeline quality, it is necessary to reduce the water content to less than approximately 1.0%.

The separated saltwater still contains certain amounts of residual oil, where now preferentially oil-in-water emulsions are formed. The separation of the residual oil is necessary in oil field water purification and treatment for ecologic and technical reasons, because the water is used for secondary production by waterflooding, and residual oil volumes in the water would increase the injection pressure.

The presence of water-in-oil emulsions often leads to corrosion and to the growth of microorganisms in the water-wetted parts of the pipelines and storage tanks. At the refinery, before distillation, the salt content is often further reduced by a second emulsification with freshwater, followed by demulsification. Crude oils with high salt contents could lead to breakdowns and corrosion at the refinery. The object of using an emulsion breaker, or demulsifier, is to break the emulsion at the lowest possible concentration and, with little or no additional consumption of heat, to bring about a complete separation of the water and reduce the salt content to a minimum.

There are oil-soluble demulsifiers and water-soluble demulsiflers, the latter being widely used. Emulsions are variable in stability. This variability is largely dependent on oil type and degree of weathering. Emulsions that have a low stability will break easily with chemical emulsion breakers. Broken emulsions will form a foamlike material, called rag, which retains water that is not part of the stable emulsions. The most effective demulsifier must always be determined for the particular emulsion. Demulsifiers are often added to the emulsion at the wellhead to take advantage of the temperature of the freshly raised emulsion to hasten the demulsification step.

5.7.1 Mechanisms of demulsification

The stabilization of water-oil emulsions happens as a result of the interfacial layers, which mainly consist of colloids present in the crude oil—asphaltenes and resins. By adding demulsifiers, the emulsion breaks up. With water-soluble demulsifiers, the emulsion stabilizers originally in the system will be displaced from the interface. In addition, a change in wetting by the formation of inactive complexes may occur. Conversely, using oil-soluble demulsifiers, the mechanism, in addition to the displacement of crude colloids, is based on neutralizing the stabilization effect by additional emulsion breakers and the breakup resulting from interface eruptions.

The effectiveness of a crude oil demulsifier is correlated with the lowering of the shear viscosity and the dynamic tension gradient of the oil-water interface. The interfacial tension relaxation occurs faster with an effective demulsifier. Short relaxation times imply that interfacial tension gradients at slow film thinning are suppressed. Electron spin resonance experiments with labeled demulsifiers indicate that the demulsifiers form reverse micelle like clusters in the bulk oil. The slow unclustering of the demulsifier at the interface appears to be the rate-determining step in the tension relaxation process.

5.7.2 Performance testing

The trial-and-error method of choosing an optimal demulsifier from a wide variety of demulsifiers to effectively treat a given oil field water-in-oil emulsion is time-consuming. However, there are methods to correlate and predict the performance of demulsifiers.

The performance of demulsifiers can be predicted by the relationship between the film pressure of the demulsifier and the normalized area and the solvent properties of the demulsifier. The surfactant activity of the demulsifier is dependent on the bulk phase behavior of the chemical when dispersed in the crude oil emulsions. This behavior can be monitored by determining the demulsifier pressure-area isotherms for adsorption at the crude oil-water interface.

In addition, the dielectric constant can be used as a criterion for screening, ranking, and selecting demulsifiers for emulsion breaking. In a study, the dielectric constants of emulsions and demulsifiers were measured using a portable capacitance meter, and bottle tests were conducted according to the API specification. The results showed that the dielectric constants can be used effectively to screen and rank demulsifiers, whereas a confirmatory bottle test should be conducted on the best demulsifiers to assist in the rapid selection of the most effective demulsifier.

A study by Environment Canada and the U. S. Minerals Management Service attempted to develop a standard test for emulsion breaking agents. Nine types of shaker test methods were tried. Although the results are comparable with different tests, a stable water-in-oil emulsion must be used to yield reproducible results. Tests with unstable emulsions showed nonreproducible and inconsistent results.

Water content and viscosity measurements in certain systems show a correlation to emulsion stability. The viscosity provides a more reliable measure of emulsion stability, but measurements of the water content are more convenient. Mixing time, agent amount, settling time, and mixing energy impact the effectiveness of an emulsifier.

5.7.3 Classification of demulsifiers

The chemicals used as demulsifiers can be classified according to their chemical structure and their applications. With the latter respect, a main division for water-in-oil and oil-in-water applications exists. Furthermore, the demulsifiers can be classified according to the oil type used. From the view of chemical classification, two major groups exist: nonionic demulsifiers and ionic demulsifiers.

1. Polyalkyleneoxides

Polyalkyleneoxides are substances of the following general structure:

$$HO—(CH_2—CHR—O)_x—H$$

The most important additives are polyethyleneoxide, polypropyleneoxide, andpolybutyleneoxide. They are also referred to aspolyalkylene glycols, but this name is correct strictly for derivatives of 1, 2-diols.

Polypropyleneoxide has a molar mass of 250 to 4000 Dalton. The lowermolecular homologues are miscible with water, whereas the higher molecularpolypropyleneoxides are sparingly soluble. They are formed by the polyaddition of, for example, propylene oxide to water or propanediol. The simplest examples are di- and triutetrapropylene glykol.

There are also block copolymers from ethylene oxide propylene oxide.

$$HO—(R_1—O)_x—(R_1—O)_y—H$$

Polytetramethyleneglycol (polytetrahydrofuran) is formed by ring opening polyetherification of tetrahydrofuran. Branched polyalkyleneoxides are formed using polyfunctional alcohols such as trimethylolpropane and pentaerythrite. The products are liquids or waxes depending on the molar mass. Polyalkyleneoxides are often precursors for demulsifiers.

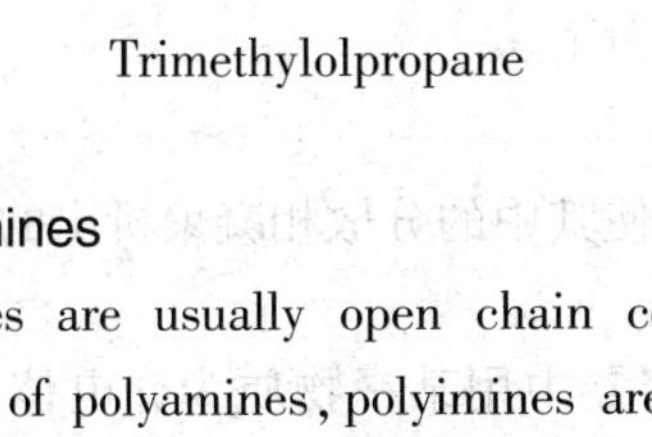

$$CH_3—CH_2—C(CH_2OH)_2—CH_2—OH$$

Trimethylolpropane

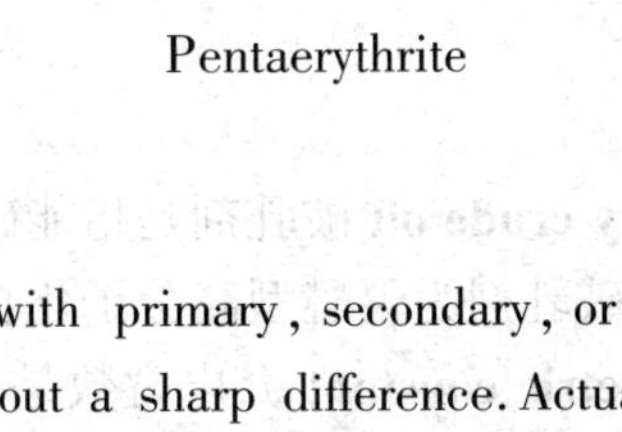

$$OH—CH_2—C(CH_2OH)_2—CH_2—OH$$

Pentaerythrite

2. Polyamines

Polyamines are usually open chain compounds with primary, secondary, or tertiary amino groups. Instead of polyamines, polyimines are used without a sharp difference. Actually, imines are compounds with the =N— group or cycles such as ethyleneimine. Examples of oligoamines and polyamines are ethylenediamine, propanediamine, and 1,4-butanediamine and the respective products of condensation such as diethyleneamine, dipropylenetriamine, and triethylenetetramine. The compounds are colorless to yellowish liquids or solids with alkaline reaction.

Polyamines can also be synthesized by cationic ring-opening polymerization of ethyleneimines, trimethyleneimines, and 2-oxazolines. Polyalkyleneimines are polyamines whose structure is classified into linear and branched types as shown below:

Linear $H_2N—(CH_2—CH_2—NH)_x—H$

Branched $H_2N—[CH_2—CH_2—N(CH_2—CH_2—NH_2]_x—H$

Linear polyalkyleneimines have amino groups only in the main chain, and branched polyalkyleneimines have amino groups in both the main and side chains. In general, nitrogen atoms are at every third or fourth atom. Linear polyethyleneimine is insoluble in benzene, diethyl ether, acetone, and water at room temperature and is soluble in hot water.

From *Oil Field Chemicals* by *Johannes Karl Fink*

New Words and Expressions

natural emulsifier systems 天然乳化体系

petroleumemulsion demulsifiers 石油乳状液破乳剂

purification [ˌpjʊərɪfɪˈkeɪʃn] 提纯

demulsification [dɪˈmʌlsɪfɪˈkeɪʃən] 破乳作用

variability[ˌveəriəˈbɪləti] 变化性,易变

foamlike material 泡沫状物质

weathering[ˈweðərɪŋ] 侵蚀,风化;雨蚀

emulsion stabilizers 乳状液稳定性

interfacial tension gradients 界面张力梯度

film pressure 膜压

nonreproducible 非再生的,不可再生的

Polyalkyleneoxides 聚氧乙烯基化合物

polyalkylene glycols 聚亚烷基二醇

pentaerythrite[ˈpenteərɪθraɪt] 季戊四醇

Notes

(1) **Heavy crude oil** 破乳剂是指能破坏乳浊液使其中的分散相凝聚析出的物质。目前油田常用的原油破乳剂为聚醚型高分子化合物。

(2) **Dielectric constant** 介电常数是指同一电容器中用某一物质为介电体与该物质在真空中的电容的比值。又称相对电容率,以 ε_r 表示。

5.8 Hydraulic Fracturing Fluids

Hydraulic fracturing is a technique to stimulate the productivity of a well. A hydraulic fracture is a superimposed structure that remains undisturbed outside the fracture. Thus the effective permeability of a reservoir remains unchanged by this process. The increased productivity results from increased wellbore radius, because in the course of hydraulic fracturing, a large contact surface between the well and the reservoir is created.

5.8.1 Characterization of fracturing fluids

Historically, viscosity measurements have been the single most important method to characterize fluids in petroleum-producing applications. Whereas the ability to measure a fluid's resistance to flow has been available in the laboratory for a long time, a need to measure the fluid properties at the well site has prompted the development of more portable and less sophisticated viscosity-measuring devices. These instruments must be durable and simple enough to be used by persons with a wide range of technical skills. As a result, the Marsh funnel and the Fann concentric cylinder, both variablespeed viscometers, have found wide use. In some instances, the Brookfield viscometer has also been used.

1. Rheologic characterization

To design a successful hydraulic fracturing treatment employing crosslinked gels, accurate measurements of rheologic properties of these fluids are required. Rheologic characterization of borate-crosslinked gels turned out to be difficult with a rotational viscometer. In a laboratory apparatus, field pumping conditions (i.e., crosslinking the fluid on the fly) and fluid flow down tubing or casing and in the fracture could be simulated. The effects of the pH and temperature of the fluid and the type and concentration of the gelling agent on the rheologic properties of fluids have been measured.

These parameters have significant effect on the final viscosity of gel in the fracture. Correlations to estimate friction pressures in field size tubulars have been developed from laboratory test data. In conjunction with field calibrations, these correlations can aid in accurate prediction of friction pressure of boratecrosslinked fluids

2. Size exclusion chromatography

Size exclusion chromatography has been used to monitor the degradation of the thickeners initiated by various oxidative and enzymatic breakers.

Maximal well production can be achieved only when the solution viscosity and the molecular weight of the gelling agent are significantly reduced after the treatment, that is, the fluid is degraded. However, the reduction of the fracturing fluid viscosity, the traditional method of evaluating these materials, does not necessarily indicate that the gelling agent has been thoroughly degraded also. The reaction between hydroxypropylguar and the oxidizing agent (ammonium peroxydisulfate) in an aqueous potassium chloride solution was studied under controlled conditions to determine changes in solution viscosity and the weight average of the molecular mass of hydroxypropylguar.

5.8.2 Water-based systems

Naturally occurring polysaccharides and their derivatives form the predominant group of water-soluble species generally used as thickeners to impart viscosity to treating fluids. Other synthetic polymers and biopolymers have found ancillary applications. Polymers increase the viscosity of the fracturing fluid in comparatively small amounts. The increase in fluid viscosity of hydraulic fracturing fluids serves for improved proppant placement and fluid loss control. Polymers suitable for fracturing fluids are hydroxypropylguar, galactomannans, hydroxyethylcellulose-modified vinyl phosphonic acid, carboxymethylcellulose, and reticulated bacterial cellulose.

Fracturing fluids have traditionally been viscosified with guar and guar derivatives. Actually, guar is a branched polysaccharide from the guar plant Cyamopsis tetragonolobus, originally from India, now also found in the southern United States, with a molar mass of approximately 220,000 Dalton. It consists of mannose in the main chain and galactose in the side chain. The ratio of mannose to galactose is 2∶1. Polysaccharides with a mannose backbone and side chains unlike mannose are referred to as heteromannans according to the nomenclature of polysaccharides, in particular as galactomannans.

Anionic galactomannans, which are derived from guar gum, in which the hydroxyl groups are partially esterified with sulfonate groups that result from 2-acrylamido-2-methylpropane sulfonic acid and l-allyloxy-2-hydroxypropyl sulfonic acid, have been claimed to be suitable as thickeners. The composition is capable of producing enhanced viscosities, when used either alone or in combination with a cationic polymer and distributed in a solvent. Boron-crosslinked galactomannan fracturing fluids have an increased temperature stability. The temperature stability of fracturing fluids containing galactomannan polymers is increased by adding a sparingly soluble borate with a slow solubility rate to the fracturing fluid. This provides a source of boron for solubilizing at elevated temperatures, thus enhancing the crosslinking of the galactomannan polymer. The polymer also improves the leak-off properties of the fracturing fluid.

5.8.3 Oil-based systems

One advantage of fracturing with hydrocarbon gels compared with waterbased gels is that some formations may tend to imbibe large quantities of water, whereas others are water-sensitive and will swell if water is introduced.

1. Organic gel aluminum phosphate ester

A gel of diesel or crude oil can be produced using a phosphate diester or an aluminum compound with phosphate diester. The metal phosphate diester may be prepared by reacting a triester with phosphorous pentoxide to produce a polyphosphate, which is then reacted with an alcohol (usually hexanol) to produce a phosphate diester. The latter diester is then added to the organic liquid along with a nonaqueous source of aluminum, such as aluminum isopropoxide (aluminum-triisopropylate) in diesel oil, to produce the metal phosphate diester. The conditions in the previous reaction steps are controlled to provide a gel with good viscosity versus temperature and time characteristics.

Enhancers for phosphate esters are amino compounds. The 2-ethylhexanoic acid trialuminum salt has been suggested with fatty acids as an activator. Another method to produce oil-based hydrocarbon gels is to use ferric salts rather than aluminum compounds for combination with orthophosphate esters. The ferric salt has the advantage of being usable in the presence of large amounts of water, up to 20%. Ferric salts can be applied in wide ranges of pH. The linkages that are formed can still be broken with gel breaking additives conventionally used for that purpose.

2. Gel breakers

Gel breakers used in nonaqueous systems have a completely different chemistry than those used in aqueous systems. A mixture of hydrated lime and sodium bicarbonate is useful in breaking nonaqueous gels. Sodium bicarbonate used by itself is totally ineffective for breaking the fracturing fluid for aluminum phosphate or aluminum phosphate ester-based gellants. An alternative is to use sodium acetate as a gel breaker for nonaqueous gels.

5.8.4 Foam-based fracturing fluids

Foam fluids can be used in many fracturing jobs, especially when environmental sensitivity is a concern. Foam-fluid formulations are reusable, are shear stable, and form stable foams over a wide temperature range. They exhibit high viscosities even at relatively high temperatures.

In addition to the normal additives, foam-fluid formulations contain surfactants, nitrogen, and carbon dioxide as essential components. Cocobetaine and aolefin sulfonate have been proposed as foamers. The content of the gas is called quality; therefore a 70 quality contains 70% gas. Recently, foams with 95% gas have been examined. For such foam types, only foam prepared from 2% of an anionic surfactant with plain water had uniform, fine-bubble structure.

From *Oil Field Chemicals* by *Johannes Karl Fink*

New Words and Expressions

hydraulic fracturing fluids 水力压裂液
superimposed structure 叠合结构
borate-crosslinked gels 硼酸盐交联凝胶
hydroxypropylguar 羟丙基胍胶
galactomannans 半乳甘露聚糖
hydroxyethylcellulose 羟乙基纤维素
nonacetylated xanthan 非乙酰基黄原胶
phosphorous pentoxide 五氧化二磷
aluminum isopropoxide 异丙醇铝,三异丙氧基铝

Notes

(1) **Brookfield viscometer** 布氏粘度计或 B 型粘度计,是美国 Brookfield 家族开创的旋转

粘度测量法,利用独特的转子与流体之间产生的剪切和阻力之间的关系而得出的全新的粘度值,正式开创了动力粘度的测量先河。

(2) **Size exclusion chromatography** 尺寸排阻色谱法是 20 世纪 60 年代发展起来的一种色谱分离方法。又称为凝胶色谱法、分子排阻色谱法、分子筛色谱法和凝胶渗透色谱法,是液相色谱的一种。

5.9 Drag-Reducing Agents

5.9.1 Introduction

The flow of liquid in a conduit, such as a pipeline, results in frictional energy losses. As a result of this energy loss, the pressure of the liquid in the conduit decreases along the conduit in thedirection of the flow. For a conduit of fixed diameter, this pressure drop increases with increasing flow rate. The effect of a drag reducer added to a liquid is to reduce the frictional resistance in turbulent flow (Reynold's number greater than about 2100) compared with that of the pure liquid. This is sometimes called the Toms effect after the inventor of the definition. Drag-reducing agents (DRAs) are sometimes known as friction reducers or flow improvers, although the latter term can be confused with wax inhibitors/pour-point depressants such as poly(meth)acrylic esters. DRAs interact with the turbulent flow processes and reduce frictional pressure losses such that the pressure drop for a given flow rate is less, or the flow rate for a given pressure drop is greater. In most petroleum pipelines, the liquid flows through the pipeline in a turbulent regime. Therefore, DRAs can perform very well in most pipelines. Because DRAs reduce frictional energy losses, increase in the flow capability of pipelines, hoses, and other conduits in which liquids flow can be achieved. DRAs can also decrease the cost of pumping fluids, the cost of equipment used to pump fluids, and provide for the use of a smaller pipe diameter for a given flow capacity.

5.9.2 Drag-reducing Agent Mechanisms

Despite considerable research in the areas of DRAs, there is no universally accepted model, which explains the mechanism by which polymers, or surfactants, reduce friction in turbulent flow. One early theory states that the stretchingof randomly coiled polymers increases the effective viscosity. By consequence, small eddies are damped, which leads to a thickening of the viscous sublayer and, thus, drag reduction. A more recent theory proposed that drag reduction is caused by elastic rather than viscous properties. This conclusion was reached by observing drag reduction in experiments where polymers were active at the center of the pipe, where viscous forces do not play a role. Another group observed that the amount of drag reduction is limited by an empirical asymptote, called the "Virk asymptote," although others have found conflicting results. Polymer elasticity has been used to scale up the laboratory characteristics to field application of DRAs. The friction pressure drop gradient is determined for a DRA by measuring the pressure drop, velocity, and relaxation time of the polymer as it passes through a narrow straight tubing at multiple injection rates.

A later qualitative theory discusses turbulent flow in a pipeline as having three parts. In the very center of the pipe is a turbulent core where one finds the eddy currents. It is the largest region and includes most of the fluid in the pipe. Nearest to the pipeline wall is the laminar sublayer. In this zone, the fluid moves laterally in sheets. Between the laminar layer and the turbulent core lies the buffer zone where turbulence is first formed. A portion of the laminar sublayer, called a "streak," occasionally will move to the buffer region. There, the streak begins to vortex and oscillate, moving faster as it gets closer to the turbulent core. Finally, the streak becomes unstable and breaks up as it throws fluid into the core of the flow. This ejection of fluid into the turbulent core is called a "turbulent burst." This bursting motion and growth of the bursts in the turbulence core results in wasted energy. Drag-reducing polymers interfere with the bursting process and reduce the turbulence in the core. The polymers absorb the energy in the streak, rather like a shock absorber, thereby reducing subsequent turbulent bursts. As such, drag-reducing polymers are most active in the buffer zone. The overall effect may be to increase the thickness of the laminar sublayer, thereby reducing convective heat transfer.

Using somewhat different terminology, another flow loop study carried out with a water-soluble polyethylene oxide (PEO) indicates that drag reduction is accompanied by the appearance of "shear layers" (i.e., thin filament-like regions of high spatial velocity gradients) that act as interfaces separating low-momentum flow regions near the pipe wall and high-momentum flow regions closer to the centerline. The shear layers are not stationary and their mean thickness correlates with the measured level of drag reduction.

Clearly, molecular weight, aggregation, and chain flexibility (versus polymer rigidity) are all important factors affecting the performance of polymer DRAs. The theory regarding molecular weight is that the longer polymers will be best suited to break up turbulence bursts or eddies in the flow. The hydrodynamic volume (coil volume) of the polymer has been proposed as a better critical factor than molecular weight by some workers based on laboratory studies. This is illustrated by oil-soluble DRAs, many of which have long alkyl side chains to increase the polymer volume. The polymer volume varies with the solvent and, if aqueous, sometimes with the pH and ionicity. Surprisingly, a recent study showed that an anionic polymer of 95% acrylamide and 5% sodium acrylate gave better DRA performance than a nonionic acrylamide homopolymer of similar molecular weights in a flow loop in low-salinity water. The ionic polymer was suggested to be the better DRA owing to added "molar masses" arising from charge repulsion.

5.9.3 Oil-soluble DRAs

(1) Oil-soluble Polymer DRAs

In their extended configuration, polymers have a size that is smaller than the smallest length scale of the turbulence. A well-known effect is the increase of the shear viscosity of a fluid due to polymers, which gives reason to suspect that polymers can affect turbulence on a microscale. However, UHMW polymers are active on both the microscale and macroscale of the turbulence. Therefore, it is a key feature of DRAs that they are as long as possible; that is, have as high a molecular weight as possible.

Oil-soluble DRAs have been tested in a rheometer and flow loop on various fluids, including a paraffinic crude oil. The experimental results have shown that either the presence of waxy crystals or emulsified water droplets can alter the DRA efficiency, while they are not affected by waxy deposits or low temperature.

Commercial DRA polymers for oil transportation that can be produced with UHMW are mostly based on Ziegler-Natta organometallic polymerization of alkenes (olefins). Examples of cheap monomers that can be used include: Isobutylene; Isoprene; Styrene; Hexene; Octene; Decene; Tetradecene.

(2) Oil-soluble surfactant DRAs

Surfactant DRAs have been more extensively explored for water or multiphase flow applications. However, there are a few reports of nonpolymeric oil-soluble DRA studies. In single-phase hexane flow tests, an alkyl phosphate ester performed better than three UHMW polymers (20~100 ppm) at concentrations >200 ppm and friction velocities <0.3 ft/s. For these conditions, the ester yielded drag reduction levels up to 85% and exhibited negligible shear degradation. For concentrations less than 100 ppm and friction velocities above 0.3 ft/s, a polymer product was the superior drag reducer in the absence of degradation. However, in a two-phase flow with hexane and natural gas, the polymer products were superior and at a lower concentration.

Aluminum carboxylate DRAs have been described. These additives are not subject to permanent sheardegradation and do not cause undesirable changes in the emulsion or fluid quality of the fluid being treated, or undesirable foam. In addition, they are claimed to be easy to inject. Examples are aluminum dioctoate, aluminum distearate, and various mixtures. Another variation on this theme is to blend an aluminum monocarboxylate with at least one carboxylic acid on site to produce an aluminum dicarboxylate DRA. This avoids handling, transportation, and injection difficulties with very viscous solutions. Some of the classes of surfactants described for drag reduction aqueous solutions (see below) can be hydrophobic enough for use as DRAs in hydrocarbon liquids or multiphase transportation.

5.9.4 Water-soluble DRAs

Water-soluble DRAs can be divided into two categories: High molecular weight linear polymers; Surfactants. Both of these categories are discussed below as well as the relationship between drag reduction and corrosion inhibition.

1. Water-soluble polymer DRAs

Many classes of water-soluble polymer exhibit drag-reducing properties, including cationic, anionic, or nonionic polymers. Water-soluble polymers that have been deployed as DRAs include PAM and partially hydrolyzed polyacrylamide (PHPA); Other copolymers of acrylamide and acrylamide derivatives and acrylates; PEO; Polyvinyl alcohols; Polysaccharides and derivatives, such as Guar gum, Hydroxypropylguar, Xanthan, Carboxymethylcellulose (CMC); Hydroxyethylcellulose (HEC).

Of these categories, UHMW acrylamide polymers and copolymers are the most-used DRAs in water injection projects in the oil industry. The other class of water-soluble synthetic polymeric DRAs that has been well researched is the UHMW PEOs.

The use of polymeric DRAs in water injection presents the problem of absorption in reservoirs, causing a decrease in the flow of natural gas or oil into the well. This is especially undesirable in shale gas reservoirs where permeability is extremely low. Moreover, it has been shown that there is potentially a higher risk of polymer adsorption when associating polymers are used compared with non-associating homologues. A potential solution that is being investigated is to use stimuli-sensitive associating polymers, which can be switched between the states, associated and dissociated.

A drag reduction method to decrease the injection pressure of flooding significantly by decreasing the drag of laminar flow of water through microchannels in reservoir rock has been proposed. The solution containing hydrophobic nanoparticles (HNPs) of SiO_2 is injected into the microchannels of the reservoir, and the HNPs are adsorbed to the wall to form a strong or superhydrophobic layer, which can lead to a slip boundary condition and decrease drag on the fluid.

2. Water-soluble surfactant DRAs

Surfactants have been known for a long time to be capable of reducing drag in turbulent flowing liquids. Surfactant additives have dual effects on frictional drag. First, they introduce viscoelastic shear stress, which increases frictional drag. Second, they dampen the turbulent vortical structures, decrease the turbulent shear stress, and then decrease the frictional drag. Since the second effect is greater than the first one, drag reduction occurs.

Surfactant DRAs have been proposed as alternatives to polymer DRAs for water injection. Surfactant DRAs are only active at higher concentrations than UHMW polymeric DRAs. For example, 200 ppm of a good surfactant DRA may be needed for the same drag performance as 20 ppm of a UHMW polymeric DRA. Up to 80% drag reduction has been observed with surfactant DRAs. The reason for the high concentration with surfactant DRAs is that the surfactants, or blends or surfactants, need to be above a critical micelle concentration so they can associate into large-enough micelles. For some surfactants, these micelles can take on a rodlike nature. It is only these associated surfactant rodlike micelles and not individual surfactant molecules or spheroidal micelles that are capable of reducing eddies and bursts in turbulent flow. Some water-soluble surfactants, which show no drag-reducing properties at ambient temperature, may perform as DRAs at elevated temperature just below their cloud point. This may be due to aggregation of the surfactants but not total collapse of the surfactant-water interactions.

Cooperative drag-reducing effects between polymers and surfactants have been reported. For example, results with PEO mixed with a homologous series of carboxylate soaps suggest a cooperative micelle formed between soap and polymer. Furthermore, theenhanced drag reduction obtained is consistent with the model of surfactant molecules hydrophobically bonded to the polymer chain in which repulsion between adjacent polar surfactant groups promotes expansion of the polymer coil.

From *Production chemicals for the oil and Gas industry by Malcolm A. Kelland*

New Words and Expressions

turbulent flow [流] 湍流;[流] 紊流

Reynold's number 雷诺数

asymptote [ˈæsɪm(p)təʊt] [数] 渐近线

eddy current 涡流

laminar layer 层流层

vortex [ˈvɔːteks] [航] [流] 涡流;漩涡

convective heat transfer 对流传热

velocity gradient 速度梯度

hydrodynamic volume [高分子] [流] 流体力学体积

dissipated [ˈdɪsɪpeɪtɪd] *adj.* 消散的;沉迷于酒色的;闲游浪荡的;放荡的

shear viscosity [物] 剪切黏度

isobutylene. [有化] 异丁烯

partially hydrolyzed polyacrylamide 部分水解聚丙烯酰胺

Polyvinyl alcohol [高分子] 聚乙烯醇

guar gum 瓜尔豆胶

carboxymethylcellulose 羧甲基纤维素

Notes

(1) **Drag-Reducing Agents**, 减阻剂,是一种能减少流体在输送时所受阻力的试剂。多为水溶性或油溶性的高分子聚合物。20 世纪 60 年代末,美国 Conoco 公司研制成 CDR-101 型减阻剂,1972 年取得专利,1977~1979 年间首次商业化应用于横贯阿斯加的原油管道的越站输送及提高输量方面,并取得巨大成功。70 年代中期,美国 Shellco 公司和加拿大 Shell Inc 公司提出申请减阻剂专利。1983 年,美国 Atlantic Richfield co 公司研制出 Arcoflo 减阻剂产品,加入 5ppm 即可达到 20%的减阻效果。中国石油管道公司管道科技研究中心开展了减阻剂的研究工作,其 EP 系列减阻剂产品的性能已经达到国际同类产品的水平。

(2) **Reynold's number**,雷诺数,是流体惯性力与黏性力比值的量度。雷诺数较小时,黏滞力对流场的影响大于惯性力,流场中流速的扰动会因黏滞力而衰减,流体流动稳定,为层流;反之,若雷诺数较大时,惯性力对流场的影响大于黏滞力,流体流动较不稳定,流速的微小变化容易发展、增强,形成紊乱、不规则的紊流流场。

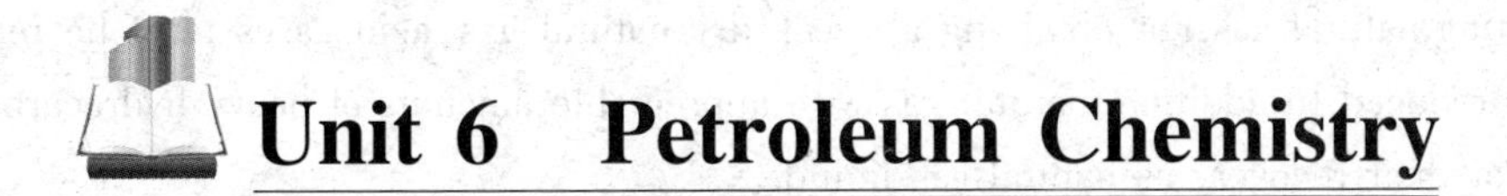

Unit 6 Petroleum Chemistry

6.1 Primary Raw Materials for Petrochemicals

In general, primary raw materials are naturally occurring substances that have not been subjected to chemical changes after being recovered. Natural gas and crude oils are the basic raw materials for the manufacture of petrochemicals.

Secondary raw materials, or intermediates, are obtained from natural gas and crude oils through different processing schemes. The intermediates may be light hydrocarbon compounds such as methane and ethane, or heavier hydrocarbon mixtures such as naphtha or gas oil. Both naphtha and gas oil are crude oil fractions with different boiling ranges.

6.1.1 Natural gas

Natural gas is a naturally occurring mixture of light hydrocarbons accompanied by some non-hydrocarbon compounds. Non-associated natural gas is found in reservoirs containing no oil (dry wells). Associated gas, on the other hand, is present in contact with and/or dissolved in crude oil and is coproduced with it. The principal component of most natural gases is methane. Higher molecular weight paraffinic hydrocarbons ($C_2 \sim C_7$) are usually present in smaller amounts with the natural gas mixture, and their ratios vary considerably from one gas field to another.

Non-associated gas normally contains a higher methane ratio than associated gas, while the latter contains a higher ratio of heavier hydrocarbons. In our discussion, both non-associated and associated gases will be referred to as natural gas. However, important differences will be noted.

The non-hydrocarbon constituents in natural gas vary appreciably from one gas field to another. Some of these compounds are weak acids, such as hydrogen sulfide and carbon dioxide. Others are inert, such as nitrogen, helium and argon. Some natural gas reservoirs contain enough helium for commercial production.

Higher molecular weight hydrocarbons present in natural gases are important fuels as well as chemical feedstocks and are normally recovered as natural gas liquids. For example, ethane may be separated for use as a feedstock for steam cracking for the production of ethylene. Propane and butane are recovered from natural gas and sold as liquefied petroleum gas (LPG). Before natural gas is used it must be processed or treated to remove the impurities and to recover the heavier hydrocarbons (heavier than methane). The 1998 U.S. gas consumption was approximately 22.5 trillion ft^3.

Raw natural gases contain variable amounts of carbon dioxide, hydrogen sulfide, and water vapor. The presence of hydrogen sulfide in natural gas for domestic consumption cannot be tolerated because it is poisonous. It also corrodes metallic equipment. Carbon dioxide is undesirable, because it reduces the heating value of the gas and solidifies under the high pressure and low temperatures used for transporting natural gas. For obtaining a sweet, dry natural gas, acid gases must be removed and water vapor reduced. In addition, natural gas with appreciable amounts of heavy hydrocarbons should be treated for their recovery as natural gas liquids.

1. Acid Gas Treatment

Acid gases can be reduced or removed by one or more of the following methods:

(1) Physical absorption using a selective absorption solvent.

(2) Physical adsorption using a solid adsorbent.

(3) Chemical absorption where a solvent (a chemical) capable of reacting reversibly with the acid gases is used.

2. Physical Absorption

Important processes commercially used are the Selexol, the Sulfinol, and the Rectisol processes. In these processes, no chemical reaction occurs between the acid gas and the solvent. The solvent, or absorbent, is a liquid that selectively absorbs the acid gases and leaves out the hydrocarbons. In the Selexol process for example, the solvent is dimethyl ether of polyethylene glycol. Raw natural gas passes countercurrently to the descending solvent. When the solvent becomes saturated with the acid gases, the pressure is reduced, and hydrogen sulfide and carbon dioxide are desorbed. The solvent is then recycled to the absorption tower.

In these processes, a solid with a high surface area is used. Molecular sieves (zeolites) are widely used and are capable of adsorbing large amounts of gases. In practice, more than one adsorption bed is used for continuous operation. One bed is in use while the other is being regenerated.

Regeneration is accomplished by passing hot dry fuel gas through the bed. Molecular sieves are competitive only when the quantities of hydrogen sulfide and carbon disulfide are low. Molecular sieves are also capable of adsorbing water in addition to the acid gases.

3. Chemical Absorption (Chemisorption)

These processes are characterized by a high capability of absorbing large amounts of acid gases. They use a solution of a relatively weak base, such as monoethanolamine. The acid gas forms a weak bond with the base which can be regenerated easily. Mono- and diethanolamines are frequently used for this purpose. The amine concentration normally ranges between 15% and 30%. Natural gas is passed through the amine solution where sulfides, carbonates, and bicarbonates are formed. Diethanolamine is a favored absorbent due to its lower corrosion rate, smaller amine loss potential, fewer utility requirements, and minimal reclaiming needs. Diethanolamine also reacts reversibly with 75% of carbonyl sulfides (COS), while the mono- reacts irreversibly with 95% of the COS and forms a degradation product that must be disposed of.

Strong basic solutions are effective solvents for acid gases. However, these solutions are not normally used for treating large volumes of natural gas because the acid gases form stable salts, which are not easily regenerated. For example, carbon dioxide and hydrogen sulfide react with aqueous sodium hydroxide to yield sodium carbonate and sodium sulfide, respectively.

6.1.2 Crude oils

Crude oil (petroleum) is a naturally occurring brown to black flammable liquid. Crude oils are principally found in oil reservoirs associated with sedimentary rocks beneath the earth's surface. Although exactly how crude oils originated is not established, it is generally agreed that crude oils derived from marine animal and plant debris subjected to high temperatures and pressures. It is also suspected that the transformation may have been catalyzed by rock constituents. Regardless of their origins, all crude oils are mainly constituted of hydrocarbons mixed with variable amounts of sulfur, nitrogen, and oxygen compounds.

Metals in the forms of inorganic salts or organometallic compounds are present in the crude mixture in trace amounts. The ratio of the different constituents in crude oils, however, varies appreciably from one reservoir to another.

Normally, crude oils are not used directly as fuels or as feedstocks for the production of chemicals. This is due to the complex nature of the crude oil mixture and the presence of some impurities that are corrosive or poisonous to processing catalysts.

Crude oils are refined to separate the mixture into simpler fractions that can be used as fuels, lubricants, or as intermediate feedstock to the petrochemical industries. A general knowledge of this composite mixture is essential for establishing a processing strategy.

6.1.3 Coal, oil shale, tar sand, and gas hydrates

Coal, oil shale, and tar sand are carbonaceous materials that can serve as future energy and chemical sources when oil and gas are consumed. The H/C ratio of these materials is lower than in most crude oils. As solids or semisolids, they are not easy to handle or to use as fuels, compared to crude oils. In addition, most of these materials have high sulfur and/or nitrogen contents, which require extensive processing. Changing these materials into hydrocarbon liquids or gaseous fuels is possible but expensive. The following briefly discusses these alternative energy and chemical sources.

Coal is a natural combustible rock composed of an organic heterogeneous substance contaminated with variable amounts of inorganic compounds. Most coal reserves are concentrated in North America, Europe, and China.

Oil shale is a low-permeable rock made of inorganic material interspersed with a high-molecular weight organic substance called "Kerogen". Heating the shale rock produces an oily substance with a complex structure.

Tar sands (oil sands) are large deposits of sand saturated with bitumen and water. Tar sand deposits are commonly found at or near the earth's surface entrapped in large sedimentary basins. Large accumulations of tar sand deposits are few. About 98% of all world tar sand is found in seven large tar deposits. The oil sands resources in Western Canada sedimentary basin is the largest in the world. In 1997, it produced 99% of Canada's crude oil. It is estimated to hold 1.7~2.5 trillion barrels of bitumen in place. This makes it one of the largest hydrocarbon deposits in the world. Tar sand deposits are covered by a semifloating mass of partially decayed vegetation approximately 6 meters thick.

Gas hydrates are an ice-like material which is constituted of methane molecules encaged in a cluster of water molecules and held together by hydrogen bonds. This material occurs in large underground deposits found beneath the ocean floor on continental margins and in places north of the arctic circle such as Siberia. It is estimated that gas hydrate deposits contain twice as much carbon as all other fossil fuels on earth. This source, if proven feasible for recovery, could be a future energy as well as chemical source for petrochemicals.

From *Chemistry of Petrochemical Processes by Sami Matar and Lewis F. Hatch*, 2nd *Edition*, 2001.

New Words and Expressions

naphtha [ˈnæfθə] 石脑油;挥发油;粗汽油
helium [ˈhiːlɪəm] [化学] 氦(符号为 He,2 号元素)
Sulfinol 环丁砜法
Rectisol 低温甲醇洗法
molecular sieves (zeolites) 分子筛
monoethanolamine [有化] 单乙醇胺
diethanolamine [有化] 二乙醇胺
bicarbonate [baɪˈkɑːbəneɪt] 碳酸氢盐;重碳酸盐;酸式碳酸盐
organometallic [ɔː,gænəʊmɪˈtælɪk] 有机金属的
tar [tɑː] 焦油;柏油;水手;涂以焦油;玷污
shale [ʃeɪl] [岩] 页岩;泥板岩
kerogen [ˈkerədʒən] [地质] [油气] 油母岩质
sedimentary [sedɪˈmentrɪ] 沉淀的
semifloating [seˈmaɪfələʊtɪŋ] 半浮动

Notes

(1) Associated gas, 伴生气。通常指与石油共生的天然气,主要成分是甲烷(CH_4),还含有少量的乙烷(C_2H_6)、丙烷(C_3H_8)、丁烷(C_4H_{10})。按有机成烃的生油理论,有机质演化可生成液态烃与气态烃。气态烃或溶解于液态烃中,或呈气顶状态存在于油气藏的上部。这两种气态烃均称为油田伴生气或伴生气。从采油工作角度考虑,指开采油田或油藏时采出的天然气。

(2) Selexol process, 天然气脱硫。在天然气中常含有 H_2S、CO_2 和有机硫化合物,这三者通称为酸性组分(或酸性气体)。这些气相杂质的存在会造成金属材料腐蚀,并污染环境。当天然气作为化工原料时,它们还会导致催化剂中毒,影响产品质量;而 CO_2 含量过高则使气体的热值达不到要求。鉴此,天然气脱硫的目的是按不同用途把气体中的上述杂质组分脱除到要求的规格。天然气脱硫包含多种化学应用程序,如胺处理或膜吸附,即从天然气去除硫磺,从而净化天然气和其他污染物。回收的硫可用于各种化学工业应用中。

6.2 Composition and Classification of Crude Oils

6.2.1 Composition of crude oils

The crude oil mixture is composed of the following groups: Hydrocarbon compounds (compounds made of carbon and hydrogen); Non-hydrocarbon compounds; Organometallic compounds and inorganic salts (metallic com-pounds).

1.Hydrocarbon compounds

The principal constituents of most crude oils are hydrocarbon compounds. All hydrocarbon classes are present in the crude mixture, except alkenes and alkynes. This may indicate that crude oils originated under a reducing atmosphere. The following is a brief description of the different hydrocarbon classes found in all crude oils (Table 6.1).

Table 6.1 Elemental composition of crude oils

Element	Composition/wt%	Element	Composition/wt%
Carbon	83.0~87.0	Oxygen	0.05~2.0
Hydrogen	10.0~14.0	Ni	<120ppm
Sulphur	0.05~6.0	V	<1200ppm
Nitrogen	0.1~0.2		

There are threemain classes of hydrocarbons. These are based on the type. of carbon-carbon bonds present. These classes are: ①Saturated hydrocarbons contain only carbon-carbon single bonds. They are known as *paraffins* (or alkanes) if they are acyclic, or *naphthenes* (or cycloalkanes) if they are cyclic. ② Unsaturated hydrocarbons contain carbon-carbon multiple bonds (double, triple or both). These are unsaturated because they contain fewer hydrogens per carbon than paraffins. Unsaturated hydrocarbons are known as olefins. Those that contain a carbon-carbon double bond are called alkenes, while those with carbon-carbon triple bond are alkyenes. ③ Aromatic hydrocarbons are special class of cyclic compounds related in structure to benzene.

2.Non-hydrocarbon compounds

Various types of non-hydrocarbon compounds occur in crude oils and refinery streams. The most important are the organic sulfur, nitrogen, and oxygen compounds. Traces of metallic compounds are also found in all crudes. The presence of these impurities is harmful and may cause problems to certain catalytic processes. Fuels having high sulfur and nitrogen levels cause pollution problems in addition to the corrosive nature of their oxidization products.

① Sulphur compounds. Thesulphur content of crude oils varies from less than 0.05 to more than 10 wt% but generally falls in the range 1 wt%~4 wt%. Crude oil with less than 1 wt % sulphur is referred to as low sulphur or sweet, and that with more than 1 wt% sulphur is referred to as high sulphur or sour.

Sulphur containing constituents of crude oils vary from simple mercaptans, also known as thiols, to sulphides and polycyclic sulphides. Mercaptans are made of an alkyl chain with —SH group at the end (R—SH). Examples of *mercaptans* and sulphides are as follows:

CH_3SH	$CH_3CH_2CH_2SH$	C_6H_5—SH
methyl mercaptan (methanethiol)	*n*-butyl mercaptan (1-butanethiol)	phenyl mercaptan (thiophenol)

In *sulphides* and *disulphides*, the sulphur atom replaces one or two carbon atoms in the chain (R-S-R' or R-S-S-R'). These compounds are often present in light fractions. Sulphides and disulphides may also be cyclic or aromatic.

② Oxygen compounds The oxygen content of crude oil is usually less than 2 wt%. A phenomenally high oxygen content indicates that the oil has suffered prolonged exposure to the atmosphere. Oxygen in crude oil can occur in a variety of forms. These include alcohols, ethers, carboxylic acids, phenolic compounds, ketones, esters and anhydrides. The presence of such compounds causes the crude to be acidic with consequent processing problems such as corrosion.

③ Nitrogen compounds Crude oils contain very low amounts of nitrogen compounds. In general, the more asphaltic the oil, the higher its nitrogen content. Nitrogen compounds are more stable than sulphur compounds and therefore are harder to remove. Even though they are present at very low concentrations, nitrogen compounds have great significance in refinery operations. They can be responsible for the poisoning of a cracking catalyst, and they also contribute to gum formation in finished products.

The nitrogen compounds in crude oils may be classified as basic or non-basic. Basic nitrogen compounds consist of pyridines. The greater part of the nitrogen in crude oils is the non-basic nitrogen compounds, which are generally of pyrrole types.

Pyridines are six-membered heteroaromatic compounds containing one nitrogen atom. When fused with benzene rings, pyridines are converted to the polycyclic heteroaromatic compounds quinolines and isoquinolines.

pyridine	quinoline	isoquinoline
(C_5H_5N)	(C_9H_7N)	(C_9H_7N)

In non-basic nitrogen compounds, *pyrroles* are five-membered heteroaromatic compounds containing one nitrogen atom. When fused with benzene ring, pyrrole is converted to the polycyclic heteroaromatic compounds *indole* and *carbazole*.

④ Asphaltenes and resins The physical properties of crude oils, such as the specific gravity (or API), are considerably influenced by high-boiling constituents, in which the heteroatoms (sulphur, nitrogen and metals) concentrate. It is therefore important to characterize the heaviest fractions of crude oils in order to determine their properties and ease of processing. This calls for determining the percentage of two generally defined classes of compounds, namely asphaltenes and resins.

3.Metallic compounds

Many metals occur in crude oils.Some of the more abundant are sodium, calcium, magnesium, aluminium, iron, vanadium, and nickel.They are present either as inorganic salts, such as sodium and magnesium chlorides, or in the form of organometallic compounds, such as those of nickel and vanadium (as in porphyrins).Calcium and magnesium can form salts or soaps with carboxylic acids.These compounds act as emulsifiers, and their presence is undesirable.

6.2.2 Classification of crude oils

The original methods of classification arose because of commercial interest in crude oil type and were a means of providing refinery operators with a rough guide to processing conditions.It is therefore not surprising that systems based on a superficial inspection of a physical property, such as specific gravity or API gravity, are easily applied, and are actually used to a large extent in expressing the quality of crude oils.Such a system is approximately indicative of the general character of a crude oil as long as materials of one general type are under consideration.For example, among crude oils from a particular area, an oil of 40°API (specific gravity = 0.825) is usually more valuable than one of 20° API (specific gravity = 0.934), because it contains more light fractions (e.g., gasoline) and fewer heavy, undesirable asphaltic constituents.

(1) By density

Density (specific gravity) has been, since the early years of the industry, the principal, and often the only specification of crude oil products and was taken as an index of the proportion of gasoline and, particularly, kerosene present.As long as only one kind of crude oil was in use, the relations were approximately true, but as crude oils having other properties were discovered and came into use, the significance of density measurements disappeared.Nevertheless, crude oils of particular types are still rated by gravity, as are gasoline and naphtha within certain limits of other properties.The use of density values has been advocated for quantitative applications using a scheme based on the American Crude oil Institute (API) gravity of the 250℃ to 275℃ and the 275℃ to 300℃ distillation fractions.

It has also been proposed to classify heavy oils according to characterization gravity.This is defined as the arithmetic average of the instantaneous specific gravity of the distillates boiling at 177℃, 232℃, and 288℃ vapor line temperature at 25 mmHg pressure in a true boiling point distillation.

(2) By API gravity

Conventional crude oil and heavy oil have also been defined very generally in terms of physical properties.For example, heavy oils were considered those crude oil-type materials that had gravity somewhat less than 20°API, with the heavy oils falling into the API gravity range of 10° to 15° (e.g., Cold Lake crude oil = 12°API) and bitumen falling into the 5° to 10°API range (e.g., Athabasca bitumen = 8°API).Residua vary depending on the temperature at which distillation is terminated.Atmospheric residua are usually in the 10° to 15°API range of, and vacuum residua are in the range of 2° to 8°API.

There have been several recent noteworthy attempts to classify crude oil using one or more of the general physical properties of crude oils. One method uses divisions by API gravity, which is already accepted by most workers, it also uses viscosity data. This method is essentially a more formal attempt in which specific numbers are applied without recognition of the implications of these numbers.

(3) By viscosity

At the same time, and in concert with the use of API gravity, the line of demarcation between crude oil and heavy oil vis-à-vis tar sand bitumen has been drawn at 10,000 centipoises (cp). Briefly, materials having viscosity less than 10,000 cp are conventional crude oil and heavy oil, whereas tar sand bitumen has a viscosity greater than 10,000 cp. Use of such a scale requires a fine line of demarcation between the various crude oils, heavy oils, and bitumen to the point where it would be confusing to differentiate between a material having a viscosity of 9950 cp and one having a viscosity of 10,050 cp. Further, the inaccuracies (i.e., the limits of experimental error) of the method of measuring viscosity also increase the potential for misclassification.

From *Fundamentals of Petroleum Refining by M. A. Fahim*, 2010; *Chemistry of Petrochemical Processes by Sami Matar and Lewis F. Hatch, 2nd Edition*, 2001.

New Words and Expressions

paraffin [ˈpærəfɪn] 石蜡;[有化] 链烷烃;硬石蜡

naphthene [ˈnæfθiːn] [有化] 环烷,[有化] 环烷属烃

cycloalkane [saikləuˈælkein] [有化] 环烷

decane [ˈdekeɪn] [有化] 癸烷;十炭矫质

triene [ˈtraii:n] [有化] 三烯

polyene [ˈpɒlɪjiːn] [有化] 多烯

cyclohexane [saɪklə(ʊ)ˈhekseɪn] [有化] 环己烷

xylene [ˈzaɪliːn] [有化] 二甲苯

asphaltenes [油气] 沥青质;沥青稀

heteroatom [ˈhetərəʊætəm] 杂原子,杂环原子

mercaptan [məˈkæpt(ə)n] [有化] 硫醇

thiophene [ˈθaɪəfiːn] [化学] 噻吩;硫(杂)茂

polynuclear [pɒlɪˈnjuːklɪə] 多核的,多环的

anhydride [ænˈhaɪdraɪd] [化学] 酸酐;脱水物

benzofuran [benzəuˈfjuəræn] 香豆酮

asphaltic [æsˈfæltik] 柏油的

gum [gʌm] 口香糖;树胶;橡皮;用胶粘,涂以树胶;使有粘性

carbazole [kɑːbəˈzəʊl] 咔唑

indole [ˈɪndəʊl] [有化] 吲哚;靛基质

liquefied [lɪkwɪfaɪd] 液化的;液化;溶解

Notes

(1) **Specific gravity**,比重。是一物体密度与水的密度之间的比值,为一个无量纲量。比重若大于1,在水中会沉下,反之,若小于1,则可以浮在水上。以水以外的物质当作参考物,通常称为相对比重(Relative density)。

(2) **API gravity**,API度,美国石油学会(American Petroleum Institute)(简称API)制订的用以表示石油及石油产品密度的一种量度。美国和中国以API度作为原油分类的基准其标准温度为15.6℃,它和15.6℃时的相对密度(与水比)的关系:API=(141.5/相对密度)-131.5。由该式可知,API度越大,相对密度越小。目前,国际上把API度作为决定原油价格的主要标准之一。它的数值越大,表示原油越轻,价格越高。

6.3 Properties of Crude Oils

Crude oils differ appreciably in their properties according to origin and the ratio of the different components in the mixture. Lighter crudes generally yield more valuable light and middle distillates and are sold at higher prices. Crudes containing a high percent of impurities, such as sulfur compounds, are less desirable than low-sulfur crudes because of their corrosivity and the extra treating cost. Corrosivity of crude oils is a function of many parameters among which are the type of sulfur compounds and their decomposition temperatures, the total acid number, the type of carboxylic and naphthenic acids in the crude and their decomposition temperatures. It was found that naphthenic acids begin to decompose at 600°F. Refinery experience has shown that above 750°F there is no naphthenic acid corrosion. The subject has been reviewed by Kane and Cayard. For a refiner, it is necessary to establish certain criteria to relate one crude to another to be able to assess crude quality and choose the best processing scheme. The following are some of the important tests used to determine the properties of crude oils.

6.3.1 Density, specific gravity and API gravity

Density is defined as the mass of unit volume of a material at a specific temperature. A more useful unit used by the petroleum industry is specific gravity, which is the ratio of the weight of a given volume of a material to the weight of the same volume of water measured at the same temperature.

Specific gravity is used to calculate the mass of crude oils and its products. Usually, crude oils and their liquid products are first measured on a volume basis, then changed to the corresponding masses using the specific gravity.

The API (American Petroleum Institute) gravity is another way to express the relative masses of crude oils. The API gravity could be calculated mathematically using the following equation:

$$^{\circ}API = \frac{141.5}{Sp.gr. 60/60^{\circ}} - 131.5$$

A low API gravity indicates a heavier crude oil or a petroleum product, while a higher API gravity means a lighter crude or product. Specific gravities of crude oils roughly range from 0.82 for lighter crudes to over 1.0 for heavier crudes (41–10°API scale).

6.3.2 Viscosity

Viscosity is a measure of a fluid's resistance to flow; the lower the viscosity of a fluid, the more easily it flows. Like density, viscosity is affected by temperature. As temperature decreases, viscosity increases. The SI unit of dynamic viscosity is the millipascal-second (mPa · s). This is equivalent to the former unit of centipoise (cP).

$$Dynamic\ Viscosity = \frac{shear\ stress}{shear\ rate}$$

Viscosity is a very important property of oils because it affects the rate at which spilled oil will spread, the degree to which it will penetrate shoreline substrates, and the selection of mechanical spill countermeasures equipment.

Viscosity measurements may be absolute orrelative (sometimes called 'apparent'). Absolute viscosities are those measured by a standard method, with the results traceable to fundamental units. "Absolute viscosities are distinguished from relative measurements made with instruments that measure viscous drag in a fluid, without known and/or uniform applied shear rates." An important benefit of absolute viscometry is that the test results are independent of the particular type or make of viscometer used. Absolute viscosity data can be compared easily between laboratories worldwide.

Modern rotational viscometers are capable of making absolute viscosity measurements for both newtonian and non-newtonian fluids at a variety of well controlled, known, and/or uniform shear rates. Unfortunately, no ASTM standard method exists that makes use of these viscometers. Nonetheless, these instruments are in widespread use in many industries.

In the dynamic viscosity data tables, NM indicates that the viscosity was 'not measurable' as it exceeded the measurement range of the instrument.

6.3.3 Salt content

The salt content expressed in milligrams of sodium chloride per liter oil (or in pounds/barrel) indicates the amount of salt dissolved in water.

Water in crudes is mainly present in an emulsified form. A high salt content in a crude oil presents serious corrosion problems during the refining process. In addition, high salt content is a major cause of plugging heat exchangers and heater pipes. A salt content higher than 10 lb/1,000 barrels (expressed as NaCl) requires desalting.

6.3.4 Sulfur content

Determining the sulfur content in crudes is important because the amount of sulfur indicates the type of treatment required for the distillates. To determine sulfur content, a weighed crude sample (or fraction) is burned in an air stream. All sulfur compounds are oxidized to sulfur dioxide, which is further oxidized to sulfur trioxide and finally titrated with a standard alkali.

Identifying sulfur compounds in crude oils and their products is of little use to a refinerbecause all sulfur compounds can easily be hydrodesulfurized to hydrogen sulfide and the corresponding hydrocarbon.

The sulfur content of crudes, however, is important and is usually considered when determining commercial values.

6.3.5 Pour point

The pourpoint of a crude oil or product is the lowest temperature at which the oil is observed to flow under the conditions of the test. Pour point data indicates the amount of long-chain paraffins (petroleum wax) found in a crude oil. Paraffinic crudes usually have higher wax content than other crude types. Handling and transporting crude oils and heavy fuels is difficult at temperatures below their pour points of ten, chemical additives known as pour point depressants are used to improve the flow properties of the fuel. Long-chain n-paraffins ranging from 16-60 carbon atoms in particular, are responsible for near-ambient temperature precipitation. In middle distillates, less than 1% wax can be sufficient to cause solidification of the fuel.

From《*Chemistry of Petrochemical Processes*》*by Sami Matar and Lewis F. Hatch*, 2nd *Edition*, 2001;《*Chemistry and Technology of Petroleum*》, *4th Edition*, *by James G. Speight*, 2006.

New Words and Expressions

corrosivity 腐蚀性, corrosive 的变形

corrosive [kəˈrəusiv] 腐蚀性的, 引起腐蚀的

viscometry [visˈkɔmitri] [物] 黏度测定法

centistoke [ˈsentəstəuk] 厘沲(动力黏度单位)[略作 cS, cSt]

kerogen [ˈkerədʒən] [地质][油气] 油母岩质

lignite [ˈlɪgnaɪt] [矿物] 褐煤

subbituminous [sʌbbiˈtjuːminəs] 亚烟煤的

bituminous [biˈtjuːminəs] 沥青的, 含沥青的

anthracite [ˈænθrəsaɪt] [矿物] 无烟煤

heavy oil 重油

misclassification [ˈmisklæsifiˈkeiʃən] [计] 误分类

Notes

(1) **SI unit**,国际单位制。来自法语(Système International d'Unités),源自公制或米制,旧称万国公制,是世界上最普遍采用的标准度量衡单位系统,采用十进制进位系统。是18世纪末科学家的

努力,最早于法国大革命时期的1799年被法国作为度量衡单位。国际单位制是在公制基础上发展起来的单位制,于1960年第十一届国际计量大会通过,推荐各国采用,其国际缩写为SI。

(2) **Newtonian and non-newtonian fluids**,牛顿流体和非牛顿流体。科学家牛顿发现,某些液体流动时切应力 τ 与切变率 D 之比为常数,即:$\eta=\tau/D$。水和油都是遵循上述规律的液体。这一公式就是牛顿粘度定律。其中,η 为液体的粘度,粘度是液体流动时内摩擦或阻力的量度,η 的单位为 Pa·s 或 mPa·s(帕斯卡.秒)。遵循牛顿粘性定律的液体称为牛顿流体,凡是流体运动时其切变率 D 与切应力 τ 不成线性关系的流体称为非牛顿流体。水、酒精等大多数纯液体、轻质油、低分子化合物溶液以及低速流动的气体等均为牛顿流体;高分子聚合物的浓溶液和悬浮液等一般为非牛顿流体。

6.4 Petroleum Refining

Petroleumrefinery processes can be conveniently divided into three different types: (1) Separation: division of the feedstock into various streams (or fractions) depending on the nature of the crude material. (2) Conversion: that is, the production of saleable materials from the feedstock by skeletal alteration, or even by alteration of the chemical type of the feedstock constituents. (3) Finishing: purification of the various product streams by a variety of processes that remove impurities from the product.

6.4.1 Physical separation processes

1. Crude distillation

Crude oils are first desalted and then introduced with steam to an atmospheric distillation column. The atmospheric residue is then introduced to a vacuum distillation tower operating at about 50 mmHg, where heavier products are obtained. Typical products from both columns and their boiling point ranges are listed in Table 6.2.

Table 6.2 Crude distillation products

		Yield/wt%	True boiling temperature/℃
Atmospheric distillation	Refinery gases (C_1-C_2)	0.1	—
	Liquid petroleum gases (LPG)	0.69	—
	Light straight run (LSR)	3.47	32~82
	Heavy straight run (HSR)	10.17	82~193
	Kerosene (Kero)	15.32	193~271
	Light gas oil (LGO)	12.21	271~321
	Heavy gas oil (HGO)	21.10	321~427
Vacuum distillation	Vacuum gas oil (VGO)	16.8	427~566
	Vacuum residue (VR)	20.3	+566

2. Solvent deasphalting

This is the only physical process where carbon is rejected from heavy petroleum fraction such as vacuum residue. Propane in liquid form (at moderate pressure) is usually used to dissolve the whole oil, leaving asphaltene to precipitate. The deasphalted oil (DAO) has low sulphur and metal contents since these are removed with asphaltene. This oil is also called "Bright Stock" and is used as feedstock for lube oil plant. The DAO can also be sent to cracking units to increase light oil production.

3. Solvent extraction

In this process, lube oil stock is treated by a solvent, such as N-methyl pyrrolidone (NMP), which can dissolve the aromatic components in one phase (extract) and the rest of the oil in another phase (raffinate). The solvent is removed from both phases and the raffinate is dewaxed.

4. Solvent dewaxing

The raffinate is dissolved in a solvent (methyl ethyl ketone, MEK) and the solution is gradually chilled, during which high molecular weight paraffin (wax) is crystallized, and the remaining solution is filtered. The extracted and dewaxed resulting oil is called "lube oil". In some modern refineries removal of aromatics and waxes is carried out by catalytic processes in "all hydrogenation process".

6.4.2 Chemical catalytic conversion processes

1. Cracking

(1) Thermal cracking The term cracking applies to the decomposition of petroleum constituents that is induced by elevated temperatures (>350℃), whereby the higher molecular weight constituents of petroleum are converted to lower molecular weight products. Cracking reactions involve carbon-carbon bond rupture and are thermodynamically favored at high temperature. Thus, cracking is a phenomenon by which higher boiling (higher molecular weight) constituents in petroleum are converted into lower boiling (lower molecular weight) products.

Two general types of reaction occur during cracking:

A. The decomposition of large molecules into small molecules (primary reactions):

$$\underset{\text{Butane}}{CH_3CH_2CH_2CH_3} \rightarrow \underset{\text{Methane}}{CH_4} + \underset{\text{Propene}}{CH_3CH=CH_2}$$

$$\underset{\text{Butane}}{CH_3CH_2CH_2CH_3} \rightarrow \underset{\text{Ethane}}{CH_3CH_3} + \underset{\text{Ethylene}}{CH_2=CH_2}$$

B. Reactions by which some of the primary products interact to form higher molecular weight materials (secondary reactions):

$$CH_2=CH_2+CH_2=CH_2 \rightarrow CH_3CH_2CH=CH_2$$

or

$$R \cdot CH=CH_2+R^1 \cdot CH=CH_2 \rightarrow \text{Cracked residuum+Coke+Other products}$$

Thermal cracking is a free radical chain reaction; a free radical is an atom or group of atoms possessing an unpaired electron. Free radicals are very reactive, and it is their mode of reaction that actually determines the product distribution during thermal cracking. The free radical reacts with a hydrocarbon by abstracting a hydrogen atom to produce a stable end product and a new free radical. Free radical reactions are extremely complex, and it is hoped that these few reaction schemes illustrate potential reaction pathways. Any of the preceding reaction types are possible, but it is generally recognized that the prevailing conditions and those reaction sequences that are thermodynamically favored determine the product distribution.

One of the significant features of hydrocarbon free radicals is their resistance to isomerization, for example, migration of an alkyl group and, as a result, thermal cracking does not produce any degree of branching in the products other than that already present in the feedstock.

The aromatic ring is considered fairly stable at moderate cracking temperatures (350℃ to 500℃). Alkylated aromatics, like the alkylated naphthenes, are more prone to dealkylation that to ring destruction. However, ring destruction of the benzene derivatives occurs above 500℃, but condensed aromatics may undergo ring destruction at somewhat lower temperatures (450℃).

(2) Catalytic cracking Catalytic cracking is the thermal decomposition of petroleum constituents' hydrocarbons in the presence of a catalyst. Thermal cracking has been superseded by catalytic cracking as the process for gasoline manufacture. Indeed, gasoline produced by catalytic cracking is richer in branched paraffins, cycloparaffins, and aromatics, all of which serve to increase the quality of the gasoline. Catalytic cracking also results in production of the maximum amount of butenes and butanes (C_4H_8 and C_4H_{10}), rather than ethylene and ethane (C_2H_4 and C_2H_6).

In the 1930s, a catalytic cracking catalyst for petroleum that used solid acids as catalysts was developed using acid-treated clays. Clays are a family of crystalline aluminosilicate solids, and the acid treatment develops acidic sites by removing aluminum from the structure. The acid sites also catalyze the formation of coke, and Houdry developed a moving-bed process that continuously removed the cooked beads from the reactor for regeneration by oxidation with air.

Although thermal cracking is a free radical (neutral) process, catalytic cracking is an ionic process involving carbonium ions, which are hydrocarbon ions having a positive charge on a carbon atom. The formation of carbonium ions during catalytic cracking can occur by:

① Addition of a proton from an acid catalyst to an olefin

② Abstraction of a hydride ion (H^-) from a hydrocarbon by the acid catalyst or by another carbonium ion.

However, carbonium ions are not formed by cleavage of a carbon-carbon bond. In essence, the use of a catalyst permits alternate routes for cracking reactions, usually by lowering the free energy of activation for the reaction. The acid catalysts first used in catalytic cracking were amorphous solids composed of approximately 87% silica (SiO_2) and 13% alumina (Al_2O_3) and were designated low-alumina catalysts. However, this type of catalyst is now being replaced by crystalline aluminosilicates (zeolites) or molecular sieves.

Catalytic cracking can be represented by simple reaction schemes. However, questions have arisen as to how the cracking of paraffins is initiated. Several hypotheses for the initiation step in catalytic cracking of paraffins have been proposed. The Lewis site mechanism is the most obvious, as it proposes that a carbonium ion is formed by the abstraction of a hydride ion from a saturated hydrocarbon by a strong Lewis acid site: a tricoordinated aluminum species. On Brønsted sites a carbonium ion may be readily formed from an olefin by the addition of a proton to the double bond or, more rarely, via the abstraction of a hydride ion from a paraffin by a strong Brønsted proton. This latter process requires the formation of hydrogen as an initial product. This concept was, for various reasons that are of uncertain foundation, often neglected.

2. Hydrogenation

The purpose of hydrogenating petroleum constituents is (1) to improve existing petroleum products or develop new products or even new uses, (2) to convert inferior or low-grade materials into valuable products, and (3) to transform higher molecular weight constituents into liquid fuels.

The distinguishing feature of the hydrogenating processes is that, although the composition of the feedstock is relatively unknown and a variety of reactions may occur simultaneously, the final product may actually meet all the required specifications for its particular use.

Hydrogenation processes for the conversionof petroleum and petroleum products may be classified as destructive and nondestructive. The former (hydrogenolysis or hydrocracking) is characterized by the rupture of carbon-carbon bonds and is accompanied by hydrogen saturation of the fragments to produce lower boiling products. Such treatment requires rather high temperatures and high hydrogen pressures, the latter to minimize coke formation. Many other reactions, such as isomerization, dehydrogenation, and cyclization, can occur under these conditions.

(1) Hydrocracking Hydrocracking is a thermal process (>350℃) in which hydrogenation accompanies cracking. Relatively high pressure (100psi to 2000psi) is employed, and the overall result is usually a change in the character or quality of the products. The wide range of products possible from hydrocracking is the result of combining catalytic cracking reactions with hydrogenation. The reactions are catalyzed by dual-function catalysts in which the cracking function is provided by silica-alumina (or zeolite) catalysts, and platinum, tungsten oxide, or nickel provides the hydrogenation function.

Essentially, all the initial reactions of catalytic cracking occur, but some of the secondary reactions are inhibited or stopped by the presence of hydrogen. For example, the yields of olefins and the secondary reactions that result from the presence of these materials are substantially diminished and branched-chain paraffins undergo demethanation. The methyl groups attached to secondary carbons are more easily removed than those attached to tertiary carbon atoms, whereas methyl groups attached to quaternary carbons are the most resistant to hydrocracking.

The effect of hydrogen on naphthenic hydrocarbons is mainly that of ring scission followed by immediate saturation ofeach end of the fragment produced. The ring is preferentially broken at favored positions, although generally all the carbon-carbon bond positions are attacked to some extent. For example, methyl-cyclopentane is converted (over a platinum-carbon catalyst) to 2-methylpentane, 3-methylpentane, and n-hexane.

Aromatic hydrocarbons are resistant to hydrogenation under mild conditions, but under more severe conditions, the main reactions are conversion of the aromatic to naphthenic rings and scissions within thealkyl side chains. The naphthenes may also be converted to paraffins. However, polynuclear aromatics are more readily attacked than the single-ring compounds, the reaction proceeding by a stepwise process in which one ring at a time is saturated and then opened. For example, naphthalene is hydrocracked over a molybdenum oxide-molecular catalyst to produce a variety of low weight paraffins ($\leqslant C_6$).

(2) Hydrotreating It is generally recognized that the higher the hydrogen content of a petrole-

um product, especially the fuel products, the better is the quality of the product. This knowledge has stimulated the use of hydrogen-adding processes in the refinery. Thus, hydrotreating (i.e., hydrogenation without simultaneous cracking) is used for saturating olefins or for converting aromatics to naphthenes as well as for heteroatom removal. Under atmospheric pressure, olefins can be hydrogenated up to about 500℃, but beyond this temperature dehydrogenation commences. Application of pressure and the presence of catalysts make it possible to effect complete hydrogenation at room or even cooler temperature; the same influences are helpful in minimizing dehydrogenation at higher temperatures.

A wide variety of metals are active hydrogenation catalysts; those of most interestare nickel, palladium, platinum, cobalt, iron, nickel-promoted copper, and copper chromite. Special preparations of the first three are active at room temperature and atmospheric pressure. The metallic catalysts are easily poisoned by sulfur-containing and arsenic-containing compounds, and even by other metals. To avoid such poisoning, less effective but more resistant metal oxides or sulfides are frequently employed, generally those of tungsten, cobalt, chromium, or molybdenum.

3. Isomerization

The importance of isomerization in petroleum-refining operations is twofold. First, the process is valuable in converting n-butane into *iso*-butane, which can be alkylated to liquid hydrocarbons in the gasoline boiling range. Second, the process can be used to increase the octane number of the paraffins, boiling in the gasoline boiling range, by converting some of the n-paraffins present into *iso*-paraffins.

The process involves contact of the hydrocarbon and a catalyst under conditions favorable to good product recovery. The catalyst may be aluminum chloride promoted with hydrochloric acid or a platinum-containing catalyst. Both are very reactive and can lead to undesirable side reactions along with isomerization. These side reactions include dis-proportionation and cracking, which decrease the yield and produce olefinic fragments that may combine with the catalyst and shorten its life. These undesired reactions are controlled by such techniques as the addition of inhibitors to the hydrocarbon feed or by carrying out the reaction in the presence of hydrogen.

Paraffins are readily isomerized at room temperature, and the reaction is believed to occur by the formation and rearrangement of carbonium ions. The chain-initiating ion R^+ is formed by the addition of a proton from the acid catalyst to an olefin molecule, which may be added, present as an impurity, or formed by dehydrogenation of the paraffin.

The isomerization of alkylaromatics may involve changes in the side-chain configuration, disproportionation of the substituent groups, or their migration about the nucleus. The conditions needed for isomerization within attached long side chains of alkylbenzenes and alkylnaphthalenes are also those for the scission of such groups from the ring. Such isomerization, therefore, does not take place unless the side chains are relatively short. The isomerization of ethylbenzene to xylenes, and the reverse reaction, occurs readily.

4. Alkylation

Alkylation in the petroleum industry refers to a process for the production of high-octane motor fuel components by the combination of olefins and paraffins. The reaction of *iso*-butane with olefins, using an aluminum chloride catalyst, is a typical alkylation reaction. In acid-catalyzed alkylation reactions, only paraffins with tertiary carbon atoms, such as *iso*-butane and iso-pentane react with the olefin. Ethylene is slower to react that the higher olefins. Olefins higher than propene may complicate the products by engaging in hydrogen exchange reactions. Cycloparaffins, especially those containing tertiary carbon atoms, are alkylated with olefins in a manner similar to the *iso*-paraffins; the reaction is not as clean, and the yields are low because of the several side reactions that take place.

Aromatic hydrocarbons are more easily alkylated than the iso-paraffins by olefins. Cumene (*iso*-propylbenzene) is prepared by alkylating benzene with propene over an acid catalyst. The alkylating agent is usually an olefin, although cyclopropane, alkyl halides, aliphatic alcohols, ethers, and esters may also be used. The alkylation of aromatic hydrocarbons is presumed to occur through the agency of the carbonium ion.

Thermal alkylation is also used in some plants, but like thermal cracking, it is presumed to involve the transient formation of neutralfree radicals and therefore tends to be less specific in production distribution.

From *Fundamentals of Petroleum Refining*, *First Ed*, *by M.A. Fahim*; *Chemistry and Technology of Petroleum*, *4th Edition*, , *by James G. Speight*, 2006

New Words and Expressions

N-methyl pyrrolidone *N*-甲基吡咯烷酮
raffinate [ˈræfɪneɪt] [化工] 萃余液;残油液;剩余液
methyl ethyl ketone 甲乙酮
decomposition [diːkɒmpəˈzɪʃn] 分解,腐烂;变质
dealkylation [diːælkiˈleiʃən] 脱烷基化作用
aluminosilicate [əljuːminəuˈsilikət] [无化] 铝硅酸盐;铝矽酸盐
carbonium [kɑːˈbəʊnɪəm] 正碳;[有化] 阳碳;正电离子
zeolite [ˈziːəlaɪt] 沸石
tricoordinated aluminum 三配位铝
hydrogenolysis [haidrəudʒiˈnɔlisis] 氢解作用
cyclization [saikləˈzeiʃən] [化学] 环化;环合作用
disproportionation [ˈdisprə,pɔːʃəˈneiʃən] 歧化作用;不相均
cyclopropane [saɪklə(ʊ)ˈprəʊpeɪn] [有化] 环丙烷

Notes

(1) Vacuum distillation,减压蒸馏,也称真空蒸馏,属于最基本的石油炼制过程,是在真空条件下用蒸馏的方法将原油分离成不同沸点范围油品(称为馏分)的过程。原油中重馏分沸点约370~535℃,在常压下要蒸馏出这些馏分,需要加热到420℃以上,而在此温度下,重馏分会发生一定程度的裂化。因此,通常在常压蒸馏后再进行减压蒸馏。在约2~8kPa的绝对

压力下，使在不发生明显裂化反应的温度下蒸馏出重组分。

(2) **Catalytic cracking**，催化裂化，是石油炼制过程之一，是在热和催化剂的作用下使重质油发生裂化反应，转变为裂化气、汽油和柴油等的过程。原料采用原油蒸馏（或其他石油炼制过程）所得的重质馏分油；或重质馏分油中混入少量渣油，经溶剂脱沥青后的脱沥青渣油；或全部用常压渣油或减压渣油。催化裂化是石油炼厂从重质油生产汽油的主要过程之一。所产汽油辛烷值高（马达法 80 左右），安定性好，裂化气（一种炼厂气）含丙烯、丁烯、异构烃多。

6.5 Petroleum Products

The customary processing of petroleum does not usually involve the separation and handling of pure hydrocarbons. Indeed, petroleum-derived products are always mixtures: occasionally simple but more often very complex. Thus, such materials as the gross fractions of petroleum (e.g., gasoline, naphtha, kerosene, and the like) that are usually obtained by distillation or refining are classed as petroleum products; asphalt and other solid products (e.g., wax) are also included in this division.

This type of classification separates this group of products from those obtained as petroleum chemicals (petrochemicals), for which the emphasis is on separation and purification of single chemical compounds, which are in fact starting materials for a host of other chemical products.

6.5.1 Gaseous fuels

The principal constituent of natural gas is methane (CH_4). Other constituents are paraffinic hydrocarbons such as ethane (CH_3CH_3), propane ($CH_3CH_2CH_3$), and the butanes [$CH_3CH_2CH_2CH_3$ and $(CH_3)_3CH$]. Many natural gases contain nitrogen (N_2) as well as carbon dioxide (CO_2) and hydrogen sulfide (H_2S). Trace quantities of argon, hydrogen, and helium may also be present. Generally, the hydrocarbons having a higher molecular weight than methane, carbon dioxide, and hydrogen sulfide are removed from natural gas before to its use as a fuel. Gases produced in a refinery contain methane, ethane, ethylene, propylene, hydrogen, carbon monoxide, carbon dioxide, and nitrogen, with low concentrations of water vapor, oxygen, and other gases.

As already noted, the compositions of natural, manufactured, and mixed gases can vary so widely, no single set of specifications could cover all situations. The requirements are usually based on performances in burners and equipment, on minimum heat content, and on maximum sulfur content. Gas utilities in most states come under the supervision of state commissions or regulatory bodies and the utilities must provide a gas that is acceptable to all types of consumers and that will give satisfactory performance in all kinds of consuming equipment. However, there are specifications for LPG that depend upon the required volatility.

As natural gas as delivered to pipelines has practically no odor, the addition of an odorant is required by most regulations so that the presence of the gas can be detected readily in case of accidents and leaks. This odorization is provided by the addition of trace amounts of some organic sulfur compounds to the gas before it reaches the consumer. The standard requirement is that a user will be able to detect the presence of the gas by odor when the concentration reaches 1% of gas in air. As the lower limit of flammability of natural gas is approximately 5%, this 1% requirement is essentially equivalent to one-fifth the lower limit of flammability. The combustion of these trace amounts of odorant does not create any serious problems of sulfur content or toxicity.

6.5.2 Gasoline

Gasoline is manufactured to meet specifications and regulations and not to achieve a specific distribution of hydrocarbons by class and size. However, chemical composition often defines properties. For example, volatility is defined by the individual hydrocarbon constituents and the lowest-boiling constituent(s) defines the volatile as determined by certain test methods.

Automotive gasoline typically contains about almost 200 (if not several hundred) hydrocarbon compounds. The relative concentrations of the compounds vary considerably depending on the source of crude oil, refinery process, and product specifications. Typical hydrocarbon chain lengths range from C_4 through C_{12} with a general hydrocarbon distribution consisting of alkanes (4% ~ 8%), alkenes (2% ~ 5%), iso-alkanes 25% ~ 40%, cycloalkanes (3% ~ 7%), cycloalkenes (1% ~ 4%), and aromatics (20% ~ 50%). However, these proportions vary greatly.

The majority of the members of the paraffin, olefin, and aromatic series (of which there are about 500) boiling below 200℃ have been found in the gasoline fraction of petroleum. However, it appears that the distribution of the individual members of straight-run gasoline (i.e., distilled from petroleum without thermal alteration) is not even.

The reduction of the lead content of gasoline and the introduction of reformulated gasoline has-been very successful in reducing automobile emissions. Further improvements in fuel quality have been proposed for the years 2000 and beyond. These projections are accompanied by a noticeable and measurable decrease in crude oil quality and the reformulated gasoline will help meet environmental regulations for emissions for liquid fuels.

Despite the diversity of the processes within a modern petroleum refinery, no single stream meets all the requirements of gasoline. Thus, the final step in gasoline manufacture is blending the various streams into a finished product. It is not uncommon for the finished gasoline to be made up of six or more streams and several factors make this flexibility critical: (1) the requirements of the gasoline specification and the regulatory requirements and (2) performance specifications that are subject to local climatic conditions and regulations.

The early criterion for gasoline quality was Baume (or API) gravity. For example, a 70°API gravity gasoline contained fewer, if any, of the heavier gasoline constituents than a 60°API gasoline. Therefore, the 70°API gasoline was of a higher quality and, hence, economically more valuable gasoline. However, apart from being used as a rough estimation of quality (not only for petroleum products but also for crude petroleum), specific gravity is no longer of any significance as a true indicator of gasoline quality.

6.5.3 Solvents (naphtha)

Naphtha is dividedinto two main types, aliphatic and aromatic. The two types differ in two ways:

first, in the kind of hydrocarbons making up the solvent, and second, in the methods used for their manufacture. Aliphatic solvents are composed of paraffinic hydrocarbons and cycloparaffins (naphthenes), and may be obtained directly from crude petroleum by distillation. The second type of naphtha contains aromatics, usually alkyl-substituted benzene, and is very rarely, if at all, obtained from petroleum as straight-run materials.

Stoddard solvent is a petroleum distillate widely used as a dry cleaning solvent and as a general cleaner and degreaser. It may also be used as a paint thinner, as a solvent in some types of photocopier toners, in some types of printing inks, and in some adhesives. Stoddard solvent is considered to be a form of mineral spirits, white spirits, and naphtha but not all forms of mineral spirits, white spirits, and naphtha are considered to be Stoddard solvent.

Stoddard solvent consists of linear alkanes (30% ~ 50%), branched alkanes (20% ~ 40%), cycloalkanes (30% ~ 40%), and aromatic hydrocarbons (10% ~ 20%). The typical hydrocarbon chain ranges from C_7 through C_{12} in length.

The main uses of naphtha fall into the general areas of (1) solvents (diluents) for paints, for example; (2) dry-cleaning solvents; (3) solvents for cutback asphalt; (4) solvents inthe rubber industry; and (5) solvents for industrial extraction processes.

Turpentine, the older more conventional solvent for paints, has now been almost completely replaced with the discovery that the cheaper and more abundant petroleum naphtha is equally satisfactory. The differences in application are slight: naphtha causes a slightly greater decrease in viscosity when added to some paints than does turpentine, and depending on the boiling range, may also show difference in evaporation rate.

The boiling ranges of fractions that evaporate at rates permitting the deposition of good films have been fairly well established. Depending on conditions, products are employed as light as those boiling from 38℃ to 150℃ and as heavy as those boiling between 150℃ and 230℃. The latter are used mainly in the manufacture of backed and forced-drying products.

6.5.4 Kerosene

(1) Composition

Chemically, kerosene is a mixture of hydrocarbons; the chemical composition depends on its source, but it usually consists of about 10 different hydrocarbons, each containing from 10 to 16 carbon atoms per molecule; the constituents include n-dodecane (n-$C_{12}H_{26}$), alkyl benzenes, and naphthalene and its derivatives. Kerosene is less volatile than gasoline; it boils between about 140℃ and 320℃.

Kerosene, because of its use as a burning oil, must be free of aromatic and unsaturated hydrocarbons, as well as free of the more obnoxious sulfur compounds. The desirable constituents of kerosene are saturated hydrocarbons, and it is for this reason that kerosene is manufactured as a straight-run fraction, not by a cracking process.

Although the kerosene constituents are predominantly saturated materials, there is evidence for the presence of substituted tetrahydronaphthalene. Dicycloparaffins also occur in substantial amounts in kerosene. Other hydrocarbons with both aromatic and cycloparaffin rings in the same molecule, such as substituted indan, also occur in kerosene. The predominant structure of the dinuclear aromatics appears to be that in which the aromatic rings are condensed, such as naphthalene whereas the isolated two-ring compounds, such as biphenyl, are only present in traces, if at all.

(2) Properties and uses

Kerosene is by nature a fraction distilled from petroleum that has been used as a fuel oil from the beginning of the petroleum-refining industry. As such, low proportions of aromatic and unsaturated hydrocarbons are desirable to maintain the lowest possible level of smoke during burning. Although some aromatics may occur within the boiling range assigned to kerosene, excessive amounts can be removed by extraction; that kerosene is not usually prepared from cracked products almost certainly excludes the presence of unsaturated hydrocarbons.

The essential properties of kerosene are flash point, fire point, distillation range, burning, sulfur content, color, and cloud point. In the case of the flash point, the minimum flash temperature is generally placed above the prevailing ambient temperature; the fire point determines the fire hazard associated with its handling and use.

The boiling range is of lesser importance for kerosene than for gasoline, but it can be taken as an indication of the viscosity of the product, for which there is no requirement for kerosene. The ability of kerosene to burn steadily and cleanly over an extended period is an important property and gives some indication of the purity or composition of the product.

6.5.5 Fuel oil

Fuel oil is classified in several ways but generally may be divided into two main types: distillate fuel oil and residual fuel oil. Distillate fuel oil is vaporized and condensed during a distillation process and thus has a definite boiling range and does not contain high-boiling constituents. A fuel oil that contains any amount of the residue from crude distillation of thermal cracking is a residual fuel oil. The terms distillate fuel oil and residual fuel oil are losing their significance, as fuel oil is now made for specific uses and may be either distillates or residuals or mixtures of the two. The terms domestic fuel oil, diesel fuel oil, and heavy fuel oil are more indicative of the uses of fuel oils.

Domestic fuel oil is fuel oil that is used primarily in the home. This category of fuel oil includes kerosene, stove oil, and furnace fuel oil; they are distillate fuel oils.

Diesel fuel oil is also a distillate fuel oil that distills between 180℃ and 380℃. Several grades are available depending on their uses: diesel oil for diesel compression ignition (cars, trucks, and marine engines) and light heating oil for industrial and commercial uses.

Heavy fuel oil comprises all residual fuel oils (including those obtained by blending). Heavy fuel oil constituents range from distillable constituents to residual (non-distillable) constituents that must beheated to 260℃ or more before they can be used. The kinematic viscosity is above 10 cSt at 80℃. The flash point is always above 50℃ and the density is always higher than 0.900. In general, heavy fuel oil usually contains cracked residua, reduced crude, or cracking coil heavy product, which is mixed (cut back) to a specified viscosity with cracked gas oils and fractionator bottoms. For some industrial purposes in which flames or flue gases contact the product (ceramics, glass, heat treating, and open hearth furnaces) fuel oils must be blended to contain minimum sulfur contents, and hence low-sulfur residues are preferable for these fuels.

6.5.6 Lubricating oil

(1) Composition

Lubricating oil is distinguished from other fractions of crude oil by their usuallyhigh (>400℃) boiling point, as well as their high viscosity. Materials suitable for the production of lubricating oils are comprised principally of hydrocarbons containing from 25 to 35 or even 40 carbon atoms per molecule, whereas residual stocks may contain hydrocarbons with 50 or more (up to 80 or so) carbon atoms per molecule. The composition of lubricating oil may be substantially different from the lubricant fraction from which it was derived, as wax (normal paraffins) is removed by distillation or refining by solvent extraction and adsorption preferentially removes nonhydrocarbon constituents as well as polynuclear aromatic compounds and the multi-ring cycloparaffins.

Mono-, di-, and tri-nuclear aromatic compounds appear to be the main constituents of the aromatic portion, but material with more aromatic nuclei per molecule may also be present. For the dinuclear aromatics, most of the material consists of naphthalene types. For the trinuclear aromatics, the phenanthrene type of structure predominates over the anthracene type. There are also indications that the greater part of the aromatic compounds occurs as mixed aromatic-cycloparaffin compounds.

(2) Properties and uses

Lubricating oil may be divided into many categories according to the types of service they are intended to perform. However, there are two main groups: (1) oils used in intermittent service, such as motor and aviation oils, and (2) oils designed for continuous service, such as turbine oils. Lubricating oil is distinguished from other fractions of crude oil by a high (>400℃) boiling point, as well as a high viscosity and, in fact, lubricating oil is identified by viscosity.

Oils used in intermittent service must show the least possible change in viscosity with temperature; that is, their viscosity indices must be high. These oils must be changed at frequent intervals to remove the foreign matter collected during service. The stability of such oils is therefore of less importance than the stability of oils used in continuous service for prolonged periods without renewal.

Oils used in continuous service must be extremely stable, but their viscosity indices may be low because the engines operate at fairly constant temperature without frequent shutdown.

6.5.7 Grease

Grease is lubricating oil to which a thickening agent has been added for the purpose of holding the oil to surfaces that must be lubricated. The most widely used thickening agents are soaps of various kinds, and grease manufacture is essentially the mixing of soaps with lubricating oils. Until a relatively short time ago, grease making was considered an art. To stir hot soap into hot oil is a simple business, but to do so in such a manner as to form a grease is much more difficult, and the early grease maker needed much experience to learn the essentials of the trade. Therefore, it is not surprising that grease making is still a complex operation. The signs that told the grease maker that the soap was cooked and that the batch of grease was ready to run have been replaced by scientific tests that follow the process of manufacture precisely.

Oil and soap are mixed in kettles that have double walls between which steam and water may be circulated to maintain the desired temperature. When temperatures higher than 150℃ are required, a kettle heated by a ring of gas burners is used. Mixing is usually accomplished in each kettle by horizontal paddles radiating from a central shaft.

The soaps used in grease making are usually made in the grease plant, usually in a grease making kettle. Soap ismade by chemically combining a metal hydroxide with a fat or fatty acid:

$$R—COOH+NaOH \longrightarrow R—COO^-Na^++H_2O$$

The most common metal hydroxides used for this purpose are calcium hydroxide, lye, lithium hydroxide, and barium hydroxide. Fats are chemical combinationsof fatty acids and glycerin.

The soaps may be combined with any lubricating oil from a light distillate to a heavy residual oil. The lubricating value of the grease is chiefly dependent on the quality and viscosity of the oil. In addition to soap and oil, greases may also contain various additives that are used to improve the ability of the grease to stand up under extreme bearing pressures, to act as a rust preventive, and to reduce the tendency of oil to seep or bleed from a grease.

6.5.8 Wax

Paraffin wax is a solid crystalline mixture of straight-chain (normal) hydrocarbons ranging from C_{20} to C_{30} and possibly higher, that is, $CH_3(CH_2)_nCH_3$ where $n \geqslant 18$.

It is distinguished by its solid state at ordinary temperatures (25℃) and low viscosity when melted. However, in contrast to petroleum wax, petrolatum (petroleum jelly), although solid at ordinary temperatures, does in fact contain both solid and liquid hydrocarbons. It is essentially a low-melting, ductile, microcrystalline wax.

The melting point of paraffin wax has both direct and indirect significance in most wax utilization. All wax grades are commercially indicated in a range of melting temperatures rather than at a single value, and a range of 1℃ usually indicates a good degree of refinement. Other common physical properties that help to illustrate the degree of refinement of the wax are color, oil content, API gravity, flash point, and viscosity, although the last three properties are not usually given by the producer unless specifically requested.

6.5.9 Coke

Coke is the residue left by the destructive distillation of petroleum residua. That formed in catalytic cracking operations is usually nonrecoverable, as it is often employed as fuel for the process.

The composition of petroleum coke varies with thesource of the crude oil, but in general, large amounts of high-molecular-weight complex hydrocarbons (rich in carbon but correspondingly poor in hydrogen) make up a high proportion. The solubility of petroleum coke in carbon disulfide has been reported to be as high as 50% to 80%, but this is in fact a misnomer, as the coke is the insoluble, honeycomb material that is the end product of thermal processes.

Petroleum coke is employed for a number of purposes, but its chief use is in the manufacture of carbonelectrodes for aluminum refining, which requires a high-purity carbon low in ash and sulfur free; the volatile matter must be removed by calcining. In addition to its use as a metallurgical reducing agent, petroleum coke is employed in the manufacture of carbon brushes, silicon carbide abrasives, and structural carbon (e.g., pipes and rashig rings), as well as calcium carbide manufacture from which acetylene is produced:

$$\text{Coke} \longrightarrow \text{CaC}_2$$

$$\text{CaC}_2 + \text{H}_2\text{O} \longrightarrow \text{HC}\equiv\text{CH}$$

From *Chemistry and Technology of Petroleum*, *4th Edition*, , *by James G. Speight.*

New Words and Expressions

odorization [əʊdəraɪˈzeɪʃn] 加臭
flammability [flæməˈbɪlətɪ] 可燃性,易燃性
gasoline [ˈgæsəliːn] 汽油
criterion [kraɪˈtɪərɪən] (批评判断的)标准;准则;规范;准据
n-heptane [ˈhepteɪn] 正庚烷
iso-octane 标准异辛烷;等辛烷值
cutback [ˈkʌtbæk] 减少,削减;情节倒叙
turpentine [ˈtɜːp(ə)ntaɪn] 松节油;松脂
kerosene [ˈkerəsiːn] 煤油,火油
tetrahydronaphthalene [ˈtetrə,haidrəuˈnæfθəliːn] 四氢化萘
dinuclear [daiˈnjukliə] 双核的,两环的
indan [ˈin,dæn] [有化] 茚满
anthracene [ˈænθrəsiːn] 蒽(一种炭氢化合物)
calcine [ˈkælsaɪn] [化工] 烧成石灰;锻烧;煅烧

Notes

(1) **Octane number**,辛烷值。是交通工具所使用的燃料抵抗震爆的指标。辛烷值越高表示抗震爆的能力越好。汽油中的辛烷值直接取决定汽油内各种碳氢化合物的成分比例。辛烷值一般是区分不同等级汽油的关键标准,车用汽油的牌号是按照辛烷值区分的。共有66、70、76、80、85等号。例如,70号车用汽油即表明该汽油辛烷值不低于70。根据辛烷值的实测结果可判定属哪一牌号的车用汽油;各地区汽油牌号的标示方法并不统一,比如台湾的95无铅汽油,表示其辛烷值为95,中国大陆的97号汽油,表示其辛烷值为97。提高汽油辛烷值的方法包括加入四乙基铅及甲基叔丁基醚、碳酸二甲酯、MMT(甲基环戊二烯三羰基锰)等。由于铅是有毒物质,因此许多国家已经不再提供含铅汽油,仅航空用油仍含有铅。

(2) **Flash point**,闪点。是指可燃性液体挥发出的蒸汽在与空气混合形成可燃性混合物并达到一定浓度之后,遇火源时能够闪烁起火的最低温度。在这温度下燃烧无法持续,但如果温度继续攀升则可能引发大火。和着火点(Fire Point)不同的是,着火点是指可燃性混合物能够持续燃烧的最低温度,高于闪点。闪点的高低也是染液是否安全的重要指标。

Unit 7 Modern Organic Synthesis in the Laboratory

7.1 Laboratory Safety Guidelines

The organic chemistry laboratory is potentially one of the most dangerous of undergraduate laboratories.That is why you must have a set of safety guidelines.It is a very good idea to pay close attention to these rules,for one very good reason.

7.1.1 Laboratory apparel

(1) Safety goggles are required in the laboratory at all times! Splash hazards are perhaps the most significant danger present in the lab,and eyes are extremely sensitive.

(2) Contact lenses are not permitted in the lab.Your goggles will protect your eyes from spill hazards,but do nothing to protect you from fumes,which can dry your contacts out and may result in the necessity of an operation for their removal.Contact lenses can also absorb chemicals from the air (especially the new "breathable" lenses),concentrate and hold them against the eye,and prevent proper flushing of the eye should a chemical be splashed into the eyes.

(3) Laboratory aprons must be donned at all times.In the event of a spill,these aprons are chemical and flame resistant,and could save you from scar tissue!

(4) Saddals,open-toed shoes and high heels are not permitted in the lab.This is to protect your feet from splashes and spills.The restriction on high heels is for balance.

(5) Shorts or skirts cut above the knee are not permitted in the lab.Again,should a spill occur, it will be your clothing that will be your protection from direct exposure of the skin to that chemical. The idea is to put as many layers of clothing as possible between you and a chemical spill.The more clothing,the more diffuse the chemical will be by the time it reaches the skin,and the greater the chance to remove the chemical before it reaches your skin.

(6) Careful consideration should be given before wearing any jewelry into the lab.Some chemicals evaporate very quickly and therefore pose relatively little danger should they get onto your skin. Hower,if they get beneath a ring,watch or some other form of jewelry,they can be prevented from evaporating,held against the skin longer and greatly increase the risk of injury.Should you decide to wear jewelry to the lab (as I will be wearing my watch),be particularly mindful of itching,burning or any other irritation under or around you jewelry.(By the way,never wear opals,pearls,or other "soft gems" in the lab.The harsh laboratory environment may dry them out or otherwise damage them,and neither your instructor nor DSU will replace or repair such items)

(7) Never wear clothes that hang, such as loose sleeves. Be sure ties and scarves are tucked well inside your laboratory apron. These pose fire hazards (if you are reaching or bending down near an open flame) as well as chemical hazards (if they accidentally get dragged through a chemical, they can transport that chemical directly to your skin). In fact, you may want to give very serious consideration to wearing only old clothes. Some of my students have, in the past, brought old clothes with them in a gym bag and changed right before and after lab. Be especially careful of sleeves around open flames.

(8) Long hair is to be constrained. Like hanging clothes, long hair is subject to fire and contact with chemicals. A rubber band will be used to constrain particularly long hair if necessary.

(9) No radios, tape players, CD players or any other devices of this type will be permitted in the laboratory at any time. Loud music is distracting, and headphones prevent you from hearing announcements or verbal warning given in the lab.

7.1.2 Safety equipment

(1) Take the time to identify all of the laboratory safety equipment, and keep their location in your mind at all times. You should be able to close your eyes any time during a lab and point to such safety equipment as the fire extinguisher, the emergency eyewash stations, the fire blankets, the safety shower, etc. If you were to spash a chemical in your eyes, you had better be albe to find that eyewash station without your eyes well before permanent damage can occur (which can be seconds depending on the nature of the chemical).

(2) Check all safety equipment. I will keep as close an eye on it as possible, but I need your help as well. Is the fire extinguisher charged? Does it have the plastic "seal"? Is there enough sodium bicarbonate in case there is a chemical spill? If anything does not look right to you, report it to your lab instructor immediately!

(3) Material Data Safety Sheets are available to you on request only. Basic safety information will be given during the safety lecture before each lab.

7.1.3 General behavior

(1) Absolutely no horseplay will be tolerated in the laboratory! Offenders of this one will be unceremoniously cast out with a zero resulting for that day' work. I realize that at times it is awfully tempting to grab that water bottle and squirt your friends, but many hazardous chemicals look like water. The humor will be lost if something other than water is in that bottle.

(2) Always read the upcoming experiments carefully and thoroughly, being sure to understand all of the directions before entering the lab. This will help you to be prepared to handle any hazards of the experiment, and will also help you to perform the experiment more quickly resulting in less "fumbling around" and reckless work as you rush to finish on time. To ensure that you have read the upcoming experiment, you are required to complete the pre-lab assignment before entering the lab. If your fail to complete the pre-lab assignment on time, you will not be allowed to perform the experiment.

(3) Be in the lab and ready promptly when the lab begins. The safety lecture (specific to that day's experiment) will be the first item of business each day. If you are not present to get this important information, you will not be allowed to do the experiment.

(4) Absolutely no food or beverages will be permitted inside the lab. They can absorb chemicals from the air (and concentrate them), or can pick them up from the bench, causing ingestion of these chemicals. Everything possible will be done to be sure the laboratory air is safe for working in without the use of special respiratory equipment. Please do not complicate the issue by eating these chemicals as well!

(5) Wash your hands! Wash your hands frequently during lab, and definitely wash you hands twice at the end of the lab, once in the lab itself, and again outside of lab (as in a public rest room), especially before eating. Once you get home, you should wash your face as well. You do not want to drag too many chemicals around with you on your skin.

(6) Do not apply makeup (including Chapstick and other lip balms) in the lab. In fact, you may want to seriously consider not wearing makeup to the lab at all. Makeup can also pick up and concentrate fumes from the air, and hold them against the skin causing irritation. Perfumes, colognes or other fragrances may also interfere with the olfactory senses when an experiment calls for "smell" something.

(7) Should an injury occur, regardless of how minor it is, report it immediately to the lab supervisor. The smallest puncture wound allows for chemicals to enter the blood stream directly. By notifying your supervisor, even if no action is taken, the incident will be reported to the student health center. In the event that this wound should become infected later, having this information on file may prove to be of extreme importance for prompt treatment.

(8) Never pick up broken glassware with your bare hands, regardless of the size of the pieces. Typically, puncture wounds occur with the largest pieces in such a situation, because they look to be the most harmless. A brush and dust pan is provided for broken glassware. Please place all broken glassware in the appropriate broken glassware container, and never put caps, paper or other waste in this same container. Very small bits of broken glassware (as in the bottom of a drawer) can be picked up with a damp paper towel.

(9) Never put broken a glass in a regular garbage can. A container is provided that is especially designed for broken glassware.

(10) Always read the labels to reagents (chemicals used in an experiment) twice! Many chemicals look identical on first glance, and may differ only slightly in their spelling or concentration. Sodium sulfate may look similar to sodium sulfite, but they are most certainly different and confusing them in the lab may result in dire consequences. Therefore, read the label as you grab the bottle, and holding it in your hand, look carefully at the label a second time and verify that it is exactly what you want.

(11) Never make unauthorized substitutions. If you are wondering what would happen if you used this instead of that, ask me. If it is safe, I may let you try it. If not, I will let you know what would have happened if you tried it.

(12) Never use reagents from an unmarked bottle. All reagent bottles will have proper labels, so

if a reagent bottles is unlabeled, it is the incorrect reagent.

(13) In any emergency, the fastest way to get the lab supervisors attention is to scream.

(14) If you are not feeling well, report it to the laboratory supervisor immediately. If your supervisor should lose consciousness during a lab period, it may be due to chemical fumes. Evacuate the lab immediately and seek another professor for help. Should anybody else lose consciousness in the lab, the lab supervisor will determine whether or not evacuation of the lab is warranted (it probably will not be)

(15) Avoid bringing excess coats, books, backpacks or other personal items to the laboratory. There is always the danger of spilling chemicals on them, and they create a fire hazard if left in the isles. In the general chemistry lab, you may use the small cabinets underneath each drawer to store personal items during an experiment.

(16) Close your lab drawer! Once you have retrieved the equipment you need from your equipment drawer, be sure to close it again. Open drawers can pose tripping hazards (especially bottom drawers) and obstruction of walkways, Thump! OUCH!! The reason we do not have stools in the lab is to avoid similar obstruction.

(17) Never smell a chemical straight out of a container. Some chemicals are extremely caustic (fumes severely irritate delicate tissue) and the fumes should be avoided. To safely smell a chemical, hold it two to three feet from your nose, and withyour other hand cupped, waft the fumes towards you. You may slowly move the chemical closer to your nose if you cannot smell it all the while taking only small sniffs.

7.1.4 Laboratory equipment

(1) Never heat a piece of glassware (beaker, flasks, etc) that is chipped or cracked unless otherwise told to do so by your lab supervisor. Heating defective glassware can cause that glassware to break (or explode!), resulting in a spill.

(2) If you have chipped or cracked glassware, or glassware with sharpor jagged edges, inform your lab supervisor immediately. The equipment will probably be replaced, or you may simply be given special instructions on using that bit of equipment.

7.1.5 General guidelines

(1) Epilepsy, pregnance, dyslexia as well as other medical conditions can be hazardous in the laboratory. Every effort will be made to keep you safe, but I will need some help. If you have any medical condition which you think may adversely affect your ability to safely perform in the laboratory, or that makes you particularly at risk to be in the laboratory, please inform me as soon as possible! Many such conditions may be deeded personal, but the chemicals themselves cannot tell the difference. Therefore, please feel free to stop in my office as soon as possible sot you can to tell me in private, and, of course, anything you do tell me will be kept in the strictest confidence.

(2) To turn on a Bunsen burner, first turn the nozzle on the bottom of the burner all the way off, then turn it back on about 2 turns. With a lit match in one hand, slowly turn on the gas at the

spigot.Hold the match near the edge of the burner as you do so the air being pushed out by the propane does not blow it out.Such a procedure will avoid "explosions" when lighting the burner.

(3) Before using a burner,be sure nobody else on the bench has any organic solvents.Organic solvents are flammable, and heavier than air, meaning that as they evaporate, they creep down the edge of their container to the bench top, whereupon they spread out horizontally. Once these fumes reach an open flame, they can ignite causing "flashback", thereby causing the beaker of solvent to catch fire four feet or more away!

(4) Before getting any organic solvent, be sure nobody on your entire lab bench has an open flame.

(5) Never take more of a reagent than you need.This means that if you need about 5 mL of a solvent, use your 10 mL beaker to get it, Not your 600 mL beaker.

(6) Never return an unused portion of a reagent to its originalcontainer.See if anybody else at your bench, or in the lab, need it. If not, give it to your instructor, who will look at you in a forlorn and sullen manner but will appreciate that you did not put it back in the original container.Returning unused portions of reagent greatly increase the odds of cross contamination, that is, getting the reagent contaminated with an unwanted chemical.

(7) Never pour a waste chemical in the drain, or put it in the garbage, unless otherwise instructed to do so by your lab supervisor. Waste bottles will be provided. Always pour waste into the appropriate and labeled waste bottle (reading the waste bottle label twice).

(8) If you have glass stirring rods or glass tubes with sharp or jagged edges, fire polish them. This means holding the sharp end in a Bunsen burner flame and rotating the rod or tube until a another minute or so, effectively melting that end a little bit. Be sure to let it cool completely before attempting to fire polish the other end.

(9) Many items (glass, metal, etc) look exactly the same hot they do cool. Be very careful whenever working with flames that all of your equipment (beakers, flasks, ring stands, etc.) are cool before handling them.

(10) If you are inserting glass tubing into a rubber stopper, use the following technique to avoid jamming a jagged piece of glass through your hand; ① use glycerol or water to lubricate either the end of the glass tubing being inserted, the hole in the stopper the tubing will be inserted into, or both; ② protect your hands by using a paper towel to hold both the glass tubing as well as the rubber stopper; ③ hold the rubber stopper in such a way that the tubing cannot go through the hole and into your palm (your fingers should actually curve, holding the edge of the stopper, as if to make the letter "C"); ④ hold the glass tubing, also with your palm away from the end, near the end being inserted into the rubber stopper; ⑤ insert the glass tubing with a twisting motion; ⑥ clean up any excess glycerol; and ⑦ live your life free scar tissue on your palms that everybody for the rest of your life will ask about by saying "how did that happen?", to which you will have to reply that did not listen to your dedicated and caring chemistry professor.

(11) Improper heating of a test tube can result in the chemicals within the test tube shooting out, possibly resulting in injury to anybody in the path. When heating a test tube, use the following procedure; ① unless directed otherwise, always place a few (five or six) boiling chips in the test tube; ② use a test tube clamp to hold the test tube; ③ hold the test tube at about a 45° angle; ④ be

sure the opening of test tube is pointing away from anybody else (preferably towards a wall in a low-traffic area of the lab); ⑤ Never heat the bottom of the test tube (unless otherwise directed); instead heat the middle of the test tube just at the level of the liquid in the test tube; ⑥ move the test tube horizontally back and forth across the flame to prevent the liquid from heating too quickly; ⑦ should the liquid begin to overheat (heat too rapidly), remove the test tube from the flame and allow the contents to cool for a minute or so.

(12) Never look down the opening of any container, including beakers, flasks, and test tubes. Should something happen to cause the chemicals to "blast out" of the container, they will go directly into your face if you are looking down the opening at the time.

(13) Do not use graduated cylinders for any purpose other than to measure a volume of as liquid. Graduated cylinders should vat he used to get reagent for an experiment (use a beaker for this) or to run reactions (use a test tube for this).

From *Organic Chemistry Laboratory Manual* by *Richard E. Bleil*

New Words and Expressions

laboratory apparel 实验室服装
Safety goggle 安全眼罩;护目镜
Contact lense 隐形眼镜
scar tissue 疤痕组织
fire extinguisher 灭火器
horseplay[ˈhɔːspleɪ] 动手脚和大声欢笑的玩闹;恶作剧
respiratory equipment 呼吸器
bunsen burner 本生灯,煤气喷灯
lip balm 润唇膏
olfactory [ɒlˈfækt(ə)rɪ] 嗅觉的;味道的;嗅觉器官
epilepsy[ˈepɪlepsɪ] 癫痫;羊角风
boiling chip 沸腾石;防止突沸石
graduated cylinder 量筒;刻度量筒

7.2 Purification

When a product has been isolated from a reaction the next step is to purify it. The degree of purity required will depend on the use for which the sample is intended, a synthetic intermediate might only require rough purification, whereas a product for elemental analysis would require rigorous purification. This section describes the most important purification techniques, crystallization, distillation, sublimation, and chromatography. It is assumed that the reader is familiar with the basic principles of these methods, so the emphasis is on more demanding applications such as the purification of air-sensitive materials, and purifications on a micro-scale.

7.2.1 Crystallization

Simple crystallization of an impure solid is a routine operation, which nevertheless requires care and good judgement if good results are to be obtained. The basic procedure can be broken down intofive steps, which are listed below, together with somc tips on how to overcome common problems.

1. Select a suitable solvent

Find a suitable solvent by carrying out small scale tests. Remember that 'like dissolves like'. The most commonly used solvents in order of increasing polarity are petroleum ether, toluene, chloroform, acetone, ethyl acetate, ethanol, and water. Chloroform and dichloromethane are rarely useful on their own because they are good solvents for the great majority of organic compounds. It is preferable to use a solvent with a boiling point in excess of 60℃, but the b.p. should be at least 10℃ lower than the m.p. of the compound to be crystallized, in order to prevent the solute from 'oiling out' of solution. In many cases a mixed solvent must be used, and combinations of toluene, chloroform, or ethyl acetate, with the petroleum ether fraction of similar boiling point are particularly useful. Consult Appendix 1 for boiling points, polarity (dielectric constant), and toxicity of common solvents.

2. Dissolve the compound in the minimum volume of hot solvent

Remember that most organic solvents are extremelyflammable and that many produce very toxic vapour.

Place the crude compound (always keep a few 'seed' crystals) in a conical flask fitted with a reflux condenser, add boiling chips and a small portion of solvent, and heat in a water bath. Continue to add portions of solvent at intervals until all of the crude has dissolved in the hot/refluxing solvent. If you are using a mixed solvent, dissolve the crude in a small volume of the good solvent, heat to reflux, add the poor solvent in portions until the compound just begins to precipitate (cloudiness), add a few drops of the good solvent to re-dissolve the compound, and allow to cool. When adding the solvent it is very easy to be misled into adding far too much if the crude is contaminated with an insoluble material such as silica or magnesium sulphate.

3. Filter the hot solution to remove insoluble impurities

This step is often problematic and should NOT be carried out unless an unacceptable (use your judgement) amount of insoluble material is suspended in the solution. The difficulty here is that the compound tends to crystallize during the filtration so an excess of solvent (ca. 5%) should be added, and the apparatus used for the filtration should be preheated to about the boiling point of the solvent. Use a clean sintered funnel of porosity 2 or 3, or a Hirsch or Büchner funnel, and use the minimum suction needed to draw the solution rapidly through the funnel. If the solution is very dark and/or contains small amounts of tarry impurities, allow it to cool for a few moments, add ca. 2% by weight of decolorising charcoal, reflux for a few minutes, and filter off the charcoal. Charcoal is very finely divided so it is essential to put a 1 cm layer of a filter aid such as Celite on the funnel before filtering the suspension. Observe the usual precautions for preventing crystallization in the funnel. Very dark or tarry products should be chromatographed through a short (2~3 cm) plug of silica before attempted crystallization.

4. Allow the solution to cool and the crystals to form

This is usually straightforward except when the material is very impure or has a low m. p. (<40℃) in which case it sometimes precipitates as an oil. If an oil forms it is best to reheat the solution and then to allow it to cool slowly. Try scratching the flask with a glass rod or adding a few 'seed' crystals to induce crystallization, and if this fails try adding some more solvent so that precipitation occurs at a lower temperature. If nothing at all precipitates from the solution, try scratching with a glass rod, seeding, or cooling the solution in ice-water. If all these fail, stopper the flask and set it aside for a few days, patience is sometimes the best policy.

5. Filter off and dry the crystals

When crystallization appears to be complete filter off the crystals using an appropriately sized sintered glass funnel. It is very important to wash the crystals carefully. As soon as all of the mother liquor has drained through the funnel, remove the suction and pour some cold solvent over the crystals, stirring them if necessary, in order to ensure that they are thoroughly washed. Drain off the washing under suction and repeat once or twice more. After careful washing allow the crystals to dry briefly in the air and then remove the last traces of solvent under vacuum, in a vacuum oven, in a drying pistol, or on a vacuum line. Take care to protect your crystals against accidental spillage or contamination. If they are placed in a dish, a beaker, or a sample vial, cover with aluminium foil, secure with wire or an elastic band, and punch a few small holes in the foil. If the crystals are in a flask connected directly to a vacuum line, use a tubing adapter with a tap and put a plug of glass wool in the upper neck of the adapter so that the crystals are not blown about or contaminated with rubbish from the tubing, when the air/inert gas is allowed in. If a relatively high boiling solvent such as toluene was used for the crystallization it is essential to heat the sample under vacuum for several hours to ensure that all of the solvent is removed.

7.2.2 Distillation

Distillation is the most useful method for purifying liquids, and is used routinely for purifying solvents and reagents. Using appropriate apparatus, and some care it can be possible to separate liquids whose boiling points are less than 5℃ apart. We will assume that the reader is familiar with the fundamentals of the theory and practice of distillation but it is appropriate to begin by reiterating some basic safety rules.

① Never heat a closed system.

② Remember that most organic liquids are extremely flammable so great care must be taken to ensure that the vapor does not come into contact with flames, sources of sparks (electrical motors), or very hot surfaces (hot plates).

③ Never allow a distillation pot to boil dry. The residues may ignite or explode with great violence.

④ Beware of the possibility that ethers and hydrocarbons may be contaminated with peroxides. Be particularly careful when distilling compounds prepared by peroxide and peracid oxidations and always take precautions to remove peroxide residues prior to distillation!

⑤ Carry out a safety audit on the compound you plan to distil to check that it is not thermally unstable.Some types of compounds, e.g.azides, should never be distilled.

Distillation under reduced pressure

Many compounds decompose when heated to their boiling points so theycannot be distilled at atmospheric pressure.In this situation it may be possible to avoid thermal decomposition by carrying out the distillation at reduced pressure.The reduction in the boiling point will depend on the reduction in pressure and it can be estimated from a pressure-temperature nomograph.

To find the approximate boiling point at any pressure simply place a ruler on the central line at the atmospheric boiling point of the compound, pivot it to line up with the appropriate pressure marking on the right-hand line, and read off the predicted boiling point from the left-hand line. You can also use the nomograph to find the b.p.at any pressure if you know the b.p.at some other pressure, by first using the known data to arrive at an estimate of the atmospheric boiling point. Note that although pressure is usually measured in millimetres of mercury (mmHg) it is often quoted in different units, especially Torr. Happily 1 Torr = 1mmHg. Boiling points measured at reduced pressure may be expressed in several ways, e.g.57℃/25mmHg or b.p.$_{25}$ 57℃. As a very rough guide a water pump (ca. 15mmHg) will give a 125℃ reduction in boiling point and an oil pump (ca.0.1mmHg) will give a reduction of 200~250℃.

A procedure for carrying out a distillation under reduced pressure is as follows:

① Place the sample in the distillation flask (no more than two thirds full) and add a stirring bar. Note: Anti-bumping granules are not effective at reduced pressure and so an alternative must be used. A very narrow capillary which allows a slow stream of air or nitrogen bubbles to pass through the solution is effective, but brisk stirring using a magnetic follower is much more convenient.

② Assemble the (oven-dried) apparatus, putting a little high vacuum grease on the outer edge of each joint. Ensure that the receiver adapter and the collection flasks are secured using clips, and connect the assembly to a vacuum pump. One convenient method of doing this is to connect it to a vacuum/inert gas double manifold. The pump must be protected with a cold finger trap and the line should incorporate a vacuum gauge for monitoring the pressure.

③ Stir the liquid rapidly and carefully open the apparatus to the vacuum. Some bumping and frothing may occur as air and volatile components are evacuated. If necessary adjust the pressure to the required value by allowing inert gas into the system via a needle valve.

④ Heat the flask slowly to drive off any volatile impurities and then to distil the product. Monitor the stillhead temperature and collect a forerun and a main fraction, which should distil at a fairly constant temperature. If fractionation is required you may have to collect several fractions and it is very important to distil the mixture slowly and steadily.

⑤ Stop the distillation when the level of liquid in the pot is running low, by removing the heating bath.

⑥ Isolate the apparatus from the vacuum and carefully fill with inert gas. If you are using a double manifold, this can be done by simply turning the vacuum/inert gas tap. The flask containing the distillate will be under a dry, inert atmosphere and should be quickly removed and fitted with a tightly fitting septum.

⑦ Switch off the pump and clean the cold trap.

7.2.3 Sublimation

Sublimation is an excellent method for purifying relatively volatile organic solids on scales ranging from a few milligrams to tens of grams. At reduced pressure many compounds, especially those of low polarity, have a sufficiently high vapour pressure that they can be sublimed, i.e. converted directly from the solid phase into the vapour phase without melting. Condensation of the vapour then gives purified solid product provided, as it is often the case, that the original impurities were much less volatile. The larger sublirnator consists of a tube with a side arm which is fitted with a cold-finger condenser, and it is used as follows:

① If the crude material is a solid, powder it and place it in the bottom of the outer vessel. If it is waxy or oily, wash it into the tube with a small amount of solvent, cover the side arm with a septum, and remove the solvent on a rotary evaporator.

② Put some vacuum grease on the joint of the cold-finger condenser and fit it into the sublimator (there should be a gap of approximately 1 cm between the solid and the condenser).

③ Evacuate the apparatus slowly to prevent any spattering of the solid. Turn on the condenser water and slowly heat the base of the sublimator.

④ A fine mist of sublimed material on the condenser indicates that sublimation is beginning and the temperature should then be held fairly constant until the process is complete.

⑤ When sublimation is complete product may be clinging precariously to the cold finger so proceed with great care. Turn off the water, carefully allow air/inert gas into the sublimator, and very carefully remove the cold-finger. Scrape off the product with a microspatula.

From *Advanced Practical Organic Chemistry* by *J. Beonard*

New Words and Expressions

dielectric constant 电容率;介电常数
conical flask 锥形烧瓶;锥形瓶
reflux condenser 回流冷凝器
magnesium sulphate 硫酸镁
sintered funnel 烧结玻璃漏斗
peroxide [pəˈrɒksaɪd] 过氧化物
peracid[ˈpɜːæsɪd] 过酸;高酸
azide [ˈeɪzaɪd] 叠氮化物
reduced pressure 减压
nomograph 计算图表;列线图解
vacuum gauge 真空计
still head 蒸馏头
flash chromatography 急骤层析;快速色谱

Notes

(1) **Crystallization**,结晶,是指物质从液态(溶液或溶融状态)或气态形成晶体的过程。从熔融体析出晶体的过程用于单晶制备,从气体析出晶体的过程用于真空镀膜,而化工生产中常遇到的是从溶液中析出晶体。根据液固平衡的特点,结晶操作不仅能够从溶液中取得固体溶质,而且能够实现溶质与杂质的分离,借以提高产品的纯度。

(2) **Distillation**,蒸馏,是一种热力学的分离工艺,它利用混合液体或液-固体系中各组分沸点不同,使低沸点组分蒸发,再冷凝以分离整个组分的单元操作过程,是蒸发和冷凝两种单元操作的联合。与其它的分离手段,如萃取、吸附等相比,它的优点在于不需使用系统组分以外的其它溶剂,从而保证不会引入新的杂质。

7.3 Characterization

7.3.1 Introduction

This Chapter deals with the type of physical data that are required for the proper characterization of the purified product. No theory is discussed as this is well covered in other sources and, given good data, it is often possible to find a colleague (for example) who will help out if you are unable to interpret a spectrum. With inadequate data it will be difficult to be certain of the structure and purity of your product, and it will certainly be more difficult to interest the colleague referred to above!

It is important to acquire as much information as possible on your product. It might be "obvious" from the NMR that the structure is what you think that it should be, but it is still necessary to record (at least) the IR and mass spectra. These might simply confirm the NMR data, or they might raise other structural possibilities.

The full set of routine physical data which could and, ideally, should be obtained on a pure compound is as follows: IR, UV, high field NMR (^1H and ^{13}C), and low and high resolution mass spectra, m.p. or b.p., microanalysis (for a new compound). If the compound is optically active then the optical rotation must be measured. Only mass spectroscopy and microanalysis from this list are destructive techniques, but modern techniques mean that only a small amount of material need be "sacrificed". Some general points concerning these techniques and the sample requirements are given below.

7.3.2 NMR

Nowadays the ^1H NMR spectrum is often the first measurement taken. The size of sample required depends on the type of spectrometer used. A continuous wave machine operating at 90 MHz will need at least 10mg of a normal organic compound, probably more. Pulsed Fourier transform spectrometers require less, 5 mg being a normal amount, and good spectra can be obtained using much smaller quantities. Most spectra are measured in deuteriochloroform ($CDCl_3$) although other solvents will be required from time to time. A typical solvent volume would be ca.0.4~0.5mL in a 5 mm tube. Routine measurement of the ^{13}C NMR spectrum usually requires more samples (25~50mg) but good spectra can be obtained on less, it simply takes more time.

Choice of solvent for the NMR spectra is important. Relative chemical shifts are solvent dependent and ideally all spectra should be measured in a standard solvent. Deuteriochlorofom ($CDCl_3$) is the generally accepted "standard" solvent and it is advisable to use this where possible. If your compound is not sufficiently soluble in $CDCl_3$ then an alternative must be found. Lack of solubility in $CDCl_3$ is usually due to the polarity of your compound, very polar materials and those with extensive, strong, hydrogen-bonding networks often need solvents other than $CDCl_3$. Such materials usually dissolve sufficiently well in hexadeuteri odimethylsulphoxide (DMSO-d_6), tetradeuteriomethanol (MeOH-d_4), or deuterium oxide (D_2O). All these solvents will absorb moisture readily from the air, and it is advisable to purchase small ampoules and protect them from air once opened. In this way, if the "water" signal should become too great then a new ampoule can be opened and relatively little solvent wasted.

Which solvent to choose depends on several factors? The most obvious of these is solubility, sufficient compound must dissolve to provide a spectrum. Given this then the nature of your compound, and what you need from the spectrum, will be important. Both D_2O and MeOH-d_4 will "exchange out" exchangeable protons such as hydroxyl, amine, or amide protons. If you wish to observe the positions of such protons then it is advisable to try DMSO-d_6. It is also wise to check in the literature for similar types of compounds and see which solvent has been used in the past for running NMR spectra.

If you are seeking to make comparisons of the NMR spectrum of your compound with a known compound, or class of compounds, then it is imperative that NMR spectra recorded in the same solvents must be compared. This also applies to solvent mixtures. Sometimes it is suggested that a small quantity of DMSO-d_6 be added to $CDCl_3$ if your compound is not sufficiently soluble in $CDCl_3$ alone. This is inadvisable if you wish to make comparisons, unless you are careful to note the exact ratio of the two solvents. Addition of DMSO-d_6 to $CDCl_3$ results in a change of chemical shift of the resonance for residual $CDCl_3$, a peak which is often used as a reference point in ^1H NMR spectra. The chemical shift of the proton of $CHC1_3$ in $CDCl_3$ is 7.27 δ, in MeOH-d_4 it is 7.88 δ, and 8.35 δ in DMSO-d_6. Be wary of using mixtures of solvents.

If a polarsolvent is needed for NMR spectroscopy, then consider also how you will recover the sample if you need to, as DMSO-d_6 is rather difficult to remove without extensive exposure to high vacuum (but prolonged exposure will usually succeed). In this case MeOH-d_4 might be preferable, provided that it is acceptable in other ways.

The solution used must be free of paramagnetic metal ions (it usually is) and particles (it usually is not). Place sufficient sample into a clean, small, vial (weigh it in if you are unsure how much to use), add the solvent (ca.0.5ml) to dissolve the sample. After taking up this solution into a Pasteur pipette, filtration through a small wad of cotton wool forced into a Pasteur pipette, directly into the NMR tube will usually remove sufficient particles to allow a good spectrum to be obtained. Be very careful that no fragments of glass are broken off the pipette during this filtration, and on removal of the pipette being used as a filter.

The high field ^{1}H NMR spectrum will show up impurities containing protons. Given that the compound has been purified, the most common impurity peaks observed in the ^{1}H NMR spectrum are those from the last solvent used (for example, solvents used in crystallization or chromatography). This should be avoided, and it is always possible unless the boiling point of your product is close to the solvent (which it should not be) or your product is a crystalline solvate (not that common). Thorough exposure to high vacuum should suffice but some very viscous oils and gums will "hold on" to solvent due to the very slow rate of diffusion. If warming in high vacuum fails to remove all solvent, and a "clean" ^{1}H NMR spectrum cannot be obtained, then dissolve the sample in a small amount of $CDCl_3$ and evaporate. Repeat once or twice and most of the residual solvent should be $CDCl_3$ rather than (say) ethyl acetate. This will improve matters, but do not forget that your sample will still be impure on evaporation as it will contain residual $CDCl_3$.

7.3.3 IR

The sample again needs to be free of impurities and solvents for infrared spectroscopy. There are various methods for sample preparation and which you choose will depend largely on the type of compound. The amount required is no more than a few milligrams. For a liquid sample the spectrum can be obtained using a thin film obtained by compressing a small drop between sodium chloride plates, or as a solution (usually in chloroform) using solution cells. The spectra of solids can be recorded either as mulls with a hydrocarbon ("Nujol", for example), or by mixing with KBr and compressing to form a thin disk. Which you use will depend upon the facilities available, and often on the usual working practice of your department.

7.3.4 UV

Ultraviolet spectroscopy is only of use if your compound has a characteristic chromophore. There is little point trying to measure weak bands which will provide no information. However, it is of considerable value in several areas of research; for example, natural product isolation, heteroaromatic chemistry, porphyrin and related chemistry, and in the study of dyestuffs. The amount of material required is usually very small (fractions of a mg) since the extinction coefficients are usually large. The sample must be as pure as possible and is dissolved in the solvent of choice (usually spectroscopically pure ethanol). The concentration must be known accurately before extinction coefficients can be calculated, and will vary depending upon the type of chromophore. An estimate of the concentration to used can be made if the extinction coefficients of compounds similar to that being studied are available. If this data is not available make up a solution accurately and dilute it (accurately!) until a reasonable spectrum is obtained.

7.3.5 Mass Spectra

Thereare three pieces of useful information which can be obtained from mass spectroscopy; the molecular mass, composition, and the fragmentation pattern of your compound. The accurate molecular mass is of primary importance since this will confirm the composition of your compound. Fragmentation information might be of value for supporting the proposed structure, possibly by comparison with known compounds. The amount required is minimal (a few mgs at most), and the material should be reasonably pure. If you are unable to obtain good microanalytical data the accurate mass measurement may provide an acceptable alternative.

7.3.6 M.p. and b.p.

These are usually straightforward. There are various forms of melting point apparatus in widespread use, so check carefully on the procedure appropriate for the apparatus available to you. Always obtain a "rough" melting point before attempting to make accurate measurements and it is often useful to "calibrate" the apparatus by measuring a known (pure!) compound with a similar mp to the product. If there is a significant discrepancy then a more reliable apparatus must be used. Do not forget to get the inaccurate apparatus repaired, and discard it if necessary. The compound needs to be pure and free from dust, and the temperature must be raised very slowly as you near the melting point. If there is a range over which the compound melts (there usually is) then record it; do not estimate an "average" reading. If a capillary tube is used, it is sometimes useful to examine the upper part of the tube for sublimate or distilled decomposition products.

If you have distilled your product to isolate and purify it then you should already have the information required for reporting the boiling point. It is important to quote the range of temperature (if observed) over which the compound distils, the pressure (measured as it is distilling), the vapor temperature (if measured), and the bath temperature. All these will be useful when you or anyone else come to repeat the work, and most of this information will be required at some time for a publication, report, or thesis.

7.3.7 Optical rotation

If your product is, or should be, optically active then the specific rotation will need to be measured and recorded. The precise value of this property is dependent on the wavelength of the light used, solvent, concentration, and temperature. Moreover, great care should be taken to exclude any by-products from the reaction since, although these might be present in small quantities, they might have very large rotations and make your measurements quite misleading. Clearly then, it is important to be sure of the purity of your product, and to make up the solution carefully and accurately. If you are unsure then make a measurement using a known compound before you try to measure the rotation of your product (assuming that the specific rotation of your product is not known). Usually you will use the sodium-D line and measure at ambient temperature, but be sure to record the concentration, solvent, wavelength, and temperature along with the actual value of the measured specific rotation. Occasionally optical rotatory dispersion and/or circular dichroism spectra will be required. These measurements will usually made by specialists and the specific requirements for your particular type of compound are best discussed with them.

From *Advanced Practical Organic Chemistry* by *J. Beonard*

New Words and Expressions

microanalysis [maɪkrəʊə'nælɪsɪs] [分化] 微量分析
deuteriochloroform[dju:tr'ɪəʊklɒrəfɔ:m] 氘氯仿
deuterium oxide 氧化氘
ampoule ['æmpu:l][医]安瓿(等于 ampul)
paramagnetic [pærəmæg'netɪk] 顺磁性的;常磁性的
pasteur pipette 巴斯德吸管
ethyl acetate 乙酸乙酯
nujol['nju:dʒɒl] 液体石蜡
chromophore ['krəʊməfɔ:] [化学] 发色团
porphyrin ['pɔ:fɪrɪn] [生化]卟啉
extinction coefficient 消光系数
hygroscopic [haɪgrə(ʊ)'skɒpɪk] 吸湿的;湿度计的;易潮湿的
cross-referencing 交叉引用
pulsed Fourier transform spectrometer 脉冲傅里叶变换核磁共振(波谱)仪

Notes

(1) **Nuclear Magnetic Resonance**,核磁共振,是通过利用原子核在磁场中的能量变化来获得关于原子核的信息,具有迅速、准确、分辨率高等优点。最常见的有^{1}H氢谱和^{13}C碳谱,通过碳谱不仅可了解分子中碳的种数,而且还可提供分子中各碳所处环境信息。氢谱可给出有机分子中不同环境氢核的信息,根据谱图中各峰的化学位移、峰的分裂情况和峰面积可判定不同种氢的个数,从而推导出分子的可能结构。

(2) **Ultraviolet spectroscopy**,紫外光谱,是利用分子中某些价电子吸收一定波长的电磁波,由低能级跃迁到高能级而产生的一种光谱,具有设备简单、操作方便、灵敏高的特点,已广泛用于有机化合物的定性、定量和结构鉴定。

(3) **Mass spectra**,质谱,是一种测量离子荷质比的分析方法,其基本原理是使试样中各组分在离子源中发生电离,生成不同荷质比的带电离子,经加速电场的作用,形成离子束,然后利用电磁学原理使离子按不同的质荷比分离并测量各种离子的强度,从而确定被测物质的分子量和结构。

(4) **Opticl rotation**,旋光性,当光通过含有某物质的溶液时,使经过此物质的偏振光平面发生旋转的现象。可通过存在镜像形式的物质显示出来,这是由于物质内存在不对称碳原子或整个分子不对称的结果。由于这种不对称性,物质对偏振光平面有不同的折射率,因此表现出向左或向右的旋光性。利用旋光性可以对物质(如某些糖类)进行定性或定量分析。

7.4 Acetylsalicylic Acid (Aspirin)

Aspirin is among the most fascinating and versatile drugs known to medicine, and it is among the oldest—the first known use of an aspirinlike preparation can be traced to ancient Greece and Rome. Salicigen, an extract of willow and poplar bark, has been used as a pain reliever (analgesic) for centuries. In the middle of the last century it was found that salicigen is a glycoside formed from a molecule of salicylic acid and a sugar molecule. Salicylic acid is easily synthesized on a large scale by heating sodium phenoxide with carbon dioxide at 150℃ under slight pressure (the Kolbe synthesis):

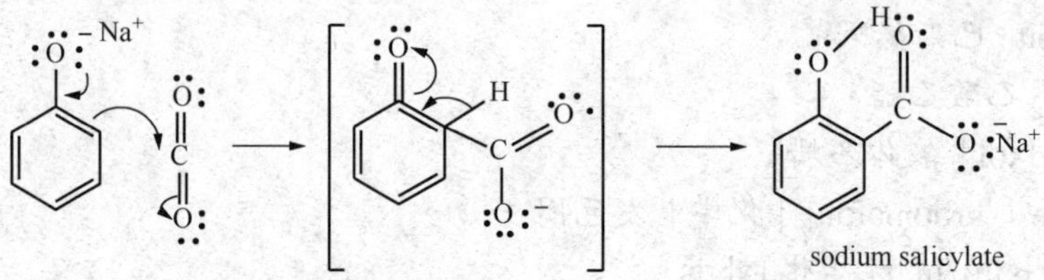

sodium salicylate

But unfortunately salicylic acid attacks the mucous membranes of the mouth and esophagus and causes gastric pain that may be worse than the discomfort it was meant to cure. Felix Hoffmann, a chemist for Friedrich Bayer, a German dye company, reasoned that the corrosive nature of salicylic acid could be altered by addition of an acetyl group; and in 1893 the Bayer Company obtained a patent on acetylsalicylic acid, despite the fact that it had been synthesized some forty years previously by Charles Gerhardt. Bayer coined the name Aspirin for their new product to reflect its acetyl nature and its natural occurrence in the Spiraea plant. Over the years they have allowed the term aspirin to fall into the public domain so it is no longer capitalized. The manufacturers of Coke and Sanka work hard to prevent a similar fate befalling their products.

In 1904 the head of Bayer, Carl Duisberg, decided to emulate John D. Rockefeller's Standard Oil Company and formed an "interessen gemeinschaft" (I.G.) of the dye industry (Farbenindustrie). This cartel completely dominated the world dye industry before World War I and it continued to prosper between the wars even though some of their assets were seized and sold after World War I. After World War I an American company, Sterling Drug, bought the rights to aspirin. The company's Glenbrook Laboratories division still is the major manufacturer of aspirin in the United States (Bayer Aspirin).

Because of their involvement at Auschwitz the top management of IG Farbenindustrie was tried and convicted at the Nuremberg trials after World War II and the cartel broken into three large branches—Bayer, Hoechst, and BASF—each of which does more business than DuPont, the largest American chemical company.

By law all drugs sold in the United States must meet purity standards set by the Food and Drug Administration, and so all aspirin is essentially the same. Each 5 grain tablet contains 0.325 g of acetylsalicylic acid held together with a binder. The remarkable difference in price for aspirin is primarily a reflection of the advertising budget of the company that sells it.

Aspirin is an analgesic (painkiller), an antipyretic (fever reducer), and an anti-inflammatory agent. It is the premier drug for reducing fever, a role for which it is uniquely suited. As an anti-inflammatory, it has become the most widely effective treatment for arthritis. Patients suffering from arthritis must take so much aspirin (several grams per day) that gastric problems may result. For this reason aspirin is often combined with a buffering agent. Bufferin is an example of such a preparation.

The ability of aspirin to diminish inflammation is apparently due to its inhibition of the synthesis of prostaglandins, a group of C20 molecules that enhance inflammation. Aspirin alters the oxygenase activity of prostaglandin synthetase by moving the acetyl group to a terminal amine group of the enzyme.

If aspirin were a new invention, the U.S. Food and Drug Administration (FDA) would place many hurdles in the path of its approval. It has been implicated, for example, in Reyes syndrome, a brain disorder that strikes children and young people under 18. It has an effect on platelets, which play a vital role in blood clotting. In newborn babies and their mothers, aspirin can lead to uncontrolled bleeding and problems of circulation for the baby—even brain hemorrhage in extreme cases. This same effect can be turned into an advantage, however. Heart specialists urge potential stroke victims to take aspirin regularly to inhibit clotting in their arteries, and it has recently been shown that one-half tablet per day will help prevent heart attacks in healthy men.

Aspirin is found in more than 100 common medications, including Alka-Seltzer, Anacin ("contains the pain reliever doctors recommend most"), Coricidin, Excedrin, Midol, and Vanquish. Despite its side effects, aspirin remains the safest, cheapest, and most effective nonprescription drug. It is made commercially, employing the same synthesis used here. The mechanism for the acetylation of salicylic acid is as follows:

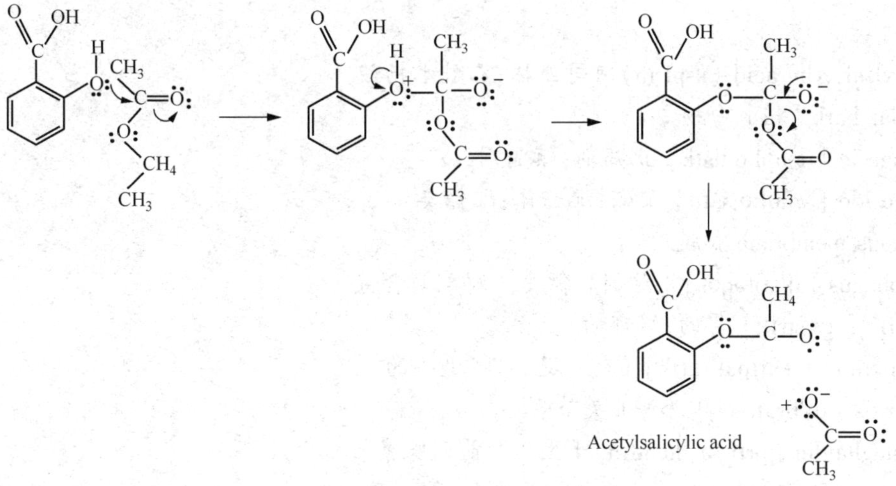

Acetylsalicylic acid

Place 1 g of salicylic acid in each of four 13 mm×100 mm test tubes and add to each tube 2 mL of acetic anhydride. To the first tube add 0.2 g of anhydrous sodium acetate, note the time, stir with a thermometer, and record the time required for a 4℃ rise in temperature. Replace the thermometer and continue to stir occasionally while starting the next acetylation. Obtain a clean thermometer, put it in the second tube, add 5 drops of pyridine, observe as before, and compare with the first results. To the third and fourth tubes add 5 drops of boron trifluoride etherate and 5 drops of concentrated sulfuric acid, respectively. What is the order of activity of the four catalysts as judged by the rates of the reactions?

Put all tubes in hot water (beaker) for 5 min to dissolve solid material and complete the reactions, and then pour all the solutions into a 125 mL Erlenmeyer flask containing 50mL of water and rinse the tubes with water. Swirl to aid hydrolysis of excess acetic anhydride and then cool thoroughly in ice, scratch the side of the flask with a stirring rod to induce crystallization, and collect the crystalline solid; yield is 4g.

Acetylsalicyclic acid melts with decomposition at temperatures reported from 128 to 137℃. It can be crystallized by dissolving it in ether, adding an equal volume of petroleum ether, and letting the solution stand undisturbed in an ice bath.

Test the solubility of your sample in toluene and in hot water and note the peculiar character of the aqueous solution when it is cooled and when it is then rubbed against the tube with a stirring rod. Note also that the substance dissolves in cold sodium bicarbonate solution and is precipitated by addition of an acid. Compare a tablet of commercial aspirin with your sample. Test the solubility of the tablet in water and in toluene and observe if it dissolves completely. Compare its behavior when heated in a melting point capillary with the behavior of your sample. If an impurity is found, it is probably some substance used as binder for the tablets. Is it organic or inorganic? To interpret your results, consider the mechanism whereby salicylic acid is acetylated.

From *Organic Experiments* by *Louis F. Fieser*

New Words and Expressions

acetylsalicylic acid (aspirin) 阿司匹林;乙酰水杨酸
poplar bark [植] 白杨
analgesic [ˌænəlˈdʒisɪk] 止痛剂;镇痛剂
glycoside [ˈglaɪkəˌsaɪd] 配醣;配糖体;配糖类
mucous membrane 黏膜
esophagus [iːˈsɒfəgəs] [解剖] 食管;[解剖] 食道
gastric [ˈgæstrɪk] 胃的;胃部的
antipyretic [æntɪpaɪˈrɛtɪk] [药] 退热剂;退热的
arthritis [ɑːˈθraɪtɪs] [外科] 关节炎
prostaglandin [prɒstəˈglændɪn] [生化] 前列腺素
oxygenase [ˈɒksɪdʒəneɪs] [生化] 加氧酶,氧合酶
platelet [ˈpleɪtlɪt] [组织] 血小板;薄片
hemorrhage [ˈhemərɪdʒ] [病理] 出血
acetylation [əsetɪˈleɪʃən] 乙酰化作用
pyridine [ˈpɪrɪdin] 吡啶;氮(杂)苯
erlenmeyer flask 锥形瓶,爱伦美氏烧瓶
boron trifluoride etherate 三氟化硼醚化物

Questions

(1) Hydrochloric acid is about as stronga mineral acid as sulfuric acid. Why would it not be a satisfactory catalyst in this reaction?

(2) How do you account for the smell of vinegar when an old bottle of aspirin is opened?

7.5 Ferrocene [Bis(cyclopentadienyl)] iron

The Grignard reagent is a classical organometallic compound. The magnesium ion in Group IIA of the periodic table needs to lose two and only two electrons to achieve the inert gas configuration. This metal has a strong tendency to form ionic bonds by electron transfer. Among the transition elements the situation is not so simple. Consider the bonding between iron and carbon monoxide in $Fe(CO)_5$:

$$Fe \longleftarrow :\overset{-}{C}\equiv\overset{+}{O}: \longrightarrow Fe=C=\ddot{O}:$$

The pair of electrons on the carbon atom is shared with iron to form a σ bond between the carbon and iron. The π bond between iron and carbon is formed from a pair of electrons in the *d*-orbital of iron. The π bond is thus formed by the overlap of a *d* orbital of iron with the p-π bond of the carbonyl group. This mutual sharing of electrons results in a relatively nonpolar bond.

Iron has 6 electrons in the 3*d* orbital, 2 in the 4*s*, and none in the 4*p* orbital. The inert gas configuration requires 18 electrons—ten 3*d*, two 4*s*, and six 4*p* electrons. Iron pentacarbonyl enters this configuration by accepting two electrons from each of the five carbonyl groups, a total of 18 electrons. Back-bonding of the d-π type distributes the excess electrons among the five carbon monoxide molecules.

Early attempts to form σ-bonded derivatives linking alkyl carbon atoms to iron were unsuccessful, but P.L.Pauson in 1951 succeeded in preparing a very stable substance, ferrocene, $C_{10}H_{10}Fe$, by reacting two moles of cyclopentadienylmagnesium bromide with anhydrous ferrous chloride. Another group of chemists—Wilkinson, Rosenblum, Whiting, and Woodward—recognized that the properties of ferrocene (remarkable stability to water, acids, and air and its ease of sublimation) could only be explained if it had the structure depicted and that the bonding of the ferrous iron with its six electrons must involve all twelve of the π-electrons on the two cyclopentadiene rings, with a stable 18-electron inert gas structure as the result.

In the present experiment ferrocene is prepared by reaction of the anion of cyclopentadiene with iron(Ⅱ) chloride. Abstraction of one of the acidic allylic protons of cyclopentadiene with base gives the aromatic cyclopentadienyl anion. It is considered aromatic because it conforms to the Hückel rule in having $4n+2\pi$ electrons (where n is 1). Two molecules of this anion will react with iron(Ⅱ) to give ferrocene, the most common member of the class of metal-organic compounds referred to as metallocenes. In this centrosymmetric sandwich-type π complex, all carbon atoms are equidistant from the iron atom, and the two cyclopentadienyl rings rotate more or less freely with respect to each other. The extraordinary stability of ferrocene (stable to 500℃) can be attributed to the sharing of the 12 π electrons of the two cyclopentadienyl rings with the six outer shell electrons of iron(Ⅱ) to give the iron a stable 18-electron inert gas configuration. Ferrocene is soluble in organic solvents, can be dissolved in concentrated sulfuric acid and recovered unchanged, and is resistant to other acids and bases as well (in the absence of oxygen). This behavior is consistent with that of an aromatic compound; ferrocene is found to undergo electrophilic aromatic substitution reactions with ease.

Cyclopentadiene readily dimerizes at room temperature by a Diels-Alder reaction to give dicyclopentadiene. This dimer can be "cracked" by heating (an example of the reversibility of the Diels-Alder reaction) to give low-boiling cyclopentadiene. In most syntheses of ferrocene the anion of cyclopentadiene is prepared by reaction of the diene with metallic sodium. Subsequently, this anion is allowed to react with anhydrous iron(Ⅱ) chloride. In the present experiment the anion is generated using powdered potassium hydroxide, which functions as both a base and a dehydrating agent.

The anion of cyclopentadiene rapidly decomposes in air, and iron(Ⅱ) chloride, although reasonably stable in the solid state, is readily oxidized to the iron(Ⅲ) (ferric) state in solution. Consequently this reaction must be carried out in the absence of oxygen, accomplished by bubbling nitrogen gas through the solutions to displace dissolved oxygen and to flush air from the apparatus. In research laboratories rather elaborate apparatus is used to carry out an experiment in the absence of oxygen. In the present experiment, because no gases are evolved, no heating is necessary, and the reaction is only mildly exothermic, very simple apparatus is used.

2 (cyclopentadienyl) $\overset{+}{K}H + FeCl_2 \cdot 4H_2O \longrightarrow$ Fe(C5H5)2 (ferrocene) $+ 2KCl + 4H_2O$

Following the procedure described in Chapter 28, prepare 6 mL of cyclopentadiene. It need not be dry. While this distillation is taking place, rapidly weigh 25 g of finely powdered potassium hydroxide into a 125 mL erlenmeyer flask, add 60 mL of dimethoxyethane ($CH_3OCH_2CH_2OCH_3$), and immediately cool the mixture in an ice bath. Swirl the mixture in the ice bath for a minute or two, then bubble nitrogen through the solution for about 2min. Quickly cork the flask and shake the mixture to dislodge the cake of potassium hydroxide from the bottom of the flask and to dissolve as much of the base as possible (much will remain undissolved).

Grind 7g of iron(Ⅱ) chloride tetrahydrate to a fine powder and then add 6.5 g of the green salt to 25 mL of dimethyl sulfoxide (DMSO) in a 50 mL Erlenmeyer flask. Pass nitrogen through the DMSO mixture for about 2 min, cork the flask, and shake it vigorously to dissolve all the iron(Ⅱ) chloride. Gentle warming of the flask on a steam bath may be necessary to dissolve the last traces of iron (Ⅱ) chloride. Transfer the solution rapidly to a 60 mL separatory funnel equipped with a cork to fit the 25 mL Erlenmeyer flask, flush air from the funnel with a stream of nitrogen, and stopper it. Transfer 5.5 mL of the freshly distilled cyclopentadiene to the slurry of potassium hydroxide in dimethoxyethane. Shake the flask vigorously and note the color change as the potassium cyclopentadienide is formed. After waiting about 5 min for the anion to form, replace the cork on the Erlenmeyer flask with the separatory funnel quickly (to avoid admission of air to the flask). Add the iron(Ⅱ) chloride solution to the base dropwise over a period of 20 min with vigorous swirling and shaking. Dislodge the potassium hydroxide should it cake on the bottom of the flask. The shaking will allow nitrogen to pass from the Erlenmeyer flask into the separatory funnel as the solution leaves the funnel. Continue to shake and swirl the solution for 10min after all the iron(Ⅱ) chloride is added, then pour the dark slurry onto a mixture of 90 mL of 6 M hydrochloric acid and 100 g of ice in a 500 mL beaker. Stir the contents of the beaker thoroughly to dissolve and neutralize all the potassium hydroxide. Collect the crystalline orange ferrocene on a Büchner funnel, wash the crystals with water, press out excess water, and allow the product to dry on a watch glass overnight.

Recrystallize the ferrocene from methanol or, better, from ligroin. It is also very easily sublimed. In a hood place about 0.5 g of crude ferrocene on a watch glass on a hot plate set to about 150℃. Invert a glass funnel over the watch glass. Ferrocene will sublime in about one hour, leaving nonvolatile impurities behind. Pure ferrocene melts at 172～174℃. Determine the melting point in an evacuated capillary since the product sublimes at the melting point. Compare the melting points of your sublimed and recrystallized materials.

The filtrate from the reaction mixture should be slightly acidic. Neutralize it with sodium carbonate, dilute it with water, and flush it down the drain. Place any unused cyclopentadiene in the recovered dicyclopentadiene or the organic solvents container. If the ferrocene has been crystallized from methanol or ligroin, place the mother liquor in the organic solvents container.

From *Organic Experiments* by *Louis F. Fieser*

New Words and Expressions

ferrocene [ˈferəusiːn] [有化] 二茂铁，环戊二烯铁
cyclopentadiene [saɪkləʊpenˈteɪdɪən] 环戊二烯
allylic [əˈlilik] 烯丙基的
metallocene [miˈtæləusiːn] 茂(合)[有化] 金属；金属茂络合物
centrosymmetric [sentrəʊsɪˈmetrɪk] 中心对称的
dimerize [ˈdaiməraiz] 使二聚；使成为二聚物
dehydrating agent 脱水剂；去水剂
potassium hydroxide 氢氧化钾
dimethyl sulfoxide 二甲基亚砜
watch glass 表面皿
sublime [səˈblaɪm] 使…纯化；使…升华
Büchner funnel 瓷漏斗；布氏漏斗

ligroin [ˈlɪgrəʊɪn] 轻石油,粗汽油,石油醚

sodium carbonate 碳酸钠

Questions

(1) If ferrocene is stable to air and all of the reagents are stable to air before the reaction begins, why must air be so carefully excluded from this reaction?

(2) What special properties do the solvents dimethoxyethane and dimethyl sulfoxide have compared to diethyl ether, for example, that make them particularly suited for this reaction?

(3) What is there about ferrocene that allows it to sublime easily where many other compounds do not?

Unit 8 Scientific Paper and Literature

Lesson 1 How to Read a Scientific Paper

John W.Little and Roy Parker—University of Arizona

The main purpose of a scientific paper is to report new results, usually experimental, and to relate these results to previous knowledge in the field. Papers are one of the most important ways that we communicate with one another. In understanding how to read a paper, we need to start at the beginning with a few preliminaries. We then address the main questions that will enable you to understand and evaluate the paper.

① How are papers organized?

② How do I prepare to read a paper, particularly in an area not so familiar to me?

③ What difficulties can I expect?

④ How do I understand and evaluate the contents of the paper?

1.Organization of a paper

In most scientific journals, scientific papers follow a standard format. They are divided into several sections, and each section serves a specific purpose in the paper. We first describe the standard format, then some variations on that format.

A paper begins with a short Summary or Abstract. Generally, it gives a brief background to the topic, describes concisely the major findings of the paper and relates these findings to the field of study. As will be seen, this logical order is also that of the paper as a whole. The next section of the paper is the Introduction. In many journals this section is not given a title. As its name implies, this section presents the background knowledge necessary for the reader to understand why the findings of the paper are an advance on the knowledge in the field. Typically, the Introduction describes first the accepted state of knowledge in a specialized field, then it focuses more specifically on a particular aspect, usually describing a finding or set of findings that led directly to the work described in the paper. If the authors are testing a hypothesis, the source of that hypothesis is spelled out, findings are given with which it is consistent, and one or more predictions are given. In many papers, one or several major conclusions of the paper are presented at the end of this section, so that the reader knows the major answers to the questions just posed. Papers more descriptive or comparative in nature may begin with an introduction to an area which interests the authors, or the need for a broader database.

The next section of most papers is the Materials and Methods. In some journals this section is the last one. Its purpose is to describe the materials used in the experiments and the methods by which the experiments were carried out. In principle, this description should be detailed enough to allow other researchers to replicate the work. In practice, these descriptions are often highly compressed, and they often refer back to previous papers by the authors.

The third section is usually Results. This section describes the experiments and the reasons they were done. Generally, the logic of the Results section follows directly from that of the Introduction. That is, the Introduction poses the questions addressed in the early part of Results. Beyond this point, the organization of Results differs from one paper to another. In some papers, the results are presented without extensive discussion, which is reserved for the following section. This is appropriate when the data in the early parts do not need to be interpreted extensively to understand why the later experiments were done. In other papers, results are given, and then they are interpreted, perhaps taken together with other findings not in the paper, so as to give the logical basis for later experiments.

The fourth section is the Discussion. This section serves several purposes. First, the data in the paper are interpreted; that is, they are analyzed to show what the authors believe the data show. Any limitations to the interpretations should be acknowledged, and fact should clearly be separated from speculation. Second, the findings of the paper are related to other findings in the field. This serves to show how the findings contribute to knowledge, or correct the errors of previous work. As stated, some of these logical arguments are often found in the Results when it is necessary to clarify why later experiments were carried out. Although you might argue that in this case the discussion material should be presented in the Introduction, more often you cannot grasp its significance until the first part of Results is given.

Finally, papers usually have a short Acknowledgements section, in which various contributions of other workers are recognized, followed by a Reference list giving references to papers and other works cited in the text.

Papers also contain several Figures and Tables. These contain data described in the paper. The figures and tables also have legends, whose purpose is to give details of the particular experiment or experiments shown there. Typically, if a procedure is used only once in a paper, these details are described in Materials and Methods, and the Figure or Table legend refers back to that description. If a procedure is used repeatedly, however, a general description is given in Materials and Methods, and the details for a particular experiment are given in the Table or Figure legend.

2. Variations on the organization of a paper

In most scientific journals, the above format is followed. Occasionally, the Results and Discussion are combined, in cases in which the data need extensive discussion to allow the reader to follow the train of logic developed in the course of the research. As stated, in some journals, Materials and Methods follows the Discussion. In certain older papers, the Summary was given at the end of the paper.

The formats for two widely-read journals, Science and Nature, differ markedly from the above outline. These journals reach a wide audience, and many authors wish topublish in them; accordingly, the space limitations on the papers are severe, and the prose is usually highly compressed. In both journals, there are no discrete sections, except for a short abstract and a reference list. In Science, the abstract is self-contained; in Nature, the abstract also serves as a brief introduction to the paper. Experimental details are usually given either in endnotes (for Science) or Figure and Table legends and a short Methods section (in Nature). Authors often try to circumvent length limitations by putting as much material as possible in these places. In addition, an increasingly common practice is to put a substantial fraction of the less-important material, and much of the methodology, into Supplemental Data that can be accessed online.

Many other journals also have length limitations, which similarly lead to a need for conciseness. For example, the Proceedings of the National Academy of Sciences (PNAS) has a six-page limit; Cell severely edits many papers to shorten them, and has a short word limit in the abstract; and so on. In response to the pressure to edit and make the paper concise, many authors choose to condense or, more typically, omit the logical connections that would make the flow of the paper easy. In addition, much of the background that would make the paper accessible to a wider audience is condensed or omitted, so that the less-informed reader has to consult a review article or previous papers to make sense of what the issues are and why they are important. Finally, again, authors often circumvent page limitations by putting crucial details into the Figure and Table legends, especially when (as in PNAS) these are set in smaller type.

3. Reading a scientific paper

Although it is tempting to read the paper straight through as you would do with most text, it is more efficient to organize the way you read. Generally, you first read the Abstract in order to understand the major points of the work. The extent of background assumed by different authors, and allowed by the journal, also varies as just discussed.

One extremely useful habit in reading a paper is to read the Title and the Abstract and, before going on, review in your mind what you know about the topic. This serves several purposes. First, it clarifies whether you in fact know enough background to appreciate the paper. If not, you might choose to read the background in a review or textbook, as appropriate.

Second, it refreshes your memory about the topic. Third, and perhaps most importantly, it helps you as the reader integrate the new information into your previous knowledge about the topic. That is, it is used as a part of the self-education process that any professional must continue throughout his/her career.

If you are very familiar with the field, the Introduction can be skimmed or even skipped. As stated above, the logical flow of most papers goes straight from the Introduction to Results; accordingly, the paper should be read in that way as well, skipping Materials and Methods and referring back to this section as needed to clarify what was actually done. A reader familiar with the field who is interested in a particular point given in the Abstract often skips directly to the relevant section of the Results, and from there to the Discussion for interpretation of the findings. This is only easy to do if the paper is organized properly.

4. Codewords

Many papers contain shorthand phrases that we might term 'codewords', since they have connotations that are generally not explicit. In many papers, not all the experimental data are shown, but referred to by "(data not shown)". This is often for reasons of space; the practice is accepted when the authors have documented their competence to do the experiments properly (usually in previous papers). Two other codewords are "unpublished data" and "preliminary data". The former can either mean that the data are not of publishable quality or that the work is part of a larger story that will one day be published. The latter means different things to different people, but one connotation is that the experiment was done only once.

5. Difficulties in reading a paper

Several difficulties confront the reader, particularly one who is not familiar with the field. As discussed above, it maybe necessary to bring yourself up to speed before beginning a paper, no matter how well written it is. Be aware, however, that although some problems may lie in the reader, many are the fault of the writer.

One major problem is that many papers are poorly written. Some scientists are poor writers. Many others do not enjoy writing, and do not take the time or effort to ensure that the prose is clear and logical. Also, the author is typically so familiar with the material that it is difficult to step back and see it from the point of view of a reader not familiar with the topic and for whom the paper is just another of a large stack of papers that need to be read.

Bad writing has several consequences for the reader. First, the logical connections are often left out. Instead of saying why an experiment was done, or what ideas were being tested, the experiment is simply described. Second, papers are often cluttered with a great deal of jargon. Third, the authors often do not provide a clear road-map through the paper; side issues and fine points are given equal air time with the main logical thread, and the reader loses this thread. In better writing, these side issues are relegated to Figure legends or Materials and Methods or clearly identified as side issues, so as not to distract the reader.

Another major difficulty arises when the reader seeks to understand just what the experiment was. All too often, authors refer back to previous papers; these refer in turn to previous papers in a long chain. Often that chain ends in a paper that describes several methods, and it is unclear which was used. Or the chain ends in a journal with severe space limitations, and the description is so compressed as to be unclear. More often, the descriptions are simply not well-written, so that it is ambiguous what was done.

Other difficulties arise when the authors are uncritical about their experiments; if they firmly believe a particular model, they may not be open-minded about other possibilities. These may not be tested experimentally, and may even go unmentioned in the Discussion. Still another, related problem is that many authors do not clearly distinguish between fact and speculation, especially in the Discussion. This makes it difficult for the reader to know how well-established are the 'facts' under discussion.

One final problem arises from the sociology of science. Many authors are ambitious and wish to publish in trendy journals. As a consequence, they overstate the importance of their findings, or puta speculation into the title in a way that makes it sound like a well-established finding. Another example of this approach is the "Assertive Sentence Title", which presents a major conclusion of the paper as a declarative sentence (such as "LexA is a repressor of the recA and lexA genes"). This trend is becoming prevalent; look at recent issues of Cell for examples. It's not so bad when the assertive sentence is well-documented (as it was in the example given), but all too often the assertive sentence is nothing more than a speculation, and the hasty reader may well conclude that the issue is settled when it isn't.

These last factors represent the public relations side of a competitive field. This behavior is understandable, if not praiseworthy. But whenthe authors mislead the reader as to what is firmly established and what is speculation, it is hard, especially for the novice, to know what is settled and what is not. A careful evaluation is necessary, as we now discuss.

6. Evaluating a Paper

A thoroughunderstanding and evaluation of a paper involves answering several questions:

a. What question does the paper address?

b. What are the main conclusions of the paper?

c. What evidence supports those conclusions?

d. Do the data actually support the conclusions?

e. What is the quality of the evidence?

f. Why are the conclusions important?

Type of research	Question asked
Descriptive	What is there? What do we see?
Comparative	How does it compare to other organism? Are our findingsgeneral?
Analytical	How does it work? What is the mechanism?

a. What question does the paper address?

Before addressing this question, we need to be aware that research in biochemistry and molecular biology can be of several different types:

Descriptive research often takes place in the early stages of our understanding of a system. We can't formulate hypotheses about how a system works, or what its interconnections are, until we know what is there. Typical descriptive approaches in molecular biology are DNA sequencing and DNA microarray approaches. In biochemistry, one could regard x-ray crystallography as a descriptive endeavor.

Comparative research often takes place when we are asking how general a finding is. Is it specific to my particular organism, or is it broadly applicable? A typical comparative approach would be comparing the sequence of a gene from one organism with that from the other organisms in which that gene is found. One example of this is the observation that the actin genes from humans and budding yeast are 89% identical and 96% similar.

Analytical research generally takes place when we know enough to begin formulating hypotheses about how a system works, about how the parts are interconnected, and what the causal connections are. A typical analytical approach would be to devise two (or more) alternative hypotheses about how a system operates. These hypotheses would all be consistent with current knowledge about the system. Ideally, the approach would devise a set of experiments to distinguish among these hypotheses. A classic example is the Meselson-Stahl experiment.

Of course, many papers are a combination of these approaches. For instance, researchers might sequence a gene from their model organism; compare its sequence to homologous genes from other organisms; use this comparison to devise a hypothesis for the function of the gene product; and test this hypothesis by making a site-directed change in the gene and asking how that affects the phenotype of the organism and/or the biochemical function of the gene product.

Being aware that not all papers have the same approach can orient you towards recognizing the major questions that a paper addresses.

What are these questions? In a well-written paper, as described above, the Introduction generally goes from the general to the specific, eventually framing a question or set of questions. This is a good starting place. In addition, the results of experiments usually raise additional questions, which the authors may attempt to answer. These questions usually become evident only in the Results section.

b. What are the main conclusions of the paper?

This question can often be answered in a preliminary way by studying the abstract of the paper. Here the authors highlight what they think are the key points. This is not enough, because abstracts often have severe space constraints, but it can serve as a starting point. Still, you need to read the paper with this question in mind.

c. What evidence supports those conclusions?

Generally, you can get a pretty good idea about this from the Results section. The description of the findings points to the relevant tables and figures. This is easiest when there is one primary experiment to support a point. However, it is often the case that several different experiments or approaches combine to support a particular conclusion. For example, the first experiment might have several possible interpretations, and the later ones are designed to distinguish among these.

In the ideal case, the Discussion begins with a section of the form "Three lines of evidence provide support for the conclusion that... First, ... Second, ... etc." However, difficulties can arise when the paper is poorly written (see above). The authors often do not present a concise summary of this type, leaving you to make it yourself. A skeptic might argue that in such cases the logical structure of the argument is weak and is omitted on purpose! In any case, you need to be sure that you understand the relationship between the data and the conclusions.

d. Do the data actually support the conclusions?

One major advantage of doing this is that it helps you to evaluate whether the conclusion is sound. If we assume for the moment that the data are believable (see next section), it still might be the case that the data do not actually support the conclusion the authors wish to reach. There are at least two different ways this can happen:

i. The logical connection between the data and the interpretation is not sound.

ii. There might be other interpretations that might be consistent with the data. One important aspect to look for is whether the authors take multiple approaches to answering a question. Do they have multiple lines of evidence, from different directions, supporting their conclusions? If there is only one line of evidence, it is more likely that it could be interpreted in a different way; multiple approaches make the argument more persuasive.Another thing to look for is implicit or hidden assumptions used by the authors in interpreting their data.This can be hard to do,unless you understand the field thoroughly.

e.What is the quality of that evidence?

This is the hardest question to answer,for novices and experts alike.At the same time,it is one of the most important skills to learn as a young scientist.It involves a major reorientation from being a relatively passive consumer of information and ideas to an active producer and critical evaluator of them.This is not easy and takes years to master.Beginning scientists often wonder,"Who am I to question these authorities? After all the paper was published in a top journal,so the authors must have a high standing,and the work must have received a critical review by experts." Unfortunately, that's not always the case.In any case,developing your ability to evaluate evidence is one of the hardest and most important aspects of learning to be a critical scientist and reader.

How can you evaluate the evidence?

First,you need to understand thoroughly the methods used in the experiments.Often these are described poorly or not at all (see above).The details are often missing,but more importantly the authors usually assume that the reader has a general knowledge of common methods in the field (such as immunoblotting,cloning,genetic methods,or DNase I footprinting).If you lack this knowledge,as discussed above you have to make the extra effort to inform yourself about the basic methodology before you can evaluate the data.

Sometimes you have to go to the library,or to a lab that has a lot of back issues of common journals,to trace back the details of the methods if they are important.One new development that eventually will make this much easier is the increasing availability of journals on the Web.A comprehensive listing of journals relevant to this course,developed by the Science Library,allows access to most of the listed volumes from any computer at the University;a second list at the Arizona Health Sciences Library includes some other journals,again from University computers.

Second,you need to know the limitations of the methodology.Every method has limitations,and if the experiments are not done correctly they can't be interpreted.

For instance,an immunoblot is not a very quantitative method.Moreover,in a certain range of protein the signal increases (that is,the signal is at least roughly "linear"),but above a certain amount of protein the signal no longer increases.Therefore,to use this method correctly one needs a standard curve that shows that the experimental lanes are in a linear range.Often,the authors will not show this standard curve,but they should state that such curves were done.If you don't see such an assertion,it could of course result from bad writing,but it might also not have been done.If it wasn't done,a dark band might mean "there is this much protein or an indefinite amount more".

Third,you need to distinguish between what the data show and what theauthors say they show. The latter is really an interpretation on the authors' part,though it is generally not stated to be an interpretation.Papers usually state something like "the data in Fig.x show that...".This is the authors' interpretation of the data.Do you interpret it the same way? You need to look carefully at the data to ensure that they really do show what the authors say they do.You can only do this effectively if you understand the methods and their limitations.

Fourth, it is oftenhelpful to look at the original journal (or its electronic counterpart) instead of a photocopy. Particularly for half-tone figures such as photos of gels or autoradiograms, the contrast is distorted, usually increased, by photocopying, so that the data are misrepresented.

Fifth, you should ask if the proper controls are present. Controls tell us that nature is behaving the way we expect it to under the conditions of the experiment (see here for more details). If the controls are missing, it is harder to be confident that the results really show what is happening in the experiment. You should try to develop the habit of asking "where are the controls?" and looking for them.

f. Why are the conclusions important?

Do the conclusions make a significant advance in our knowledge? Do they lead to new insights, or even new research directions?

Again, answering these questions requires that you understand the field relatively well.

http://www.biochem.arizona.edu/classes/bioc568/papers.htm

Lesson 2 Writing a Paper

1. What is a scientific paper?

A paper is an organized description of hypotheses, data and conclusion, intended to instruct the reader. Papers are a central part of research. If your research does not generate papers, it might just as well not have been done. "Interesting and unpublished" is equivalent to "non-existent".

Realize that your objective in research is to formulate and test hypotheses, to draw conclusions from these tests, and to teach these conclusions to others. Your object is not to "collect data".

A paper is not just an archival device for storing a completed research program; it is also a structure for *planning* your research in progress. If you clearly understand the purpose and form of a paper, it can be immensely useful to you in *organizing* and conducting your research. A good outline for the paper is also a good plan for the research program. You should write and rewrite these plans/outlines throughout the course of the research. At the beginning, you will have mostly plan; at the end, mostly outline. The continuous effort to understand, analyze, summarize, and reformulate hypotheses on paper will be immensely more efficient for you than a process in which you collect data and only start to organize them when their collection is "complete".

2. Outlines

2.1 The reason for outlines

I emphasize the central place of an outline in writing papers, preparing seminars, and planning research. I especially believe that for you, and for me, it is most *efficient* to write papers from outlines. An outline is a written plan of the organization of a paper, including the data on which it rests. You should, in fact, think of an outlines as a carefully organized and presented set of data, with attendant objects, hypotheses, and conclusions, rather than an outline of text.

An outline itself contains little text. If you and I can agree on the details of the outline (that is, on the data and organization), the supporting text can be assembled fairly easily. If we do not agree on the outline, any text is useless. Much of the time in writing a paper goes into the text; most of the thought goes into the organization of the data and into the analysis. It can be relatively efficient in time to go through several (even many) cycles of an outline before beginning to write text; writing many versions of the full text of a paper is slow.

All writing that I do-papers, reports, proposals (and, of course, slides for seminars)—I do from outlines. I urge you to learn how to use them as well.

2.2 How should you construct an outline?

The classical approach is to start with a blank piece of paper, and write down, in any order, all important ideas that occur to you concerning the paper. Ask yourself the obvious questions: "Why did I do this work?"; "What does it mean:"; "What hypotheses did I mean to test?"; "What ones did I actually test?"; "What were the results? Did the work yield a new method of compound? What?"; "What measurements did I make?"; "What compounds? How were they characterized?". Sketch possible equations, figures, and schemes. It is essential to try to get the major ideas. If you start the research to test one hypothesis, and decide, when you see what you have, that the data really seem to test some other hypothesis better, do not worry. Write them both down, and pick the best combinations of hypotheses, objectives, and data. Often the objectives of a paper when it is finished are different from those used to justify starting the work. Much of good science is opportunistic and revisionist.

When you have written down what you can, start with another piece of paper and try to organize the jumble of the first one. Sort all of your ideas into three major heaps.

(1) Introduction

Why did I do this work? What were the central motivations and hypotheses?

(2) Results and discussion

What were the results? How were compounds made and characterized? What was measured?

(3) Conclusion

What does it all mean? What hypotheses were proved or disproved? What didI learn? Why does it make a difference?

Next, take each of these sections, and organize it on yet finer scale. Concentrate on organizing the data. Construct figures, tables, and schemes to present the data as clearly and compactly as possible. This process can be slow—I may sketch a figure five to ten times in different ways trying to decide how it is most clear (and looks best aesthetically).

Finally, put everything—outline of sections, tables, sketches of figures, equations—in good order.

When you are satisfied that you have included all the data (or that you know what additional data you intend to collect), and have a plausible organization, give the outline to me. Simply indicate where missing data will go, how you think (hypothesize) they will look, and how you will interpret them if you hypothesis is correct. I will take this outline, add my opinions, suggest changes, and return it to you. It usually takes four to five iterations (often with additional experiments) to agree on an outline. When we have agreed, the data are usually in (or close to) final form (that is, the tables, figures, etc., in the outline will be the tables, figures, in the paper).

You can then start writing, with some assurance that much of your prose will be used.

The key to efficient use of your and my time is that we start exchanging outlines and proposals as early in a project as possible. Do not, under any circumstances, wait until the collection of data is "complete" before starting to write an outline. No project is ever complete, and it saves enormous effort and much time to propose a plausible paper and outline as soon as you see the basic structure of a project. Even if we decide to do significant additional work before seriously organizing a paper, the effort of writing an outline will have helped to guide the research.

2.3 The outline

What an outline should contain: (1) Title; (2) Authors; (3) Abstract: Do not write an abstract. That can be done when the paper is complete; (4) Introduction.

The first paragraphor two should be written out completely. Pay particular attention to the opening sentence. Ideally, it should state concisely the objective of the work, and indicate why this objective is important.

- The objectives of the work.
- The justification for these objectives: Why is the work important?
- Background: Who else has done what? How? What have we done previously?
- Guidance to the reader: What should the reader watch for in the paper? What are the interesting high points? What strategy did we use?
- Summary/conclusion: What should the reader expect as conclusion? In advanced versions of the outline, you should also include all the sections that will go in the Experimental section (at the level of paragraph subheadings) and indicate what information will go in the Microfilm section.

(5) Results and discussion

The results and discussion are usually combined. This section should be organized according to major topics. The separate parts should have subheadings in boldface to make this organization clear, and to help the reader scan through the final text to find the parts of interest. The following list includes examples of phrases that might plausibly serve as section headings:

- Synthesis of Alkane Thiols
- Characterization of Monolayers
- Absolute Configuration of the Vicinal Diol Unit
- Hysteresis is Correlates with Roughness of the surface
- Dependence of the Rate Constant on Temperature
- The Rate of Self-Exchange Decreases with the Polarity of the Solvent

Try to make these section headings as specificand information-rich as possible. For example, the phrase "The Rate of Self-Exchange Decreases with the Polarity of the Solvent" is obviously longer than "Measurement of Rates", but much more useful to the reader. In general, try to cover the major common points:

- Synthesis of starting materials
- Characterization of products
- Methods of characterization

- Methods of measurement
- Results (rate constants, contact angles, whatever)

In the outline, do not write any significant amount of text, but get all the data in their proper place: Any text should simply indicate what will go in that section.

- Section Headings
- Figures (with captions)
- Schemes (with captions and footnotes)
- Equations
- Tables (correctly formatted)

Remember to think of a paper as a collection of experimental results, summarized as clearly and economically as possible in figures, tables, equations, and schemes. The text in the paper serves just to explain the data, and is secondary. The more information can be compressed into tables, equations, etc., the shorter and more readable the paper will be.

(6) Conclusion

In the outline, summarize the conclusions of the paper as a list of short phrases or sentences. Do not repeat what is in the results section, unless special emphasis is needed. The conclusions section should be just that, and not a summary. It should add a new higher lever of analysis, and should indicate explicitly the significance of the work.

(7) Experimental

Include, in the correct order to correspond to the order in the Results section, all of the paragraph subheadings of the Experimental section.

2.4 In summary

- Start writing possible outlines for papers early in a project. Do not wait until the "end". The end may never come.
- Organize the outline and the paper around easily assimilated data—tables, equations, figures, schemes—rather than around text.
- Organize in order of importance, not in chronological order. An important detail in writing papers concerns the weight to given to topics. Neophytes often organize a paper in terms of chronology: that is, they give a recitation of their experimental program, starting with their cherished initial failures and leading up to a climactic successful finale. This approach is completely wrong. Start with the most important results, and put the secondary results later, if at all. The reader usually does not care how you arrived at your big results, only what they are. Shorter papers are easier to read than longer ones.

3. Some points of style

- Do not use nouns as adjective:

Not: ATP formation; reaction product

But: Formation of ATP; product of the reaction

- The word "this" must always be followed by a noun, so that its reference is explicit.

Not: This is a fast reaction; This leads us to conclude

But: This reaction is fast; This observations leads us to conclude

- Describe experimental results uniformly in the past tense.

Not: Addition of water gives product.

But: Addition of water gave product.

- Use the active voice whenever possible

Not: It was observed that the solution turned red.

But: The solution turned red.or

We observed that the solution turned red.

- Complete all comparisons

Not: The yield was higher using bromine.

But: The yield was higher using bromine than chlorine.

- Type all paper double-spaced (not single- or one-and-a-half-spaced), and leave two spaces after colons, and after periods at the end of sentences. Leave generous margins.

Assume that we will write all papers using the style of the American Chemical Society. You can get a good idea of this style from three sources:

- The journals. Simply look at articles in the journals and copy the organization you see there.
- Previous papers from the group. By looking at previous papers, you can see exactly how a paper should "look". If what you wrote looks different, it probably is not what we want.
- *The ACS handbook for Authors.* Useful, detailed, especially the section on references, pp. 173-229.

I also suggest you read Strunk and White, The Elements of Style (Macmillan: New York, 1979, 3rd ed.) to get a sense for usage. A number of other books on scientific writing are in the group library; these books on scientific writing are in the group library; these books all contain useful advice, but are not lively reading. There are also several excellent books on the design of graphs and figures.

From DOI: 10.1002/adma.200400767 by *George M. Whitesides*

Lesson 3 Reading Material

Synthesis and performance of Novel Sulfonate Gemini surfactant with trialkyl chains

Jie Li[1], Ruihua Zhu[1], Yu Liu[1], Lu Yin[1], and Wenxiang Wu[2]

[1]College of Chemistry and Chemical Engineering, Northeast Petroleum University, Daqing 163318, P.R.China; [2]College of Petroleum Engineering, Northeast Petroleum University, Daqing 163318, P.R.China

Abstract: A new type of sulfonate gemini surfactant with three lipophilic alkyl chains ($3C_{10}$-DS) was synthesized, and the structure of the product was confirmed by using the infrared spectrum and mass spectrum. Its critical micelle concentration (CMC) is 0.41 mmol/L, one order of magnitude lower than those of convectional (single-chain) surfactants, and the minimum surface tension is 27.6 mN/m. The interfacial tension (IFT) between the compound system of $3C_{10}$-DS and petroleum sulfonate (PS) and the simulated oil reaches ultra-low levels (10^{-3} mN/m), and there exists significant synergistic effect between $3C_{10}$-DS and PS. The compound flooding system consisting of polymer and the mixture of $3C_{10}$-DS and PS can effectively improve oil recovery for high-medium permeability cores, and have a good application prospect in enhancing oil recovery.

Keywords: Sulfonate gemini surfactant, critical micelle concentration, interfacial tension, oil displacement efficiency

1. Introduction

At present, the petroleum industryis facing pressing challenges to enhance oil recovery. Surfactants of the chemical flooding formulation in the tertiary oil recovery technique can reduce the interfacial tension of the oil-water interface, make oil bead deform easily, and improve the oil and water seepage characteristics[1]. Sulfonate surfactants exhibiting good salt tolerance and temperature resistance are the major important class of anionic surfactants in terms of volumes and range of application in the chemical flooding[2]. As a new generation of surfactants[3], gemini surfactants have attracted great interest in the last two decades. Gemini surfactants are made up of two hydrophilic and two hydrophobic groups connected through a spacer at or near the polar head groups[4]. These surfactants are superior to conventional surfactants in many aspects such as lower CMC, higher effectiveness in reducing the surface/interface tension, resistance to extreme environmental conditions, which make them potentially useful in such applications as enhanced oil recovery, drug carriers, and gene therapy, antibactericides, and the construction of new materials[5-7]. To our knowledge, most of studies mainly focus on synthesis and properties of cationic gemini surfactants, while few reports deal with anionic gemini surfactants due to the difficulties in synthesis and purification[8]. Thus, it is necessary to synthesize novel anionic gemini surfactants with different structures and study their potential application.

In this work, we designed and synthesized a novel sulfonate gemini surfactant with three lipophilic alkyl chains and two sulfonate groups ($3C_{10}$-DS), and the synthetic route was shown in Fig. 8.1. The structure of the product was characterized by using the infrared spectrum and mass spectrum. The surface tension was investigated by pendant drop method, and the interfacial tension between the solution and the simulated oil was measured by spinning drop method. The oil displacement efficiencies of the product and its compound systems were evaluated in order to explore the applications in enhancing oil recovery technology.

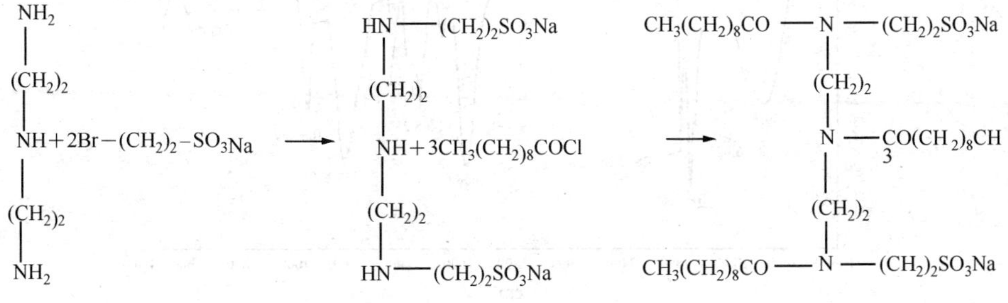

$NH_2-(CH_2)_2-NH-(CH_2)_2-NH_2 + 2Br-(CH_2)_2-SO_3Na \longrightarrow$

$NaSO_3(CH_2)_2-HN-(CH_2)_2-NH-(CH_2)_2-NH-(CH_2)_2SO_3Na + 3CH_3(CH_2)_8COCl \longrightarrow$

$CH_3(CH_2)_8CO-N[(CH_2)_2SO_3Na]-(CH_2)_2-N[CO(CH_2)_8CH_3]-(CH_2)_2-N[(CH_2)_2SO_3Na]-COCH_3(CH_2)_8CH_3$

Fig.8.1 Synthetic route of sulfonate gemini surfactant $3C_{10}$-DS.

2.Experimental Section

2.1 Materials

Diethylenetriamine (97%), Triethylamine (99%), and decanoyl chloride (98%) were purchased from Aladdin reagent Co.Ltd., China.Sodium 2-bromoethanesulfonate was obtained from Nanjing Robiot Co.Ltd., China.Petroleum sulfonates (PS), and the polymer (partially hydrolyzed polyacrylamides) were obtained from the Third Oil Production Plant of Daqing.All the solvents were of analytical grade and used without further purification.Water used in this study was redistilled water. The simulated oil was obtained by mixing kerosene and crude oil of Daqing oilfield with the mass ratio of 2:1.

2.2 Synthesis of Sulfonate Gemini Surfactants $3C_{10}$-DS

Diethylenetriamine (0.05mol) in 5mL water was placed in 500mL three-necked flask equipped with a magnetic stirrer bar and a condenser.The solution of Sodium 2-bromoethanesulfonate (0.11 mol) dissolved in 100 mL water was added dropwise into the flask under stirring.The reaction mixture was heated up to 60℃ and maintained pH 10 by adding 10% NaOH solution for 8 h.150 mL ethanol was then added to the mixture at room temperature.The resulting mixture was refrigerated, precipitated and filtered.The isolated precipitate was dried under reduced pressure to obtain the reaction intermediate as pale yellow power.The reaction intermediate (0.015mol) dissolved in 100mL mixture of water and acetone ($V:V=1:2$) was put into a 250mL three-necked flask.Decanoyl chloride (0.05mol) was slowly added dropwise to the reaction flask with vigorous stirring.The reaction mixture was stirred for 12h at room temperature and maintained pH 8 by adding dropwise the mixture of triethylamine and acetone ($V:V=2:1$).The crude product precipitated from the reaction mixture after standing 12h at 0℃.Then, the obtained precipitate was further purified by recrystallization for 3 times.The chemical structure of the synthesized compound ($3C_{10}$-DS) was confirmed by FTIR and mass spectroscopy analysis, as shown in Fig.8.2 and Fig.8.3.FTIR (Perkin Elmer Spectrum One, US;KBr):2956.8, 2919.8, 2850.3, 1638.6, 1468.2, 1377.9, 1190.8, 1060.5, 721.1, 624.9cm^{-1}.ESI-MS (microTOF-Q II, Bruker) m/z:[M-Na]$^-$, 802.49 (Calcd.803.12).

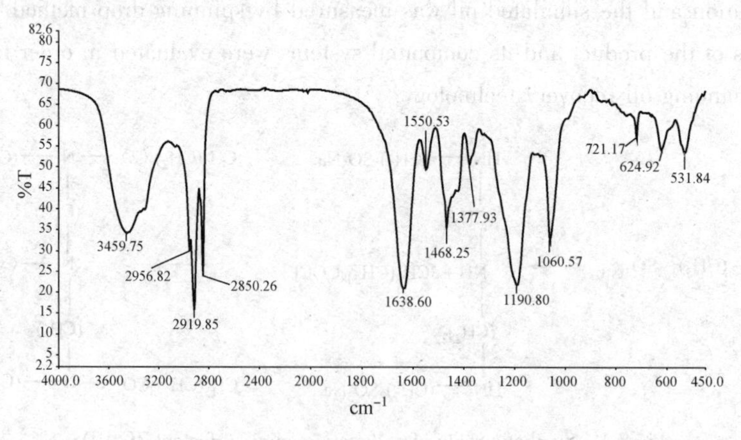

Fig.8.2 Infrared spectrum of sulfonate gemini surfactant $3C_{10}$-DS

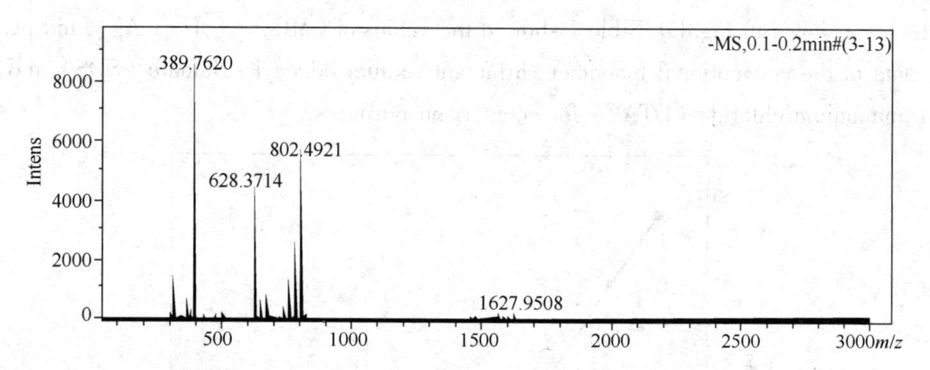

Fig.8.3 Mass spectra of sulfonate gemini surfactant $3C_{10}$-DS

2.3 Surface Tension Measurements

The surface tensions of aqueous solutions of sulfonate gemini surfactant $3C_{10}$-DS were measured with Pendant Drop method (JC2000C, China) at 25℃. The curve of surface tension versus logarithm of surfactant concentrations was plotted, and its CMC value was determined from the break point. The maximum surface excess (Γ_{max}) is calculated using the Gibbs adsorption isotherm equation of $\Gamma = -(1/2.303\ nRT)(d\gamma/d\log C)$, Where n represents the number of species at air-water interface whose concentration changes with surfactant concentration, R is the gas constant ($8.314\ J \cdot mol^{-1} \cdot K^{-1}$), T is the absolute temperature (K) and ($d\gamma/d\log C$) is the slope in the surface tension isotherm when the concentration is near the CMC. The areas per molecule (A_{min}) at air-water interface were calculated from the equation of $A_{min} = (N_A \Gamma_{max})^{-1}$, where N_A is Avogadro constant. pC_{20} is the value of the logarithm of surfactant concentration of required to reduce the surface tension of water by 20 mN/m.

2.4 Interfacial Tension Measurements

The interfacial tensions between the aqueous solution of surfactants and the simulated oil were measured by the spinning drop interface tensiometer (Model TX-500) at 45℃, corresponding to the reservoir temperature in the Daqing oil field. The spinning oil droplet in the surfactant solution was stretched until the oil-water phase reached equilibrium at a rotation speed of 6,000 r/min.

2.5 Evaluation of Oil Displacement Efficiency

The oil displacement tests were conducted in homogeneous artificial core (4.5 cm×4.5 cm×30.0 cm) at 45℃. In the experiments, the cores were saturated with the simulated formation water of 6.778g/L total salinity. Then the simulated oil was injected into the cores until the core reached irreducible water saturation. After aging 12 h, the simulated formation water was injected into the core saturated with the simulated oil at a constant speed until the water cut reached 0.98. In this process, cumulative oil production was recorded, and the recovery of the first water flooding was calculated. Following water flooding, 0.35 pore volume of chemical slug was injected into the core. Then, the second water flooding was continued to until the water cut reached 0.98 again. The recovery of the second water flooding was calculated.

3. Results and Discussion

3.1 Surface Properties

The surface tensions of $3C_{10}$-DS solutions of different concentration were measured at 25℃, and the results were shown in Fig.8.4. Table 1 showed the values of CMC, γ_{cmc}, Γ_{max}, A_{min}, and pC_{20}, along with the data of the conventional monomer surfactant sodium dodecyl sulfonate (SDS) and dodecyl trimethyl ammonium chloride (DTAC) for comparison purposes.

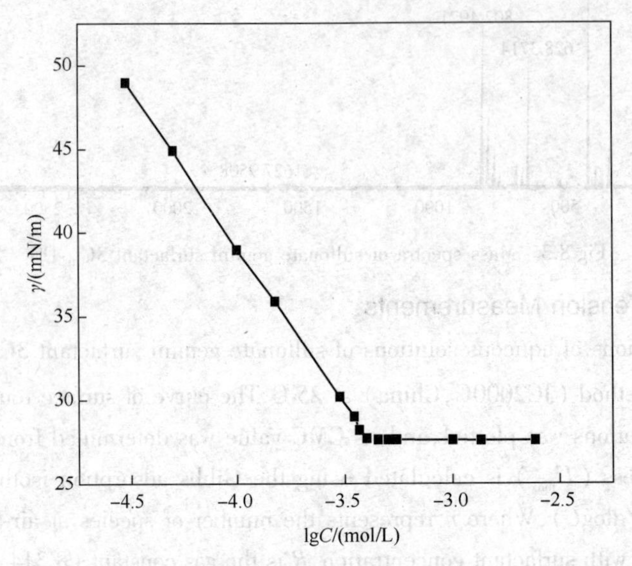

Fig.8.4 Surface tension vs concentration isotherm of $3C_{10}$-DS.

As can be seen from Figure 4, the surface tensiondecreases sharply with the increasing of $3C_{10}$-DS concentration, and then reaches clear break point, which is taken as critical micelle concentration. CMC and the surface tension at the CMC (γ_{cmc}) are 0.41mmol and 27.6 mN/m, respectively. These results indicate that CMC of $3C_{10}$-DS is about one order of magnitude lower than those of the corresponding monomeric surfactants SDS and DTAC. The low CMC of $3C_{10}$-DS with respect to the conventional surfactants arises mainly because the strong hydrophobic effect of three alkyl chains makes it easily transfer from water to the micelle pseudo-phase[11]. γ_{cmc} and pC_{20} can reflect the effectiveness and efficiency in surface tension reduction by the surfactant at the air-water surface, respectively. As expected, the synthesized gemini surfactant $3C_{10}$-DS exhibits higher effectiveness (γ_{cmc}) and efficiency (pC_{20}) of surface tension reduction than those of the conventional surfactants SDS and DTAC, and displays remarkable surface activities.

The values of Γ_{max} and A_{min} depend on molecular structure of the surfactant and its orientation at the air-water surface. The higher the effectiveness of adsorption, the greater the maximum surface excess and the smaller the minimum surface area occupied a surfactant molecule. As shown in Table 8.1, A_{min} of $3C_{10}$-DS is smaller than triple those of the conventional surfactants, as the three hydrophilic polar groups within $3C_{10}$-DS are covalently connected by a spacer group. In addition, the separation between the adjacent hydrophobic groups connecting with polar head groups is strongly decreased due to overcome the electrostatic repulsion of the head groups, and there exists strong hydrophobic effect between the adjacent hydrophobic alkyl chains. This means the packing of the hydrophobic alkyl chains in the Gemini surfactant $3C_{10}$-DS at the air-water interface is closer than that found in conventional surfactants, which may be the reason for the increased lowering of surface tension at CMC by the gemini surfactant $3C_{10}$-DS.

Table 8.1 Surface properties of $3C_{10}$-DS, SDS, and DTAC

surfactant	CMC/(mmol/L)	γ_{cmc}/(mN/m)	Γ_{max}/(μmol/m^2)	A_{min}/(nm^2)	PC_{20}
$3C_{10}$-DS	0.41	27.6	1.66	1.01	2.53
SDS[a]	8.20	32.5	3.16	0.53	2.50
DTAC[b]	12.0	39.0	3.42	0.49	2.60

[a] Reported in Ref.[9]; [b] Reported in Ref.[10]

3.2 Interfacial Tension

In many practical applications, different kinds of surfactants are deliberately mixed together to achieve properties of the mixture that are better than those of the individual surfactant. From cost and performance considerations, gemini surfactants will likely be used in mixtures with conventional surfactants in the future. In present study, the synergistic effect of the two compound method that is being used in oil field production in recent years was studied in terms of IFT reduction for $3C_{10}$-DS.

The interfacial tensions between the solutions of surfactant $3C_{10}$-DS and the simulated oil were measured at 45℃, and the results were shown in Figure 8. 5. The interfacial tensions first decrease and then increase. When the mass concentration of surfactant $3C_{10}$-DS reached 400 mg/L, there was the lowest point at which the interfacial tension was 0.343 mN/m · A possible explanation is that adsorption of the molecules of surfactant at the oil-water interface increases over time, and so the interfacial tension decreases and reaches a minimum when the surfactant molecules occupy the whole oil-water interface. Then the hydrophobic groups of surfactant molecules $3C_{10}$-DS in the bulk solution cluster together into micelles which may solubilize surfactant molecules desorbing from the oil-water interface, and consequently the interfacial tension increases.

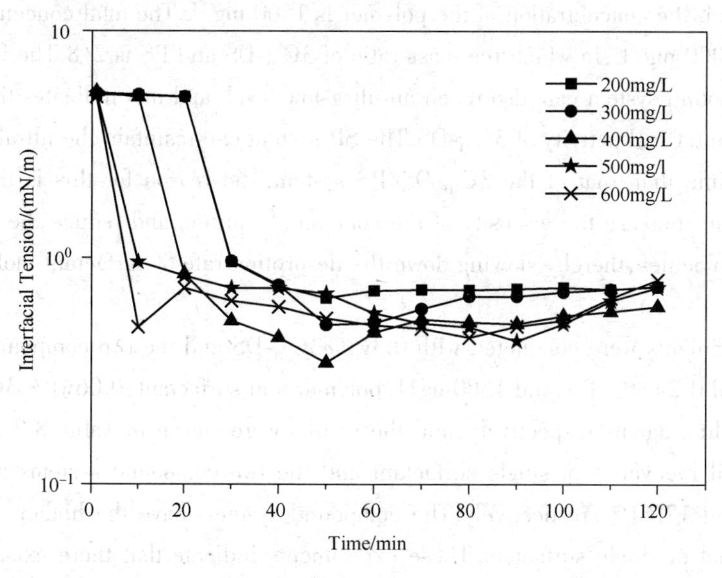

Fig.8.5 Time variation of interfacial tension between the simulated oil and $3C_{10}$-DS

The interfacial tensions between the simulated oil and the compound systems with different mass ratio of $3C_{10}$-DS and PS were measured at 45℃, in which the total concentration of the mixed surfactants is 3000 mg/L, and the results are shown in Fig.8.6.

As can be seen from Fig.8.6, the interfacial tension from the $3C_{10}$-DS/PS compound system at a

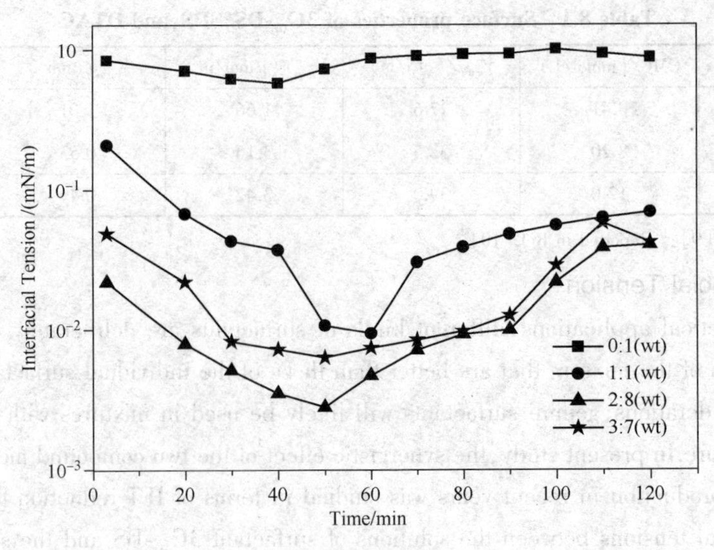

Fig.8.6 Time variation of interfacial tension between the simulated oil and the $3C_{10}$-DS/PS compound systems

proper proportion reaches ultra-low levels (10^{-3}mN/m) ,which is not achievable by either individual surfactant.This indicates that the combination of $3C_{10}$-DS and PS has significant synergistic effect. The lower concentration of $3C_{10}$-DS in the compound system is helpful to reduce the cost of crude oil production because the product cost of $3C_{10}$-DS is higher than that of petroleum sulfonate.

Thesurfactant-polymer (SP) formulations are often used to enhance oil recovery in tertiary oil recovery techniques.The interfacial tensions between the simulated oil and the SP compound system are shown in Fig.8.7, in which the concentration of the polymer is 1500 mg/L.The total concentration of the mixed surfactants is 3000 mg/L, in which the mass ratio of $3C_{10}$-DS and PS is 2:8.The interfacial tensions of the SP compound system can also reach an ultra-low level, and this indicates the polymer has little effect on the interfical activity of $3C_{10}$-DS.The SP system can maintain the ultralow interfacial tension for longer time than that of the $3C_{10}$-DS/PS system.The reason for this is that the polymer in the SP system can increase the viscosity of the compound solution and reduce the rate of diffusion of the surfactant molecules, thereby slowing down the desorption rate of surfactant molecules at the oil-water interface.

The core flooding experiments were conducted with 0.3wt%$3C_{10}$-DS and the two compound systems (0.06wt% $3C_{10}$-DS and 0.24wt% PS, and 1500mg/L polymer and surfactant; 0.06wt% $3C_{10}$-DS and 0.24 wt.% PS) as flooding agents respectively, and the results were shown in Table 8.2.As can be seen from Table 2, the oil recoveries of single surfactant and the two compound systems are improved by 2.89%, 5.54% and 17.61%, respectively.The compound systems have the higher oil displacement efficiency than that of single surfactant.These experiments indicate that there exists good synergistic effect between $3C_{10}$-DS and PS, which agrees well with the aforementioned results of the interfacial tension.The polymer in the SP compound system can increase the viscosity of water phase, lower the mobility ratio of between displacing phase and displaced phase, expand the sweep efficiency of the injected water, and thus substantially enhance the oil recovery of the high-medium permeability cores.So the sulfonate gemini surfactant $3C_{10}$-DS and its compound system have a good application prospect in oilfield development.

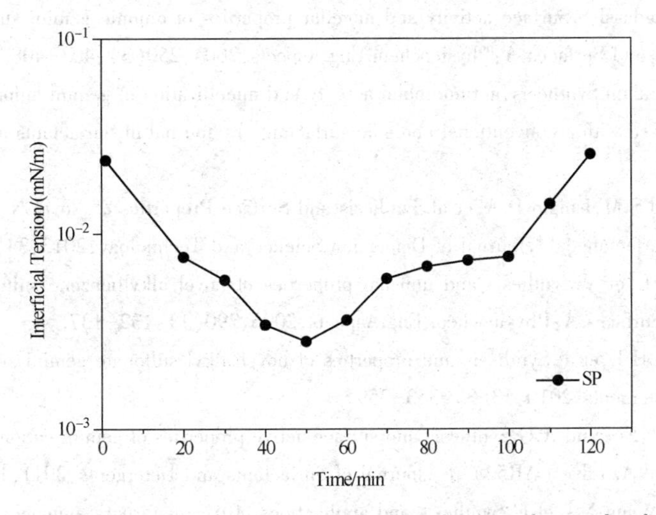

Fig.8.7 Time variation of interfacial tension between the simulated oil and the SPcompound system

Table 8.2 Data of coreflooding experiments

Core number	Chemical flooding system	Permeability/ $10^{-3}\mu m^{-2}$	Oil saturation/ %	Recovery efficiency/%		
				Water flooding	Enhancing oil recovery	Total Oil recovery
1	3C10-DS	700	71.7	45.35	2.89	48.24
2	3C10-DS/PS	735	73.2	45.08	5.54	50.62
3	SP	727	72.2	45.38	17.61	62.99

4.Conclusions

A new type of sulfonate gemini surfactant with three lipophilic alkyl chains was synthesized, which had lower critical micelle concentration, better surface/interfacial properties than those of conventional surfactants. There existed significant synergistic effect between $3C_{10}$-DS and Petroleum sulfonate. The SP compound system could effectively improve oil recovery for high-medium permeability cores, and thus they had a good application prospect in enhancing oil recovery.

The authors are grateful for the financial support from the National Natural Science Foundation of China (Grant No.51074033)

References

[1] Guo D.H, Li S, Yuan J.G. Flooding mechanism and application of surfactant flooding[J]. Advances in Fine Petrochemicals, 2002, 3(7): 36-41.

[2] Ge J.J, Zhang G.C, Jing P, et al. Development of surfactants as chemicals for EOR[J]. Oilfield Chemistry, 2007, 3(24): 287-292.

[3] Rosen M.J. Gemini: a new generation of surfactants[J]. Chemtech, 1993, 23(3): 30-33.

[4] Wu H.J, Du X.G, Zhang X, et al. Synthesis and properties of alkylbenzene sulfonate gemini surfactants with high viscosity in dilute solution without additives[J]. Journal of Dispersion Science and Technology, 2011, 32(4):

596-600.
[5] Ben-Moshe M, Magdassi S. Surface activity and micellar properties of anionic gemini surfactants and their analogues[J]. Colloids and Surfaces A: Physicochem. Eng. Aspects, 2004, 250(3): 403-408.
[6] Fatma H. Abd El-Salam. Synthesis, antimicrobial activity and micellization of gemini anionic surfactants in a pure state as well as mixed with a conventional nonionic surfactant[J]. Journal of Surfactants and Detergents, 2009, 12(4): 363-370.
[7] El-Dib F.I, Ahmed S.M, Ismaio D.A, et al. Synthesis and Surface Properties of Novel N-Alkyl Quinoline-Based Cationic Gemini Surfactants[J]. Journal of Dispersion Science and Technology, 2013, 34(4): 596-603.
[8] Du X.G, Lu Y, Li L, et al. Synthesis and unusual properties of novel alkylbenzene sulfonate gemini surfactants[J]. Colloids and Surfaces A: Physicochem. Eng. Aspects, 2006, 290(1): 132-137.
[9] Li X, Hu Z.Y, Zhu H.L, et al. Synthesis and properties of novel alkyl sulfonate gemini surfactants[J]. Journal of Surfactants and Detergents, 2010, 13(3): 353-359.
[10] Abdel-Salam F.H, Ei-Said A.G. Synthesis and surface active properties of gemini cationic surfactants and interaction with anionic Azo dye (AR52)[J]. Journal of Surfactants and Detergents, 2011, 14(3): 371-379.
[11] Wu X.J, Zhao L, Wang X.J, et al. Synthesis and applications of tri-quaternary ammonium salt gemini surfactant[J]. Journal of Dispersion Science and Technology, 2013, 34(1): 106-110.

Appendix Ⅰ Common Chemical Glassware Names

附录1 常见玻璃仪器名称

英语名称	中文名称	英语名称	中文名称
adapter	接液管	long-stem funnel medicine	长颈漏斗
air condenser	空气冷凝管	magnetic stirrer	磁力搅拌器
Balance	天平	mohr burette for use with pinchcock	碱氏滴定管
beaker	烧杯	mohr measuring pipette	量液管
boiling flask	烧瓶	mortar	研钵
boiling flask-3-neck	三口烧瓶	pestle	研杵
burette clamp	滴定管夹	pinch clamp	弹簧节流夹
burette stand	滴定架台	plastic squeeze bottle	塑料洗瓶
busher funnel	布氏漏斗	reducing bush	大变小转换接头
claisen distilling head	减压蒸馏头	rubber pipette bulb	吸耳球
condenser-Allihn type	球型冷凝管	screw clamp	螺旋夹
condenser-west tube	直型冷凝管	separatory funnel	分液漏斗
crucible tongs	坩埚钳	stemless funnel	无颈漏斗
crucible with cover	带盖的坩埚	stir bar	搅拌子
distilling head	蒸馏头	stirring rod	搅拌棒
distilling tube	蒸馏管	stopper	塞子
Dropper	滴管	test tube holder	试管夹
Dryer	干燥器	test tube	试管
erlenmeyer flask	锥型瓶	thermometer	温度计
evaporating dish (porcelain)	瓷蒸发皿	thiele melting point tube	提勒熔点管
filter flask(suction flask)	抽滤瓶	transfer pipette	移液管
florence flask	平底烧瓶	tripod	三角架
fractionating column	分馏柱	volumetric flask	容量瓶
geiser burette (stopcock)	酸氏滴定管	vacuum pump	真空泵
graduated cylinder	量筒	watch glass	表面皿
hirsch funnel	赫氏漏斗	wide-mouth bottle	广口瓶

Appendix Ⅱ Common Chemical Abbreviation

附录 2 化学常见缩略语

英文缩略语	英文全称	中文意思
ab.	Absolute	绝对的
addn.	Addition	添加
alc.	Alcohol	醇
alk.	Alkali	碱
amt.	Amount	量
A.P.	analytically pure	分析纯
app.	apparatus	装置
approx.	approximate	大约
aqu.	aqueous	水的
asym.	asymmetric	不对称的
atm.	atmospheric	大气压
av.	average	平均的
b.p.	boiling point	沸点
ca.	circa	大约
cal.	caloric	卡路里
calc.	calculate	计算
cf.	compare	比较
chem.	chemistry	化学
conc.	concentrated	浓缩的
const.	constant	常数
contg.	containing	含有…的
compd.	compound	化合物
C.P.	chemically pure	化学纯
cryst.	crystalline	晶体
decomp.	decompose	分解

续表

英文缩略语	英文全称	中文意思
deriv.	derivative	衍生
detn.	determination	测定
dil.	dilute	稀释的
distd.	distilled	蒸馏的
e.g.	for example	例如
elec.	electric	电的
eq.	equation	方程
equil.	equilibrium	平衡
equiv.	equivalent	等价的
et.al.	and others	以及其他人
etc.	et cetera	等等
evap.	evaporation	蒸发
expt.	experimental	实验的
fig.	Figure	图
hyd.	hydrous	水的
ibid.	in the same place	在同一地方
lab.	laboratory	实验室
liq.	Liquid	液体
L.R.	laboratory reagent	实验试剂
manf.	manufacture	制造
max.	maximum	最大的
min.	minute	最小的
mixt.	mixture	混合物
mol.wt.	molecular weight	分子量
m.p.	melting point	熔点
org.	organic	有机的
ppm.	parts per million	百万分之一
ppt.	precipitated	沉淀的
prep.	prepare	制备
resp.	respectively	分别地
sec.	second	第二
soln.	solution	溶液
solv.	solvent	溶剂

续表

英文缩略语	英文全称	中文意思
sp.gr.	specific gravity	比重
sq.	square	平方
sub.	sublime	升华
susp.	suspended	悬浮地
tech.	technical	技术的
Tech.P.	technically pure	技术纯
temp.	temperature	温度
vol.	volume	体积
wt.	Weight	重量

Appendix Ⅲ Common Chemical Prefix and Suffix

附录3 化学专业英语词汇常用前后缀

前/后缀	中文意思	前/后缀	中文意思
-acetal	缩醛	centi-	10^{-2}
acetal-	乙酰	chloro-	氯代
acid	酸	chromo-	铬的
-al	醛	cis-	顺式
alcohol	醇	-cide	除……剂,防……剂
aldehyde	醛	condensed	缩合的、冷凝的
alkali-	碱	cyclo-	环
allyl	丙烯基	deca-	十
alkoxy-	烷氧基	deci	10^{-1},二
-amide	酰胺	-dine	啶
amino-	氨基的	dodeca-	十二
-amidine	脒	-ene	烯
-amine	胺	epi-	表
-ane	烷	epoxy-	环氧
anhydride	酐	-ester	酯
anilino-	苯胺基	-ether	醚
aquo-	含水的	ethoxy-	乙氧基
-ase	酶	ethyl	乙基
-ate	含氧酸的盐、酯	ferro-	亚铁
-atriene	三烯	fluoro-	氟代
-atriyne	三炔	form	仿
azo-	偶氮	Formyl-	甲酰基
bi-	二,双,酸式	-glycol	二醇
bis-	双	hemi-	半
-borane	硼烷	hendeca-	十一
bromo-	溴	hepta-	七
butyl	丁基	heptadeca-	十七

前/后缀	中文意思	前/后缀	中文意思
carbonyl	羰基	hexa-	六
-caboxylic acid	羧酸	hexadeca-	十六
homo-	均,同,单	-ous	亚酸的,低价金属
hydro-	氢或水	oxa-	氧杂
hydroxyl	羟基	-oxide	氧化合物
Hyper-	高,超	-oxime	肟
hypo-	低级的,次	oxo-	酮
-ic	酸的,高价金属离子	oxy-	氧化
-ide	无氧酸盐,酐	-oyl halide	酰卤
-imine	亚胺	-oyl	酰
iodo-	碘代	para-	对位,仲
-iridine	丙啶	penta-	五
iso-	异,等,同	pentadeca-	十五
-ite	亚酸盐	per-	高,过
keto-	酮	petro-	石油
ketone	酮	phenol	苯酚
-lactone	内酯	phenyl	苯基
mega-	10^{6}	-philic	亲……的
meta-	间,偏	pico-	10^{-12}
methoxy-	甲氧基	poly-	聚,多
methyl	甲基	quadri-	四
micro-	10^{-6}	quinque-	五
milli-	10^{-3}	semi-	半
mono-	一,单	septi-	七
nano-	10^{-9}	sesqui	一个半
nitro-	硝基	sexi-	六
nitroso-	亚硝基	-side	苷
nona-	九	sub-	亚
nonadeca-	十九	sulfa-	磺胺
octa-	八	sulfo-	磺酸基
octadeca-	十八	sym-	对称
-oic	酸的	syn-	顺式,同,共
-ol	醇	ter-	三
-one	酮	tetra-	四
ortho-	邻,正,原	tetradeca-	十四
tetrakis-	四个	undeca-	十一

续表

前/后缀	中文意思	前/后缀	中文意思
thio-	硫代	uni-	单,一
trans-	反式,超,跨	unsym-	不对称的,偏位
tri-	三	-yl	基
trideca-	十三	-ylene	亚基
tris-	三个	-yne	炔

References

[1] Abdel-Raouf M E.Crude Oil Emulsions-composition Stability and Characterization[M].Croatia:InTech,2012.

[2] Alka L Gupta.Polymer chemistry[M].Meerut:Pragati Prakashan,2010.

[3] Beonard J, Lygo B, Procter G. Advanced Practical Organic Chemistry [M]. 2nd ed. London: Stanley Thornes Ltd,1998.

[4] Bleil R E.Organic Chemistry Laboratory Manual[M].Dakota State University Press,2005.

[5] Bruice P Y .Organic Chemistry[M].4th ed.New Jersey:Prentice Hall,2003.

[6] Carey F A,Sundberg R J.Advanced OrganicChemistry[M].5th ed.New York:Springer Science+Business Media, LLC,2006.

[7] European commission. Best Available Techniques Reference Document for the Manufacture of Organic Fine Chemicals[M],2006.

[8] Farn R J.Chemistry and Technology of Surfactants[M].Oxford:by Blackwell Publishing Ltd,2006.

[9] Fahim M A,Al-Sahhaf T A,Elkilani A S.Fundamentals of Petroleum Refining[M].Oxford:Elsevier B.V.,2010.

[10] Fieser L F,Williamson K L.Organic Experiments[M].7th ed.Lexington:D.C.Heath and Company,1992.

[11] Fink J K.Oil Field Chemicals[M].Houston:Gulf professional publishing,2003.

[12] Davis F J.Polymer chemistry[M].Oxford:Oxford University Press,2004.

[13] Hoggett J G,Moodie R B,Penton J R,et al.Nitration and aromatic reactivity[M].London:Cambridge University Press,1971.

[14] House J E.Inorganic Chemistry[M].2nd ed.California:Elsevier Inc.,2012.

[15] Speight J G.Chemistry and Technology of Petroleum.4th ed.New York:CRC Press,2006.

[16] Speight J G.HandbookOf Petroleum Product Analysis[M].New Jersey:John Wiley & Sons,Inc.,2002.

[17] Miessler G L,Tarr D A.Inorganic Chemistry[M].3rd ed.New Jersey:Prentice Hall,2004.

[18] Mortimer R G.Physical Chemistry[M].3rd ed.Burlington:Elsevier Academic Press,2008.

[19] Myers D.Surfactant Science and Technology[M].3rd ed.Hoboken:John Wiley & Sons,Inc.,2006.

[20] OSHA Technical Manual.U.S.Department of Labor Occupational Safety & Health Administration[M],2006.

[21] Patrick G.Organic Chemistry[M].2nd ed.New York:Taylor & Francis,2003.

[22] Peter Chave.Chemical works:fine chemicals manufacturing works[M].DOE(Department of the Environment) Industry Profile,1995.

[23] Peter Pollak.Fine Chemicals:The Industry and the Business[M].New Jersey:John Wiley & Sons,Inc.,2007.

[24] Sami Matar,Lewis F.Hatch.Chemistry of Petrochemical Processes[M].2nd ed.Houston:Gulf Publishing Company,2001.

[25] Tadros T F.Emulsion Science and Technology[M].Weinheim:WILEY-VCH Verlag GmbH & Co.KGaA,2009.

[26] Malcolm A. Kelland. Production Chemicals for the Oil and Gas Industry [M]. 2nd ed. New York: CRC Press,2014.

[27] Fred J.Davis.Polymer Chemistry[M].Oxford:Oxford University Press,2004.

[28] Heinz Heinemann.The chemistry and Technology of Petroleum[M].4th ed.New York:CRC Press,2006.